Current Progress in Petroleum Engineering

Current Progress in Petroleum Engineering

Editor: Michael Dedini

www.callistoreference.com

Callisto Reference,
118-35 Queens Blvd., Suite 400,
Forest Hills, NY 11375, USA

Visit us on the World Wide Web at:
www.callistoreference.com

ISBN: 978-1-63239-888-8 (Hardback)

The publisher's policy is to use permanent paper from mills that operate a sustainable forestry policy. Furthermore, the publisher ensures that the text paper and cover boards used have met acceptable environmental accreditation standards.

Trademark Notice: Registered trademark of products or corporate names are used only for explanation and identification without intent to infringe.

Printed in the United States of America.

Cataloging-in-Publication Data

Current progress in petroleum engineering / edited by Michael Dedini.
      p. cm.
Includes bibliographical references and index.
ISBN 978-1-63239-888-8
1. Petroleum engineering. 2. Oil wells. 3. Shale oils. 4. Petroleum--Prospecting. I. Dedini, Michael.
TN870 .C87 2017
665.5--dc23

Table of Contents

Preface

This book on petroleum engineering deals with the theoretical and technical aspects of petroleum exploration and extraction. Reservoir engineering is the primary methodology that is implemented in the extraction of crude oil. Modern technology in this field seeks to be cost-effective, environmentally friendly and efficient in petroleum recovery. The use of computer technology for reservoir design and hydraulic fracturing are other modern innovations in this field. The various advancements in petroleum engineering are glanced at and their applications as well as ramifications are looked at in detail. Different approaches, evaluations and methodologies in this field have been included in this book. Students, researchers, experts and all associated with petroleum engineering will benefit alike from this book.

After months of intensive research and writing, this book is the end result of all who devoted their time and efforts in the initiation and progress of this book. It will surely be a source of reference in enhancing the required knowledge of the new developments in the area. During the course of developing this book, certain measures such as accuracy, authenticity and research focused analytical studies were given preference in order to produce a comprehensive book in the area of study.

This book would not have been possible without the efforts of the authors and the publisher. I extend my sincere thanks to them. Secondly, I express my gratitude to my family and well-wishers. And most importantly, I thank my students for constantly expressing their willingness and curiosity in enhancing their knowledge in the field, which encourages me to take up further research projects for the advancement of the area.

Editor

Model of interval multi-attribute optimization for overseas oil–gas projects

Yong-Zhang Huang[1] · Bao-Sheng Zhang[1] · Xin-Qiang Wei[2] · Ren-Jin Sun[1]

Abstract Because of the incompleteness and uncertainty in the information on overseas oil–gas projects, project evaluation needs models able to deal with such problems. A new model is, therefore, presented in this paper based on interval multi-attribute decision-making theory. Analysis was made on the important attributes (index) and the relationships affecting the basic factors to the project economic results were described. The interval numbers are used to describe the information on overseas oil and gas projects. On these bases, an improved TOPSIS model is introduced for the evaluation and ranking of overseas oil and gas projects. The practical application of the new model was carried out for an oil company in selecting some promising blocks from 13 oil and gas blocks in eight different countries in the Middle East. Based on these innovative studies, some conclusions are given from theoretical and application aspects. The practical application shows that the introduction of interval numbers into the evaluation and ranking of the overseas oil and gas projects can lead to more reasonable decisions. The users can do the project evaluation based on the comprehensive values as well as based on some preferred index in the project evaluation and ranking.

Keywords Interval data · Improved TOPSIS model · Multiple attribute decision making · Overseas oil–gas project · Alternative ranking

1 Introduction

Up till now, the decision-making on overseas oil and gas projects is mainly based traditional economic evaluation, usually called the net present value (NPV) method. This method can, however, work with relatively complete information and accurate data. In fact, the information about oversea oil and gas projects is often not perfect. Some indices such as geological condition parameters may be missing or the values of some index can only be estimated to an interval range. It is, therefore, not easy and nor appropriate to use the traditional NPV method in such situations. For solving this problem, the multi-attribute decision-making (MADM) theory has been put forward and applied in project evaluation research over the last 10 years. Kong (2005) has proposed a decision-making model for mineral resources by a fuzzy method using MADM theory. Liu (2007) has studied project economic evaluation using a 3-dimensionality factor, namely information, space, and reference group. Li (2004) has introduced fuzzy theory into MADM and established an economic decision model for development of mineral resources. Srdjevic et al. (2004) using the traditional TOPSIS model to solve comprehensive management problems, including economic problems and water resources. Some theoretical research work and application research have also been done in the transportation field (Janic 2003; Herrera et al. 2005).

In the oil and gas industry, similar research work has been done in geological and reservoir condition analysis

✉ Bao-Sheng Zhang
  bshshysh@cup.edu.cn

[1] School of Business Administration, China University of Petroleum, Beijing 102249, China

[2] Overseas Investment Environment Research Department, CNPC Economics and Technology Research Institute, Beijing 100724, China

Edited by Xiu-Qin Zhu

and production prediction. Liu (2010) described a method based on MADM theory for evaluation of oil–gas reserves. In order to improve the forecast accuracy, Hou and Gui (2010) presented research work for oil production forecasting by combining the neural network technology with MADM theory. In view of the risk–benefit co-analysis, Wang et al. (2010) proposed a dynamic MAUT(Multi-Attribute Utility Theory) model for oil–gas project evaluation with respect to the three attributes, namely geological risk, market risk, and economic benefits. It can be seen that most existing research in the oil and gas industry was focused on the analysis of the geological or technical condition factors. Few reports have been found about similar research into direct consideration of the economic evaluation.

Some other MADM research work (Fan et al. 2002; Albayrak 2004; Dağdeviren 2008; Hladik 2007) have also been found. Most of this research was conducted based on point data. But the information on overseas oil–gas projects cannot, as referred above, be estimated accurate to a particular number. Only an interval can be given in practice. So, a logical approach is to introduce the interval number into the models based on the MADM theory. In this paper, the geological index and economic index are combined to build an index system expressed in interval numbers. In this way, the traditional technique for order of preference by similarity to ideal solution (TOPSIS) model is improved for overseas oil–gas project evaluation.

2 Analysis of the factors influencing overseas oil–gas project benefits

It is very important for decision makers to select the appropriate indices in the decision process. In this process, there are two issues worth noting, one is the type of indices, and the other is the method used to compare them. According to their nature, each index can be divided into an efficiency (positive) and cost (negative) index. For the efficiency (positive) index, such as IRR and NPV, a larger value means better. For a cost index, such as the static investment payback period (SIPP) or the total estimated investment (TEI), a smaller value means better. We may also meet another classification from the index values, namely the fixed and interval index. For fixed index, the value closer to a given fixed number suggests a better situation. By interval index, whereas all the index values falling in between a given interval can be taken as better solutions.

This paper will analyze the relationships among all the main factors affecting oil–gas project decision from two aspects, namely from the oil or gas reserve and production aspect, and from the economic value as the other aspect. For example, the reserve amount (resource scale) is

affected by many geological factors, with the affecting relationships shown in Fig. 1.

As shown in Fig. 1, the factors affecting the reserve amount (resource scale) include reservoir porosity, the density of crude oil, crude oil volume factor, original oil saturation, oil (gas) layer thickness, and oil (gas) area coefficient, and they are often affected by some other factors.

Beside such important parameters as reserve amount and oil or gas production, more attention should be given to economic benefits, because the ultimate goal of a company is to make money. The factors affecting economic benefits include geological condition, market situation, engineering technology, policy and social factors, and so on. NPV is regarded as the final economic benefit index, and its relation to the other factors is shown in Fig. 2.

The NPV is calculated directly according to the project's cash input and the cash output (cash in and cash out, as shown in Fig. 2). It is also affected directly by the benchmark discount rate and contract period, which are normally taken as fixed factors. All other factors shown in Fig. 2 will affect the NPV through influencing the cash inflow and cash outflow.

3 The index system

Based on the research work from some scholars (e.g. Liu 2010; Wang et al. 2010, etc.) and referring to the project evaluation practices of the petroleum companies, the index system was established for this work, with the index weight expressed in interval numbers. All the indices can be classified into 3 groups.

(1) The project condition index group
    The index in this group is the parameters describing the oil–gas block's basic condition information. Among these indices, the reserve scale, resource abundance and the density of crude oil are the important ones.
(2) The index related to geological conditions
    The index of this group reflects some basic geological information from the oil–gas blocks, such as trap condition, hydrocarbon source rocks condition, reservoir condition. Generally, the values of these indices have strong relations to the block exploration degree. The higher the degree of exploration of the block, the more complete the information that can be obtained and the values of the index can be more accurately estimated.
(3) The economic index
    The index of this group indicates what types of technology should be applied and what economic results can be obtained based on the geological situation of project.

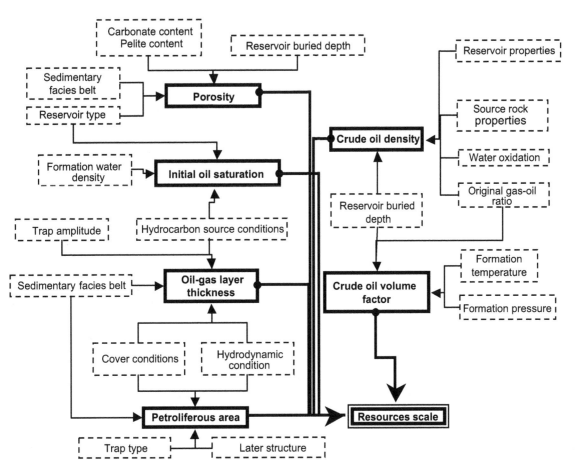

Fig. 1 The relationships among geological factors affecting the resource scale

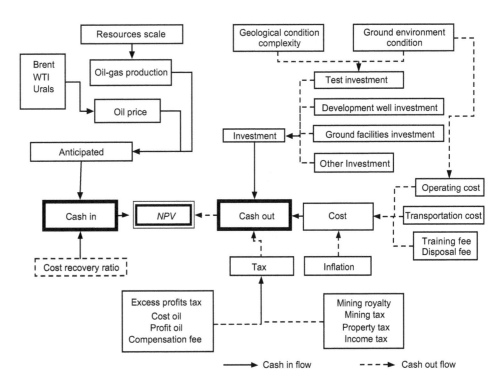

Fig. 2 The relationships among factors affecting the economic result

The established index system is shown in Table 1.

The evaluation criteria should be selected for the project evaluation based on the index system. Considering that there is no uniform global standard for judging the index value, the evaluation criteria suggested in the third resource evaluation handbook published by China National Petroleum Corporation (CNPC) are applied in our research work.

4 Method

4.1 Multiple attribute decision-making (MADM)

To carry out the MADM analysis, two important things should be prepared, that is to build the set of the decision alternatives and to build the set of the attributes (index). The following expressions are generally used:

$S = \{S_1, S_2, \ldots, S_m\}$—set of the decision alternatives,

$X = \{X_1, X_2, \ldots, X_n\}$—set of the attributes (index).

Meanwhile, the set X should be coupled with an index weights vector, $W = (W_1, W_2, \ldots, W_n)^T$.

In the traditional multi attribute decision process, a point number, a_{ij}, will be used to present the value of the alternative i evaluated under the attribute j. A comprehensive value can be calculated for every alternative according to the set X and its coupled weight vector W. The best alternative will be selected by comparison of the comprehensive values of the different alternatives.

Similar to the analysis process in the ordinary MADM process by the point data, a so called decision matrix should be also first built in this research work. The difference here in comparing to the ordinary one is that the elements of the decision matrix are all expressed in interval numbers. Generally, we use $A = \left(a_{ij}\right)_{m \times n}$ to stand for the decision matrix, where

$$a_{ij} = [a_{ij}^L, a_{ij}^U] \, (1 < i < m, \, 1 < j < n),$$

and a_{ij}^L is the lower limit of the interval value of the alternative i evaluated under the attribute j; a_{ij}^U is the upper limit of the interval value of the alternative i evaluated under the attribute j.

The coupled weight vector $w_j = \left[w_j^L, w_j^U\right]$ is also described with interval numbers, where w_j^L is the lower limit of the interval value of the weight of the attribute j; w_j^U is the upper limit of the interval value of the weight of the attribute j.

The decision matrix of the MADM under interval data is shown in Table 2.

4.2 The traditional TOPSIS model

TOPSIS is a method for ranking the considered alternatives by evaluating the similarity of each alternative to the given ideal modes. Usually, two ideal modes are given, with one as the optimal mode (positive mode) and the other as worst mode (negative mode). The evaluation is

Table 1 Indices used in this work

No.	Index	Type, unit	Attribute	Weight interval
1	Reserves	Digital, 10^4t	Efficiency	$[w_1^L, w_1^U]$
2	Reserve abundance	Digital, 10^4t/km^2	Efficiency	$[w_2^L, w_2^U]$
3	Hydrocarbon source rock thickness	Digital, m	Efficiency	$[w_3^L, w_3^U]$
4	Lithology	Discrete	–	$[w_4^L, w_4^U]$
5	Organic carbon content	Digital, %	Efficiency	$[w_5^L, w_5^U]$
6	Hydrocarbon generation peak time	Discrete	–	$[w_6^L, w_6^U]$
7	Trap type	Discrete	–	$[w_7^L, w_7^U]$
8	Reservoir thickness	Digital, m	Efficiency	$[w_8^L, w_8^U]$
9	Porosity	Digital, %	Efficiency	$[w_9^L, w_9^U]$
10	Permeability	Digital, $10^{-3}\mu$m^2	Efficiency	$[w_{10}^L, w_{10}^U]$
11	Buried depth	Digital, m	Cost	$[w_{11}^L, w_{11}^U]$
12	Cover lithology	Discrete	–	$[w_{12}^L, w_{12}^U]$
13	Cover layer thickness	Digital, m	Efficiency	$[w_{13}^L, w_{13}^U]$
14	Crude oil density	Digital, g/cm^3	Efficiency	$[w_{14}^L, w_{14}^U]$
15	Peak production	Digital, t/d	Efficiency	$[w_{15}^L, w_{15}^U]$
16	Well depth	Digital, m	Cost	$[w_{16}^L, w_{16}^U]$
17	Expected investment	Digital, 10^8\$	Cost	$[w_{17}^L, w_{17}^U]$
18	The total cost	Digital, 10^8\$	Cost	$[w_{18}^L, w_{18}^U]$
19	Expected NPV (10 %)	Digital, 10^8\$	Efficiency	$[w_{19}^L, w_{19}^U]$
20	Expected IRR	Digital, %	Efficiency	$[w_{20}^L, w_{20}^U]$

Table 2 Decision matrix of MADM under interval data

Alternative set S	Attribute set, X			
	X_1	X_2		X_n
S_1	$[a_{11}^L, a_{11}^U]$	$[a_{12}^L, a_{12}^U]$	...	$[a_{1n}^L, a_{1n}^U]$
S_2	$[a_{21}^L, a_{21}^U]$	$[a_{22}^L, a_{22}^U]$		$[a_{2n}^L, a_{2n}^U]$
$\vdots$	$\vdots$	$\vdots$	...	$\vdots$
S_m	$[a_{m1}^L, a_{m1}^U]$	$[a_{m2}^L, a_{m2}^U]$	...	$[a_{mn}^L, a_{mn}^U]$
Weight vector, W	$[w_1^L, w_1^U]$	$[w_2^L, w_2^U]$	...	$[w_n^L, w_n^U]$

conducted in such a way: firstly, each alternative will be compared with the positive mode as well as with the negative mode, and on this basis, the alternative nearest to the positive mode and furthest from the negative mode will be selected as the best alternative. As for the distance in measuring the 'nearest' or 'furthest', we mean the Euclidean geometric distance. Based on the distance, a measurement called similarity degree can be calculated for each alternative in comparing its attributes to those of the positive mode and the negative mode. And then, all the alternatives can be ranked in accordance with the similarity degrees.

The decision analysis based on the traditional TOPSIS model can be carried out in accordance to the following 7 steps:

(1) Determining the values of all the alternatives under different attributes and then, buiding the initial decision matrix $A = (a_{ij})_{m \times n}$,

(2) To turn the original decision matrix $A = (a_{ij})_{m \times n}$ into the dimensionless matrix $R = (r_{ij})_{m \times n}$ according to the given dimensionless principles,

(3) Building the weighted standard decision matrix $Z = (z_{ij})_{m \times n}$ (If z_{ij} is not in the interval [0, 1], then z_{ij} need to be further standardized),

(4) Giving the positive mode $U = (u_1^+, u_2^+, \ldots, u_n^+)$ and the negative mode $V = (v_1^-, v_2^-, \ldots, v_n^-)$

(5) Calculating the Euclidean geometric distance of each alternative to the positive and negative modes:

$$d_i^+ = \sqrt{\sum_{j=1}^n (z_{ij} - u_j^+)^2}, \quad d_i^- = \sqrt{\sum_{j=1}^n (z_{ij} - v_j^-)^2}$$

$$i = 1, 2, \ldots, m \tag{1}$$

(6) Calculating the similarity degree of each alternative to the ideal modes according to Eq. (2)

$$C_i = \frac{d_i^-}{d_i^+ + d_i^-} \quad 0 \le C_i \le 1, \quad i = 1, 2, \ldots, m \tag{2}$$

(7) Alternative ranking in accordance with the relative similarity degree C_i.

4.3 Improved TOPSIS model

Generally, the above described traditional TOPSIS model can be solved by means of the Lagrangian function. In practical application, some problems may occur if the index weights get zero values. Under such situation, the solution process cannot be realized. Some improvements should, therefore, be made to the traditional TOPSIS model.

4.3.1 Index standardization

In the translation process from the original decision matrix $A = (a_{ij})_{m \times n}$ to the dimensionless matrix $R = (r_{ij})_{m \times n}$, $r_{ij} = [r_{ij}^L, r_{ij}^U]$, a new term 'base point' is introduced which will be determined as follows:

For the efficiency index, $x_j(x_j \ne 0, j = 1, 2, 3, \ldots, n)$ are set as the 'base points' for the comparison process according to Eq. (3),

$$x_j = \max_{i=1}^m \left\{ |a_{ij}^L|, |a_{ij}^U| \right\}. \tag{3}$$

For the cost index, $y_j(y_j \ne 0, j = 1, 2, 3, \ldots, n)$ are set as the 'base points' for comparison process according to Eq. (4),

$$y_j = \min_{i=1}^m \left\{ |a_{ij}^L|, |a_{ij}^U| \right\}. \tag{4}$$

The dimensionless process can be undertaken for both the efficiency index and cost index respectively according to Eqs. (5) and (6),

$$r_{ij}^L = \frac{a_{ij}^L}{x_j}, \quad r_{ij}^U = \frac{a_{ij}^U}{x_j} \tag{5}$$

$$r_{ij}^L = \min \left(\frac{y_j}{a_{ij}^L}, \frac{y_j}{a_{ij}^U} \right), \quad r_{ij}^U = \max \left(\frac{y_j}{a_{ij}^L}, \frac{y_j}{a_{ij}^U} \right) \tag{6}$$

In the practical decision process, both the efficiency index and cost index may occur in the index set. In such a situation, all the cost index would be turned into their reciprocals according to Eq. (7),

$$r_{ij} = \frac{(x_{ij})^{-1}}{\max_{1 \le i \le m} (x_{ij})^{-1}} = \frac{\min_{1 \le i \le m} (x_{ij})}{(x_{ij})} = \frac{x_{ij}^{\min}}{x_{ij}}, \tag{7}$$

where $x_{ij}^{\min} = \min_{1 \le i \le m} (x_{ij}) \quad (1 \le j \le n)$.

And then, the whole index could be considered as 'efficiency index'.

4.3.2 Establishing the weighted normalized decision matrix

According to the information obtained from the first step, the weighted normalized decision matrix can be established as follow,

$$B = \left(b_{ij}\right)_{m \times n}, \quad b_{ij} = \left[w_j^L r_{ij}^L, w_j^U r_{ij}^U\right].$$

4.3.3 Setting the ideal model

Because the interval numbers are composed of upper- and lower-limits, we do not need both positive and negative modes together. In order to avoid redundant calculations in solving the decision problem, the positive mode and the negative mode are combined into one ideal mode as follows:

$$S^* = \left\{s_1^*, s_2^*, \ldots, s_j^*\right\} \quad (j = 1, 2, 3, \ldots, n),$$

where, $s_j^* = \left[s_j^L, s_j^U\right]$,

$$s_j^L = \max_i b_{ij}^L = \max_i w_j^L r_{ij}^L, \quad s_j^U = \max_i b_{ij}^U = \max_i w_j^U r_{ij}^U$$
(8)

Let

$$(r_j^*)^L = \max_i r_{ij}^L, \quad (r_j^*)^U = \max_i r_{ij}^U$$
(9)

and then, Eq. (8) can be expressed as follows:

$$s_j^L = w_j^L (r_j^*)^L, \quad s_j^U = w_j^U (r_j^*)^U.$$
(10)

4.3.4 Calculating the geometric distance and determining the interval weight

The Euclidean geometric distance principle is applied here as in the traditional TOPSIS model. The distance of alternative S_i to the ideal mode S^* is calculated according to the formula (11),

$$d_i = \sqrt{\sum_{j=1}^{n} (b_{ij} - s_j^*)^2}$$

$$= \sqrt{\sum_{j=1}^{n} \left\{\left[w_j^L r_{ij}^L - w_j^L (r_j^*)^L\right]^2 + \left[w_j^U r_{ij}^U - w_j^U (r_j^*)^U\right]^2\right\}}.$$
(11)

The smaller the geometric distance, the better the alternative. According to this principle, an optimization

model can be established for determining the index weights as follows:

$$\min\left[\sum_{i=1}^{m} d_i(w)\right] = \min\left[\sum_{i=1}^{m} \left(\sqrt{\sum_{j=1}^{n} (b_{ij} - s_j^*)^2}\right)\right]$$

$$s.t. \begin{cases} \sum_{j=1}^{n} w_j^L \leq 1 \\ \sum_{j=1}^{n} w_j^U \geq 1 \\ w_j > 0, \quad j = 1, 2, \ldots, n \end{cases}$$
(12)

The model (12) can be turned into its equivalent form [see model (13)].

$$\min[d_i^2(w)] = \min\left[\sum_{i=1}^{m} \left(\sum_{j=1}^{n} (b_{ij} - s_j^*)^2\right)\right]$$

$$= \min\left[\sum_{i=1}^{m} \left(\sum_{j=1}^{n} \left\{\left[w_j^L r_{ij}^L - w_j^L (r_j^*)^L\right]^2 \right.\right.\right.$$
$$\left.\left.\left. + \left[w_j^U r_{ij}^U - w_j^U (r_j^*)^U\right]^2\right\}\right)\right]$$

$$s.t. \begin{cases} \sum_{j=1}^{n} w_j^L \leq 1 \\ \sum_{j=1}^{n} w_j^U \geq 1 \\ w_j > 0, \quad j = 1, 2, \ldots, n \end{cases}$$
(13)

It can be seen that this is a nonlinear optimization model. The Lagrangian function (14) can be introduced for solving the model.

$$L(w, \lambda) = \sum_{i=1}^{m} \left(\sum_{j=1}^{n} \left[\left[r_{ij}^L - (r_j^*)^L\right]^2 + \left[r_{ij}^U - (r_j^*)^U\right]^2\right] [w_j]^2\right)$$
$$+ 2\lambda \left(\sum_{j=1}^{n} [w_j] - [1, 1]\right)$$
(14)

The partial derivative equations can be then easily obtained as follows:

$$\begin{cases} \dfrac{\partial L}{\partial [w_j]} = 2[w_j] \sum_{i=1}^{m} \left[(r_{ij} - r_j^*)^2\right] + 2[\lambda, \lambda] = 0 \\ \dfrac{\partial L}{\partial \lambda} = \sum_{j=1}^{n} [w_j] - [1, 1] = 0 \end{cases}$$
(15)

By solving the model, the following index weight vector can be obtained,

$$W = (w_1, w_2, \ldots, w_n)^T, \quad \text{where,} \ w_j = \left[w_j^L, w_j^U\right].$$

4.3.5 Calculating the comprehensive evaluation value of each alternative

By applying the above obtained index weights, the comprehensive evaluation value of each alternative can be calculated as follows:

$$Z_i(w_j) = [z_i^L, z_i^U], (i = 1, 2, 3, \ldots, m),$$

where

$$z_i^L = \sum_{j=1}^{n} w_j^L r_{ij}^L, \quad z_i^U = \sum_{j=1}^{n} w_j^U r_{ij}^U \qquad (16)$$

4.3.6 Alternative ranking

For an interval MADM problem, the optimal alternative can be found by means of comparing the comprehensive evaluation values obtained from a series of calculations described in step (5). Instead of using the similarity degree principle in the traditional TOPSIS model, the possibility degree principle is used for the alternative ranking in this paper. The ranking method based on the possibility degree is a popular method of dealing with interval number models. The 'possibility degree' is defined as follows.

Suppose that we have two interval numbers, a and b, ($a = [a^L, a^U]$ and $b = [b^L, b^U]$). The lengths of interval numbers are written as $l(a) = a^U - a^L$ and $l(b) = b^U - b^L$. The possibility degree is calculated as follows:

$$P(a \geq b) = \min\left\{\max\left(\frac{a^U - b^L}{l(a) + l(b)}, 0\right), 1\right\} \qquad (17)$$

$$P(a \geq b) = \max\left\{1 - \max\left(\frac{b^U - a^L}{l(a) + l(b)}, 0\right), 0\right\} \qquad (18)$$

The ranking and comparison among more than two interval numbers will be more complicated. Under such situation, we should, firstly, carry out the comparison between each other. A possibility degree matrix P will be, then, obtained by arranging all the results from the comparisons (ref. to formula (19)).

$$P = \begin{pmatrix} 0.5 & p_{12} & \cdots & p_{1n} \\ p_{21} & 0.5 & \cdots & p_{2n} \\ \vdots & \vdots & \vdots & \vdots \\ p_{n1} & p_{n2} & \cdots & 0.5 \end{pmatrix}, \quad \text{where } p_{ij} = P(S_i > S_j)$$
$$(19)$$

Based on the possibility degree matrix, the ranking value of an alternative is calculated according to formula (20). All the alternatives can be ranked by comparing their ranking values.

$$v_i = \frac{\sum_{j=1}^{m} p_{ij} + \frac{n}{2} - 1}{n(n - 1)}, \quad (i = 1, 2, \ldots, n) \qquad (20)$$

5 Application

An oil company had an investment opportunity for development of oil and gas blocks in the Middle East. The company could select some promising blocks from the 13 oil and gas blocks in 8 different countries in the Middle East. The MADM theory and the improved TOPSIS model described above were to be applied for supporting the company's decision.

These 13 blocks (projects)) would be taken as the alternatives and formed the alternative set. S,

S_i = {Ash Sham block, Oude block, Gbeibe block, Abu Al Bukhoosh block, Bunduq block, Darquain block, Masjid-e-Suleiman block, Majnoon block, Rumaila block, Onshore Partitioned Zone block, Block 6 block, Malik block, Mukhaizna block}. More details about the alternatives are shown in Table 3.

The interval weights of the index and the comprehensive evaluation values of the blocks were obtained using the improved TOPSIS model, with the results shown in Tables 4 and 5. As we can see from Table 4, no index has zero-weight, which proves the effectiveness of the model improvement described above in this paper.

Based on the comprehensive evaluation values above, the possibility degrees were calculated using Eqs. (17) and (18). The possibility degree matrix was obtained through arranging all the results from the comparisons between every alternative. According to Eqs. (20), the ranking values of all the alternatives were calculated, and on these bases the project ranking was given. The result is shown in Table 6.

According to the results shown in Table 6, the best five projects (blocks) are Rumaila in Iraq, Abu Al Bukhoosh in United Arab Emirates, Onshore Partitioned Zone in Saudi Arabia, Gbeibe in Syria and Majnoon in Iraq, with ranking values 0.07970, 0.07966, 0.07948, 0.07924 and 0.07889 respectively. The worst two projects (blocks) are Malik block in Yemen and Mukhaizna block in Oman.

As suggestions to the oil company, Iraq Rumaila project (block) should be chosen first, while Malik block in Yemen and Mukhaizna block in Oman could not be considered. Taking the geopolitical and other related factors into account, the company should pay more attention to the blocks in Iran, Saudi Arabia and the United Arab Emirates. In consideration of the situation in Iraq and the instability in Syria, the company might give up the blocks in these two countries, even though these blocks have better ranking values. If decision makers

Table 3 The information of projects (blocks)

No.	Name	Country	No.	Name	Country
1	Ash Sham	Syria	8	Majnoon	Iraq
2	Oude	Syria	9	Rumaila	Iraq
3	Gbeibe	Syria	10	Block 6	Qatar
4	Abu Al Bukhoosh	UAE	11	Malik	Yemen
5	Bunduq	UAE	12	Mukhaizna	Oman
6	Darquain	Iran	13	Onshore Partitioned Zone	Saudi Arabia
7	Masjid-e-Suleiman	Iran			

Table 4 The index weights

Index		w_j^L	w_j^U	Index		w_j^L	w_j^U
Reserves	w_1	0.0032	0.0088	Buried depth	w_{11}	0.1038	0.1186
Reserve abundance	w_2	0.0041	0.011	Cover lithology	w_{12}	0.0834	0.1024
Hydrocarbon source rock thickness	w_3	0.0204	0.0243	Cover layer thickness	w_{13}	0.0291	0.0743
Lithology	w_4	0.0474	0.0703	Crude oil density	w_{14}	0.0127	0.0686
Organic carbon content	w_5	0.1087	0.1226	Peak production	w_{15}	0.0034	0.0094
Hydrocarbon generation peak time	w_6	0.0201	0.0214	Well depth	w_{16}	0.01	0.017
Trap type	w_7	0.0558	0.1124	Expected investment	w_{17}	0.004	0.0104
Reservoir thickness	w_8	0.0207	0.0282	The total cost	w_{18}	0.0042	0.0112
Porosity	w_9	0.1175	0.1499	Expected NPV	w_{19}	0.0054	0.0134
Permeability	w_{10}	0.1137	0.1281	Expected IRR	w_{20}	0.0089	0.0211

Table 5 The comprehensive evaluation values of the alternatives

No.	Name	Value	No.	Name	Value
1	Ash Sham	[0.5470, 0.9803]	8	Majnoon	[0.5311, 1.0015]
2	Oude	[0.5084, 0.9331]	9	Rumaila	[0.5656, 1.0771]
3	Gbeibe	[0.4984, 0.9104]	10	Block 6	[0.5289, 0.9533]
4	Abu Al Bukhoosh	[0.5028, 0.9093]	11	Malik	[0.4922, 0.9573]
5	Bunduq	[0.4636, 0.8964]	12	Mukhaizna	[0.4412, 0.9270]
6	Darquain	[0.5340, 0.9703]	13	Onshore Partitioned Zone	[0.5128, 1.0171]
7	Masjid-e-Suleiman	[0.5660, 1.0602]			

prefer some economic index (such as NPV, etc.), they can calculate the NPVs (expressed in interval numbers) for every block separately according to the interval index weights in Table 4. They can build the possibility degree matrix based on the NPV comparisons. so as to get the final ranking results.

6 Summary and conclusions

Analysis was made on the important attributes (index) to which more attention can be given in overseas oil–gas project selection. The relationships among factors affecting the economic result were described. A new TOPSIS model was presented based on improving the traditional one so as to meet the special requirements from overseas oil and gas project selection. This model was applied in the ranking and selection decision of the oil and gas blocks (projects) in the Middle East. Some innovative work and conclusions are summarized as follows.

6.1 Improvements in theory and method aspect

(a) The index weights are expressed in interval number form instead of in point data form as in the traditional models. Because of some weakness in knowledge and experience, it is very difficult to give a relatively accurate estimate to the weights of attribute or index in the evaluation of overseas oil and gas projects. The interval number is the most suitable form for expressing such uncertainty. The index

Table 6 The result of ranking

No.	Name	Country	Rank value	Ranking
1	Ash Sham	Syria	0.07764	10
2	Oude	Syria	0.07790	8
3	Gbeibe	Syria	0.07924	4
4	Abu Al Bukhoosh	UAE	0.07966	2
5	Bunduq	UAE	0.07732	11
6	Darquain	Iran	0.07807	7
7	Masjid-e-Suleiman	Iran	0.07814	6
8	Majnoon	Iraq	0.07889	5
9	Rumaila	Iraq	0.07970	1
10	Onshore Partitioned Zone	Saudi Arabia	0.07948	3
11	Block 6	Qatar	0.07784	9
12	Malik	Yemen	0.07646	12
13	Mukhaizna	Oman	0.07424	13

weights expressed in interval number form are, therefore, introduced in this paper.

(b) In order to avoid redundant calculations in solving the decision problem, the positive mode and the negative mode are combined into one ideal mode, as the interval numbers are composed of upper- and lower-limits. The closer an alternative is to the ideal mode, the better it is. An alternative will be considered as better in view of the index, when the standardized interval value of the efficiency index is closer to the interval [1,1]. With the cost index, an alternative will be considered as better in view of the index, when the standardized interval value of the cost index is closer to the interval [0, 0].

(c) Instead of using the similarity degree principle, the possibility degree principle is applied in the comparison of the comprehensive evaluation values of the alternatives. The alternative ranking is conducted based on the possibility degree matrix.

6.2 Improvements in application aspect

(a) The application of the MADM theory is expanded from reserve evaluation to the total project economic evaluation in the oil and gas industry.

(b) The introduction of interval numbers into the evaluation and ranking of the overseas oil and gas projects can lead to more reasonable decisions, because interval data can much better accommodate the incompleteness of geological information and the uncertainty of the economic results.

(c) A new tool, the improved TOPSIS model, is provided for the evaluation of overseas oil and gas projects. The users can do the project

evaluation based on the comprehensive values as in the model application in the paper. They can also only select their preferred index in the project evaluation and ranking.

Acknowledgments This work was supported by the National Social Science Foundation key projects (13&ZD159, 11&ZD164).

References

Albayrak E. Using analytic hierarchy process (AHP) to improve human performance. J Intell Manuf. 2004;15:491–503.

Dağdeviren M. Decision making in equipment selection: an integrated approach with AHP and Prometree. J Intell Manuf. 2008;19(4): 397–06.

Fan ZP, Ma J, Zhang Q. An approach to multiple attribute decision making based on Fuzzy preference information on alternatives. Fuzzy Sets Syst. 2002;131:101–6.

Herrera F, Herrera E, Martinez L, et al. Managing non-homogeneous information in group decision making. Eur J Oper Res. 2005; 166(1):115–32.

Hladik M. Solution set characterization of linear interval systems with a specific dependence structure. Reliable Comput. 2007;13(4): 361–74.

Hou B, Gui ZX. Application of multi-attribute and neural network method to hydrocarbon reservoir prediction. Lithol Reserv. 2010;3(22):118–21.

Janic M. Multicriteria evaluation of high-speed rail, transrapid maglev, and air passenger transport in Europe. Transp Plan Technol. 2003;26(6):491–512.

Kong F. On fuzzy multi-attribute decision making theory and its applications in technical economic analysis. Thesis. North China Electric Power University; 2005.

Li DX. The application of fuzzy multi-attribute decision making method to mineral resource exploitation. Thesis. Wuhan University of Technology; 2004.

Liu XF. Research on investment project economic appraisal based on multiple attribute decision-making. Thesis. TianJin University; 2007.

Liu YA. Evaluation of undeveloped oil & gas reserves based on multi-attribute decision-making. Stat Inf Forum. 2010;9(25): 63–8.

Srdjevic B, Medeiros YDP, Faria AS. An objective multi-criteria evaluation of water management scenarios. Water Resour Manag. 2004;18:35–54.

Wang Z, Zhang SN, Kuang JC. Risk decision-making model for oil & gas exploration based on dynamic MAUT. China Min Mag. 2010;19(1):110–3.

2

Laboratory investigation of a new scale inhibitor for preventing calcium carbonate precipitation in oil reservoirs and production equipment

Azizollah Khormali[1] · Dmitry G. Petrakov[1]

Abstract The formation of mineral scale is a complex problem during the oilfield operations. Scale inhibitors are widely used to prevent salt precipitation within reservoirs, in downhole equipment, and in production facilities. The scale inhibitors not only must have high effectiveness to prevent scale formation, but also have good adsorption–desorption characteristics, which determine the operation duration of the scale inhibitors. This work is focused on the development of a new scale inhibitor for preventing calcium carbonate formation in three different synthetic formation waters. Scale inhibition efficiency, optical density of the solution, induction time of calcium carbonate formation, corrosion activity, and adsorption–desorption ability were investigated for the developed scale inhibitor. The optimum concentration of hydrochloric acid in the inhibitor was determined by surface tension measurement on the boundary layer between oil and the aqueous scale inhibitor solution. The results show that the optimum mass percentage of 5 % hydrochloric acid solution in the inhibitor was in the range of 8 % to 10 %. The new scale inhibitor had high efficiency at a concentration of 30 mg/L. The results indicate that the induction period for calcium carbonate nucleation in the presence of the new inhibitor was about 3.5 times longer than the value in the absence of the inhibitors. During the desorption process at reservoir conditions, the number of pore volumes injected into the carbonate core for the developed inhibitor was significantly greater than the volume of a tested industrial inhibitor, showing better adsorption/desorption capacity.

Keywords Scale inhibitor · Desorption · Corrosion activity · Precipitation · Optical density

1 Introduction

Huge amounts of water are injected into reservoirs to maintain the reservoir pressure at the required level, whereby salt deposition occurs as a result of water combination (Fan et al. 2012). As the reservoirs continue to deplete and more and more wells are experiencing increasing high water cut, the scaling problem is aggravated. Besides, there is a need for withdrawal of residual oil, requiring the use of modern technologies to improve oil recovery, including physical and chemical methods, which also stimulate the deposition of salts (Demadis et al. 2007; Awan and Al-Khaledi 2014). The dynamics of gas–liquid mixtures in wellbores, degassing, and various flow rates, which are determined by the flow rate of wells and the construction of the lifting equipment, affect the balance of a salt solution. Therefore, precipitation of salts can occur. This can be promoted by ingress of mechanical impurities, corrosion products as crystallization centers, various chemical treatments, and other mechanisms (Al-Tammar et al. 2014). The main reasons for deposition of salts are the changing of pressure–temperature conditions in the process of production and the incompatibility of injection water and formation water. Salt precipitation can damage the formation by permeability reduction (Moghadasi et al. 2004).

Inorganic salts may be deposited on the inner surface of the oilfield equipment during production. Salt precipitation

✉ Azizollah Khormali
aziz.khormaly.put@gmail.com; khormali@spmi.ru

[1] Department of Oil and Gas Field Development and Operation, Oil and Gas Faculty, National Mineral Resources University (Mining University), Saint Petersburg, Russia 199106

Edited by Yan-Hua Sun

occurs in all operations of wells, but the most negative consequences of scaling occur during oil production using electric submersible pumps (ESPs) (Poynton et al. 2008; Chen et al. 2013). Intense deposition of calcium carbonate on impellers of ESPs results from an increase in the produced fluid temperature, which is caused by the heat generated by the operating submersible motor. Along with salt deposition in wells, intense salt precipitation is observed in the wellhead, oil pipeline gathering, metering devices, facilities for the preparation of oil, and in reservoirs in waterflooding operations (Mackay 2005).

The process of precipitation of calcium carbonate occurs in three stages. In the first step, calcium ions combine with carbonate ions to form calcium carbonate ($CaCO_3$) microparticles. Next, $CaCO_3$ particles combine in microcrystals that serve as crystallization centers for the remainder of the solution. Crystal aggregates grow and precipitate or attach to the walls of the equipment at certain sizes (Tomson et al. 2003; El-Said et al. 2009; Kelland 2011). Calcium carbonate is found in the form of white crystals. The main factor influencing the formation of carbonate deposits is that the formation water must be supersaturated with calcium, carbonate, or bicarbonate ions (Kumar et al. 2010; Bezerra et al. 2013; Mavredaki and Neville 2014). More, inorganic calcium carbonate precipitates from the supersaturated solution of salts as a result of changes in temperature and pressure during oil–water flow in the wellbore. Temperature and pressure have a great effect on $CaCO_3$ formation and precipitation. An increase in temperature causes the outgassing of CO_2 contained in water, raising the pH, and this provokes the calcium carbonate precipitation.

All scaling control technology is divided into either prevention or removal of scaling. Chemical methods for preventing scale formation by use of scale inhibitors are more effective than other possible methods (Lakshmi et al. 2013). Basic technologies of inhibitor injection are divided as follows: reagent delivery into the wellbore and into the formation. Delivery into the wellbore is carried out by means of a dosing pump at the surface, into a given point along the well and the periodic injection into the annulus through aggregators. Scale inhibitor delivery into the formation is done through squeeze treatments via injection wells (in the pressure maintenance system), injection of the inhibitor with the fracturing fluid during fracturing (ScaleFrac), or addition of the inhibitor to the proppant (ScaleProp) (Levanyuk et al. 2012; Khormali and Petrakov 2014).

In this paper, the ongoing development of a new scale inhibitor for preventing calcium carbonate deposition during the production from oil wells is explained and a range of experimental work is described to evaluate the new inhibitor.

2 Experimental

In this study, the following experimental work has been carried out:

- Optimization of the hydrochloric acid (HCl) concentration in the developed scale inhibitor by measuring surface tension on the boundary layer between the oil and the aqueous scale inhibitor solution.
- Jar test (static test for scale inhibition efficiency).
- Compatibility of the scale inhibitors with formation waters.
- Change in absorbance of the aqueous scale inhibitor solution.
- Improvement in induction time of calcium carbonate formation in presence of scale inhibitors.
- Corrosiveness of the scale inhibitors.
- Investigation of adsorption–desorption properties of the scale inhibitors in real core samples.

2.1 Experimental materials

2.1.1 Synthetic formation waters

Three different synthetic formation waters, the ionic composition of which is similar to actual formation waters, are shown in Table 1. Table 1 indicates that the synthetic waters had high concentrations of bicarbonate, carbonate, and calcium ions, which are the main factors of $CaCO_3$ precipitation in water because the formation water must be supersaturated with these ions to precipitate this salt (Chen et al. 2005). The ions were contained in the formation waters in various concentrations and ratios of their mutual concentrations.

There are different formation water classification systems. Sulin's system, among other systems, is more descriptive of petroleum formation waters (Ostroff 1967). According to the Sulin's system, all the synthetic formation waters are a calcium chloride type.

2.1.2 Preparation of aqueous scale inhibitor solution

The following chemical reagents were used as the scale inhibitor components in the developed inhibitor: 1-hydroxyethane-1, 1-diphosphonic acid (HEDP), ammonium chloride, hydrochloric acid, isopropyl alcohol, polyethylene polyamine-N-methylphosphonic acid, and water. In addition, an industrial scale inhibitor, which is based on compounds of phosphorus, was used to compare with the developed inhibitor.

The aqueous scale inhibitor solution was prepared by dissolving the chemical reagents at different mass

Table 1 Characteristics of synthetic formation waters

Synthetic Formation water	pH	Density, mg/m^3	Ion content, mg/L							Total dissolved salts, mg/L
			HCO$_3^-$	CO$_3^{2-}$	Cl$^-$	Ca^{2+}	Mg^{2+}	Na$^+$	K$^+$	
1	6.92	1012E+06	9654	895	11,840	17,194	3489	10,753	647	54,470
2	7.34	1023E+06	1633	1547	30,558	21,469	5287	17,347	518	93,080
3	7.13	1018E+06	22,784	7871	34,772	19,836	4173	33,248	972	123,660

concentrations. Distilled water was used to prepare chemical solutions of scale inhibitor in laboratory experiments. This was done to eliminate the influence on properties of the composition and the results of experiments of ion determination. Medical syringes and high precision laboratory balances were used for the exact values of the reagents masses. This ensures the precision of the required volume of the composition, as well as the precision of concentrations of components, in the preparation of the aqueous scale inhibitor solution.

2.2 Experimental methods

2.2.1 Scale inhibitor performance measurement

The effectiveness of the inhibitor can be evaluated by its effect on real formation water or synthetic formation water. However, the use of synthetic formation water provides a more comprehensive assessment for a specific type of salt (Matty and Tomson 1988; Tortolano et al. 2014). The effectiveness of an inhibitor was evaluated by the mass change of precipitates, which were formed in mineralized water in the presence of inhibitor with respect to water with no inhibitor (Drela et al. 1998). The protective effect of an inhibitor is calculated by the following equation:

$$E = \frac{m_0 - m}{m_0} \times 100 \%, \qquad (1)$$

where E is the scale inhibitor efficiency; m and m_0 are the mass of salt precipitates in water with and without inhibitor, respectively, mg.

Working solutions were prepared for studying the spontaneous process of CaCO$_3$ precipitation in the aqueous solutions. CaCl$_2$ solutions of different concentrations have been used in the experiments. Throughout the experiment, the solutions were mixed constantly with a magnetic stirrer.

To determine the efficiency of the scale inhibitors, two narrow neck flasks of 200 mL were used. The first flask consisted of formation water without inhibitor, and the second consisted of formation water with 30 mg/L scale inhibitor. The flasks were closed by a lid and heated to 80 °C. Then, the solutions were filtered through a 0.45-μm filter paper. The weight of crystals on the filter paper was measured.

2.2.2 Optical density of the inhibitor solution and induction period of CaCO$_3$

The developed inhibitor was evaluated by determining the residual content of scale inhibitors in liquid samples. For this purpose, the concentration of the developed inhibitor in the formation water was determined based on the reaction of the ions with molybdate in an acid medium (MacAdam and Parsons 2004). The absorbance (optical density) of the obtained solutions was measured with a photo colorimeter at a wavelength $\lambda = 540$ nm because the most frequent absorbance of the inhibitor occurred at this wavelength. It was done in cells with a layer thickness of 30 mm at different concentrations of the developed scale inhibitor. The optical density should not exceed one. A control sample was taken as a standard solution. Optical density is defined as the logarithmic ratio of the intensity of the incident light to the intensity of the transmitted light through the aqueous scale inhibitor solution.

Induction time is a major factor in studying the prevention of the deposition of inorganic salts. The induction time was determined by measuring the changes in the intensity of the transmitted light through the formation waters with and without the scale inhibitor. For this purpose, a laser analyzer was used to analyze the microparticle size distribution. In addition, a photoelectric nephelometer-absorptiometer was used to obtain curves, which show the dependency of the light absorption on the testing time. The period, at which the light absorption is constant, displays the induction time of the salt formation.

2.2.3 Corrosion rate of the scale inhibitor

The corrosion activity of the inhibitors was evaluated by the gravimetric method—by measuring weight loss of the metal samples (with dimensions of 1.92 cm × 1 cm × 0.2 cm length, width, and thickness, respectively) before and after soaking in the aqueous inhibitor solution for 72 h at 80 °C. The corrosion rate of samples (density of the steel samples is 7821 kg/m^3) is calculated as follows (Oso-kogwu and Oghenekaro 2012):

$$v_c = 1.12 \times 10^{-3} \times \frac{m_1 - m_2}{St}, \qquad (2)$$

where v_c is the corrosion rate of the used metal sample, mm/year; m_1 and m_2 are the mass of the metal sample before and after the corrosion test, respectively, mg; S is the surface area of the metal sample, m^2; and t the test time, h.

2.2.4 Adsorption–desorption properties of the scale inhibitor under dynamic conditions

The adsorption and desorption ability of the inhibitors was studied by injecting aqueous scale inhibitor solution into the core samples. Figure 1 illustrates an apparatus for injecting formation waters and the aqueous scale inhibitor solution into the core samples at desired reservoir conditions. This apparatus is composed of a core holder, a pressure transducer, dosing pumps for injecting formation waters and inhibitors into the core samples, an oven, a vacuum pump, and a brine collection tank.

The adsorption–desorption characteristics of the scale inhibitors and the volume of the scale inhibitors or formation water injected were evaluated in the core samples using this apparatus. Therefore, a relative concentration of the inhibitor solution at the core outlet during adsorption and desorption processes was measured at different injection volumes. The relative concentration of the inhibitor was defined as a ratio of the inhibitor concentrations after and before injecting the inhibitor solution into the core samples.

The volume of the solution injected is defined as follows:

$$V_{inj} = 100 \times \frac{Qt}{LA\phi}, \qquad (3)$$

where V_{inj} is the volume of the solution injected into the core sample, PV (pore volume of the core sample); Q is the volumetric flow rate of the injected solution, m^3/h; t is the time, h; L is the length of the core sample, m; A is the cross-sectional areas of the core sample, m^2; and ϕ is the porosity, %.

The tests were performed at 80 °C in an open system with a jacketed crystallizer; the exposure time was 72 h. During the tests, a pressure of 8.1 MPa and concentration of 30 mg/L of scale inhibitors were applied. The pH value of the system was 6.9. The flow rate was 0.0006 m^3/h. The core samples used had an average porosity of 20 %, a length of 0.035 m, a cross-sectional area of 0.000625 m^2, and a permeability of 70 mD. The used core samples were carbonate type from an Iranian carbonate reservoir (limestone and dolomite were about 79 % and 13 %, respectively). They were saturated with oil.

3 Results and discussion

3.1 Hydrochloric acid content in the developed scale inhibitor

In the investigation of the scale inhibitor, the optimum mass concentration of 5 % hydrochloric acid (HCl) solution in the inhibitor was determined by measuring the change in the surface tension on the boundary layer between oil and the aqueous scale inhibitor solution. The used crude oil was an Iranian light type, with a density of 852 kg/m^3 under reservoir conditions and an oil viscosity of 0.89 cP at 80 °C. The content of sulfur in the oil was

Fig. 1 Apparatus for injecting the inhibitor and water into a core

about 1.08 wt%. Experimental results of this test are shown in Fig. 2. The figure shows that a significant reduction in the surface tension occurred with an addition of 5 % HCl solution up to 8 %. The surface tension reduced slightly when adding 8–10 mass percent of 5 % HCl solution, indicating a further increase in the mass fraction of 5 % HCl solution practically did not reduce this parameter. It can be concluded that the mass percentage of 5 % HCl solution in the range of 8–10 % is suitable for the preparation of an inhibitor composition for preventing the formation and deposition of $CaCO_3$.

3.2 $CaCO_3$ scaling inhibition under static conditions

3.2.1 Effectiveness of inhibitors and their compatibility with formation waters

In the laboratory, the protective performance of scale inhibitors was evaluated in synthetic formation water because using synthetic formation water, instead of actual formation water, can improve the reproducibility of test results. The synthetic formation water was prepared based on the chemical compositions of the produced water.

Experimental results of the efficiency of inhibitors are listed in Table 2, which reveals that the developed chemical compositions had the necessary protective effect (effectiveness of more than 85 %) for calcium carbonate. Inhibitor No. 1 had the highest effectiveness for preventing calcium carbonate precipitation in all formation waters. The difference between inhibitors No. 1 and No. 2 is the change in the mass fraction of inhibitor components. Inhibitor No. 3 is an industrial inhibitor for preventing calcium carbonate formation, which is commonly used in oilfields. Table 2 clearly illustrates that inhibitor No. 3 could prevent the formation of calcium carbonate up to 87 %.

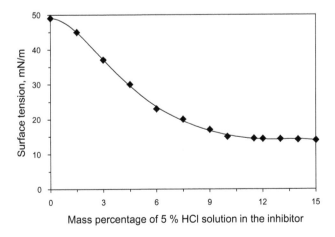

Fig. 2 Surface tension on the boundary layer between oil and the aqueous scale inhibitor solution at different mass concentrations of 5 % HCl solution in the inhibitor

The scale inhibitor should be fully compatible with formation water without any precipitate formation (Dawe and Zhang 1997). Studies have been conducted to determine the compatibility of scale inhibitors with formation waters. All inhibitors were compatible with the three synthetic formation waters, and the experimental results show that all the above chemical compositions could be prepared with formation waters. In case of formation water containing high concentrations of Ca^{2+}, the scale inhibitor is considered compatible with the formation water if turbidity is not observed within 24 h. In addition, the inhibitor effectiveness largely depends on Ca^{2+} concentration in formation waters.

3.2.2 Optical density of the aqueous scale inhibitor solution

To determine the optical density of the developed aqueous scale inhibitor solution, each tested sample was measured with a photo-electrocalorimeter two or three times, and the arithmetic mean was calculated. From the obtained data, a calibration curve was plotted, where the inhibitor concentration in mg/L was plotted on the horizontal axis, and the magnitude of its corresponding optical density on the vertical axis as shown in Fig. 3. The figure displays the high coefficient of determination for the dependency of the optical density on the inhibitor concentration. In this curve, knowing the value of the absorbance, it is possible to find the required scale inhibitor concentration for further development by initiating a new aqueous scale inhibitor solution.

3.2.3 Effect of the scale inhibitor on induction time of $CaCO_3$ scale formation

Figure 4 illustrates the curves of light transmission before and after the inhibition treatment of the salt solution with time. In these curves, transmittance (light absorption) was constant with time from the start of the test until inorganic salt is formed. After salt precipitation, the light absorption reduced with time. The period from the beginning until the salt is deposited is called the induction time. This figure indicates that the induction period of $CaCO_3$ crystallization in the supersaturated aqueous solution was increased in the presence of scale inhibitors. As shown in the figure, inhibitor No. 1 had the longest induction period of all the inhibitors.

3.2.4 Corrosion assessments of scale inhibitors

Scale inhibitor must have low corrosiveness. The corrosion activity of the used scale inhibitors was evaluated through the mass reduction of reference samples after their immersion in the aqueous scale inhibitor solution. A scale

Table 2 Effectiveness of scale inhibitors

Scale inhibitor number	Chemical composition of the scale inhibitor (in mass percent)	Scale inhibition efficiency (in, 30 mg/L of inhibitor), %		
		First water	Second water	Third water
1	HEDP 3 %, ammonium chloride 4 %, polyethylene polyamine-*N*-methylphosphonic acid 4 %, hydrochloric acid 10 %, isopropyl alcohol 2 %, water—remaining	91	90	92
2	HEDP 1 %, ammonium chloride 6 %, polyethylene polyamine-*N*-methylphosphonic acid 2 %, hydrochloric acid 5 %, isopropyl alcohol 6 %, water—remaining	89	87	88
3	Tested inhibitor, which is based on a composite reagent containing phosphorus	87	85	81

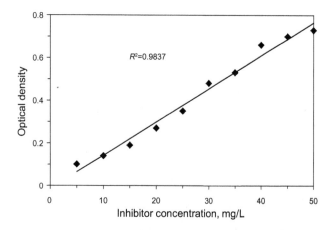

Fig. 3 Change in the optical density of the solution, depending on the inhibitor concentration in water

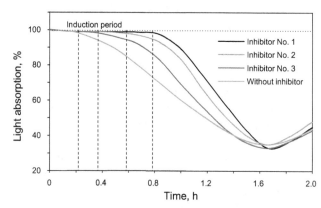

Fig. 4 Dependence of the induction period of calcium carbonate on the curves of light transmission

inhibitor is anticorrosion if pitting does not exist on the surface of the sample and the corrosion rate does not exceed 0.1 mm/year. Table 3 shows that all the chemical compositions exhibited an allowable corrosion rate (less than 0.1 mm/year). Therefore, these reagents can be considered as reagents to prevent scaling in wells.

3.3 Dynamic core testing analysis

Scale inhibitors should have good adsorption–desorption characteristics, heat resistance, and minimal toxicity (García et al. 2006). The scale inhibitor solutions were injected into the carbonate core samples to study adsorption–desorption properties of the scale inhibitors. The results of this study are shown in Figs. 5 and 6. Figure 5 shows the relative concentrations of inhibitors depending on the volume of the inhibitor solution injected into the core sample in the adsorption process at 80 °C and at 8.1 MPa. The relative inhibitor concentrations were determined by a ratio of the current (after injecting) to initial (before injecting) concentrations of the inhibitor. Laboratory studies show that the limiting adsorption value (relative concentration of the inhibitor is 1)

was achieved at 14 PV inhibitor solution injected for inhibitors No. 1 and No. 2, and at 15 PV for inhibitor No. 3. Figure 5 illustrates the achievement to the equilibrium adsorption on the surface of porous media under dynamic conditions after injecting the inhibitors into the core samples. As shown in Fig. 5, inhibitor No. 1 had a more uniform curve of changes in the relative concentration than inhibitors No. 2 and No. 3 during the adsorption process.

Once the core was left for 24 h to attain adsorption equilibrium, the formation water was pumped into the core to displace the inhibitor components. The relative concentrations of inhibitors are shown during the desorption process in Fig. 6. The desorption process can be performed for the optimal concentration of the scale inhibitor in field conditions, when it corresponds to the relative concentration of 0.0001. Figure 6 illustrates that inhibitor No. 1 could effectively protect the precipitation of $CaCO_3$ in the carbonate core sample before 37 PV water was injected into the core in the desorption process. The values for inhibitors No. 2 and No. 3 were 32 PV and 27 PV, respectively. This demonstrates that the developed inhibitor has 1.37 times greater duration of desorption than inhibitor No. 3.

Data on the removal of considered inhibitor components show that a significant portion of the free inhibitor (non-

Table 3 Corrosion rates of scale inhibitors

Scale inhibitor	Test duration, h	First water		Second water		Third water	
		Mass reduction, mg	Corrosion rate, mm/year	Mass reduction, mg	Corrosion rate, mm/year	Mass reduction, mg	Corrosion rate, mm/year
1	72	1.3	0.0404	1.5	0.0467	1.7	0.0529
2	72	1.7	0.0529	1.9	0.0591	2.0	0.0622
3	72	2.1	0.0650	2.1	0.0653	2.2	0.0684

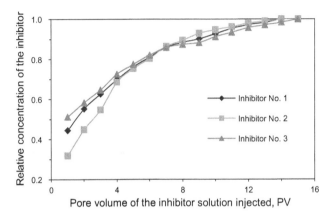

Fig. 5 Change of the inhibitor concentration at the core outlet in the adsorption process

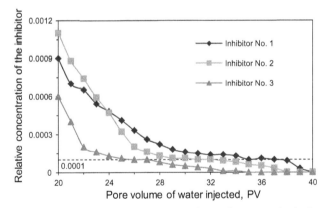

Fig. 6 Change of the inhibitor concentration at the core outlet in the desorption process

adsorbed) passed through the core during pumping the first 2 PV of the inhibitor solution. HCl solution is necessary to maintain the increased acidity of the medium in the solution. HCl solution is also capable of changing the wettability of rocks, and clearing the surface of the oil (the core samples were saturated with the oil). This ensures a uniform and more complete adsorption of the scale inhibitor. During the injection of the scale inhibitor solution into the core samples, the presence of HCl provides an increase in the desorption degree of the inhibitors on the rock surfaces. Thus, the desorption property of the new inhibitor has been

investigated by considering the influence of the acidic components of the inhibitor on the rock. The inhibitor components formed a contact layer between the scale inhibitor and mineral surfaces. Owing to this, desorption occurs slowly in reservoir rocks because the inhibitor layer on the rock surface is substantially resistant to leaching. This characteristic of the inhibitor leads to the fact that the period of desorption is increased, and from this, it can be concluded that the effectiveness of the scale inhibitor is increased. Thus, the developed inhibitor effectively protects the downhole equipment and reservoir from deposition of $CaCO_3$ for a long time.

4 Conclusions

(1) Comprehensive experimental work was carried out to study $CaCO_3$ scaling inhibition by the developed inhibitor in production equipment and carbonate oil reservoirs. Based on the experimental results, it can be concluded that the new scale inhibitor can be used in oilfields.

(2) The mass fractions of the components in the aqueous scale inhibitor solution, at which the developed scale inhibitor was effective for preventing scale formation of $CaCO_3$ up to 92 %, were determined.

(3) Surface tension between the oil and the aqueous scale inhibitor solution was reduced by increasing the mass percentage of 5 % HCl solution in the inhibitor until its value has reached 10 %, after which the surface tension remained constant.

(4) The induction period of $CaCO_3$ scale formation in the supersaturated formation brine water increased in the presence of the scale inhibitor. The corrosion rate of the developed scale inhibitor was in the range of 0.04–0.053 mm/year (less than the maximum allowable corrosion rate—0.1 mm/year).

(5) Adsorption of the developed inhibitor onto the formation rock occurred more rapidly than the used industrial inhibitor. Also, the new inhibitor was more slowly and in larger pore volume of injected synthetic formation water desorbed in comparison with the tested industrial inhibitor.

References

Al-Tammar JI, Bonis M, Salim Y, et al. Saudi Aramco downhole corrosion/scaling operational experience and challenges in HPHT gas condensate producers. In: SPE international oilfield corrosion conference and exhibition, 12–13 May, Aberdeen, Scotland; 2014. doi:10.2118/169618-MS.

Awan MA, Al-Khaledi SM. Chemical treatments practices and philosophies in oilfields. In: SPE international oilfield corrosion conference and exhibition, 12–13 May, Aberdeen, Scotland; 2014. doi:10.2118/169626-MS.

Bezerra MCM, Rosario FF, Rosa KRSA. Scale management in deep and ultradeep water fields. In: OTC Brasil, 29–31 Oct, Rio de Janeiro, Brazil; 2013. doi:10.4043/24508-MS.

Chen T, Neville A, Yuan M. Calcium carbonate scale formation—assessing the initial stages of precipitation and deposition. J Pet Sci Eng. 2005;46(3):185–94. doi:10.1016/j.petrol.2004.12.004.

Chen T, Chen P, Montgomerie H, et al. Scale squeeze treatments in short perforation and high water production ESP wells—application of oilfield scale management toolbox. In: International petroleum technology conference, 26–28 March, Beijing, China; 2013. doi:10.2523/IPTC-16844-MS.

Dawe RA, Zhang Y. Kinetics of calcium carbonate scaling using observations from glass micromodels. J Pet Sci Eng. 1997;18(3–4):179–87. doi:10.1016/S0920-4105(97)00017-X.

Demadis KD, Stathoulopoulou A, Ketsetzi A. Inhibition and control of colloidal silica: can chemical additives untie the "Gordian Knot" of scale formation?. In: Corrosion 2007, 11–15 March, Nashville, Tennessee; 2007. NACE 7058.

Drela I, Falewicz P, Kuczkowska S. New rapid test for evaluation of scale inhibitors. Water Res. 1998;32(10):3188–91. doi:10.1016/S0043-1354(98)00066-9.

El-Said M, Ramzi M, Abdel-Moghny T. Analysis of oilfield waters by ion chromatography to determine the composition of scale deposition. Desalination. 2009;249(2):748–56. doi:10.1016/j.desal.2008.12.061.

Fan C, Kan AT, Zhang P, et al. Scale prediction and inhibition for oil and gas production at high temperature/high pressure. SPE J. 2012;17(2):379–92. doi:10.2118/130690-PA.

García AV, Thomsen K, Stenby EH. Prediction of mineral scale formation in geothermal and oilfield operations using the Extended UNIQUAC model: Part II. Carbonate-scaling minerals. Geothermics. 2006;35(3):239–84. doi:10.1016/j.geothermics.2006.03.001.

Kelland MA. Effect of various cations on the formation of calcium carbonate and barium sulfate scale with and without scale inhibitors. Ind Eng Chem Res. 2011;50(9):5852–61. doi:10.1021/ie2003494.

Khormali A, Petrakov D. Scale inhibition and its effects on the demulsification and corrosion inhibition. IJPGE. 2014;2(1):22–33.

Kumar T, Vishwanatham S, Kundu SS. A laboratory study on pteroyl-L-glutamic acid as a scale prevention inhibitor of calcium carbonate in aqueous solution of synthetic produced water. J Pet Sci Eng. 2010;71(1–2):1–7. doi:10.1016/j.petrol.2009.11.014.

Lakshmi DS, Senthilmurugan B, Drioli E, et al. Application of ionic liquid polymeric microsphere in oil field scale control process. J Pet Sci Eng. 2013;112:69–77. doi:10.1016/j.petrol.2013.09.011.

Levanyuk OV, Overin AM, Sadykov A, et al. A 3-year results of application a combined scale inhibition and hydraulic fracturing treatments using a novel hydraulic fracturing fluid, Russia. In: SPE international conference on oilfield scale, 30–31 May, Aberdeen, Scotland; 2012. doi:10.2118/155243-MS.

MacAdam J, Parsons SA. Calcium carbonate scale formation and control. Rev Environ Sci Bio/Technol. 2004;3(2):159–69. doi:10.1007/s11157-004-3849-1.

Mackay EJ. Scale inhibitor application in injection wells to protect against damage to production wells: when does it work?. In: SPE European formation damage conference, 25–27 May, Scheveningen, The Netherlands; 2005. doi:10.2118/95022-MS.

Matty JM, Tomson MB. Effect of multiple precipitation inhibitors on calcium carbonate nucleation. Appl Geochem. 1988;3(5):549–56. doi:10.1016/0883-2927(88)90026-1.

Mavredaki E, Neville A. Prediction and evaluation of calcium carbonate deposition at surfaces. In: SPE international oilfield scale conference and exhibition, 14–15 May, Aberdeen, Scotland; 2014. doi:10.2118/169796-MS.

Moghadasi J, Müller-Steinhagen H, Jamialahmadi M, et al. Model study on the kinetics of oil field formation damage due to salt precipitation from injection. J Pet Sci Eng. 2004;43(3–4):201–17. doi:10.1016/j.petrol.2004.02.014.

Osokogwu U, Oghenekaro E. Evaluation of corrosion inhibitors effectiveness in oil field production operation. IJSTR. 2012;1(4):19–23.

Ostroff AG. Comparison of some formation water classification systems. AAPG Bull. 1967;51(3):404–16.

Poynton N, Miller A, Konyukhov D, et al. Squeezing scale inhibitors to protect electric submersible pumps in highly fractured, calcium carbonate scaling reservoirs. In: SPE Russian oil and gas technical conference and exhibition, 28–30 Oct, Moscow, Russia; 2008. doi:10.2118/115195-MS.

Tomson MB, Fu G, Watson MA, et al. Mechanisms of mineral scale inhibition. SPE Prod Facil. 2003;18(3):192–99. doi:10.2118/84958-PA.

Tortolano C, Chen T, Chen P, et al. Mechanisms, new test methodology and environmentally acceptable inhibitors for co-deposition of zinc sulfide and calcium carbonate scales for high temperature application. In: SPE international oilfield scale conference and exhibition, 14–15 May, Aberdeen, Scotland; 2014. doi:10.2118/169810-MS.

3

Burial and thermal maturity modeling of the Middle Cretaceous–Early Miocene petroleum system, Iranian sector of the Persian Gulf

Zahra Sadat Mashhadi[1] · Ahmad Reza Rabbani[1] · Mohammad Reza Kamali[2] ·
Maryam Mirshahani[2] · Ahmad Khajehzadeh[2]

Abstract The Cretaceous Kazhdumi and Gurpi formations, Ahmadi Member of the Sarvak Formation, and Paleogene Pabdeh Formation are important source rock candidates of the Middle Cretaceous–Early Miocene petroleum system in the Persian Gulf. This study characterizes generation potential, type of organic matter, and thermal maturity of 262 cutting samples (marls and argillaceous limestones) from these rock units taken from 16 fields in the Iranian sector of the Persian Gulf. In addition, the burial and thermal histories of these source rocks were analyzed by one-dimensional basin modeling. Based on the total organic carbon and genetic potential values, fair hydrocarbon generation potential is suggested for the studied samples. Based on T_{max} and vitrinite reflectance values, the studied samples are thermally immature to mature for hydrocarbon generation. The generated models indicate that studied source rocks are immature in central wells. The Gurpi and Pabdeh formations are immature and the Ahmadi Member and Kazhdumi Formation are early mature in the western wells. The Pabdeh Formation is within the main oil window and other source rocks are at the late oil window in the eastern wells. The hydrocarbon expulsion from the source rocks began after deposition of related caprocks which ensures entrapment and preservation of migrated hydrocarbon.

Keywords Persian Gulf · Kazhdumi Formation · Ahmadi Member · Gurpi Formation · Pabdeh Formation · Middle Cretaceous–Early Miocene petroleum system

1 Introduction

The Persian Gulf and its coastal areas (Fig. 1) contain the largest occurrence of crude oil in the world (Haghi et al. 2013) accounting for two-thirds of the world's proven oil reserves and approximately more than one-third of total proven world gas reserves (Rabbani 2007). Existence of repeated and extensive source rock beds, substantial carbonate and some sandstone reservoirs, excellent regional caprocks, huge anticlinal traps, and continuous sedimentation are the major factors making this region a remarkable area for hydrocarbon accumulations (Rabbani 2008).

The Middle Cretaceous–Early Miocene petroleum system is one of the five petroleum systems of the Zagros foldbelt and the Persian Gulf area (Bordenave and Hegre 2010). The Oligo-Miocene Asmari and Cretaceous Bangestan are the main reservoirs and the Cretaceous Kazhdumi Formation, Ahmadi Member of the Sarvak Formation, Gurpi Formation, and Paleogene Pabdeh Formation are important source rock candidates of this petroleum system. The evaporites of the Gachsaran Formation are cap rocks of this petroleum system. The Kazhdumi and Pabdeh formations are excellent source rocks and the Ahmadi Member and Gurpi Formation have been identified as marginal source rocks in the Dezful embayment (Bordenave and Burwood 1990; Bordenave and Huc 1995; Bordenave 2002; Bordenave and Hegre 2010; Rabbani and Tirtashi 2010; Alizadeh et al. 2012; Opera et al. 2013).

Despite the significant hydrocarbon accumulation in the Middle Cretaceous–Early Miocene petroleum system

✉ Ahmad Reza Rabbani
rabbani@aut.ac.ir

[1] Petroleum Engineering Department, Amirkabir University of Technology, Hafez Street, 15875-4413 Tehran, Iran

[2] Research Institute of Petroleum Industry (RIPI), West Blvd. Azadi Sport Complex, 14665-37 Tehran, Iran

Edited by Jie Hao

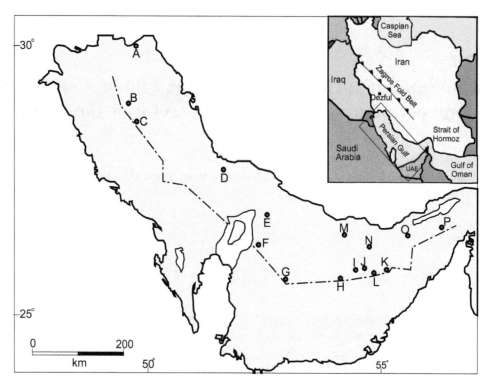

Fig. 1 Location of the studied fields in the Iranian sector of the Persian Gulf

within the Persian Gulf, little is known about the quality and maturity of the potential source rocks of this petroleum system in this area. This study tries to investigate hydrocarbon generation potential, depositional environment, and thermal maturity of the Kazhdumi Formation, Ahmadi Member, Gurpi Formation, and Pabdeh Formation in 16 fields located in the Iranian sector of the Persian Gulf (Fig. 1) by using Rock-Eval pyrolysis, molecular composition, and vitrinite reflectance measurement. Also, 1D basin modeling, a very useful tool in exploration-related studies, was applied to investigate the thermal maturity evolution and timing of hydrocarbon generation of these source rock candidates in the study area. The integration of the results of source rock characterization and basin modeling provides more detailed information to answer exploration questions. Accurate identification of a source rock helps to characterize the petroleum system and predict the location of future prospects charged by that source rock.

2 Geological setting

The Persian Gulf forms the northeast portion of the anti-clockwise-moving Arabian Plate and formed during the Late Miocene (Alavi 2004). The Persian Gulf is situated at the junction of the Arabian and Eurasian lithospheric plates. It is structurally a foreland basin filled by

terrigenous clastics transported from adjacent regions and carbonate sediments generated across the ramp surface (Ghazban 2009). Figure 2 shows the general lithostratigraphic column for the Iranian sector of the Persian Gulf.

During the Paleozoic, the Arabian Plate including the Persian Gulf region was located in the southern hemisphere with predominantly clastic sedimentation (Konert et al. 2001). Afterwards, during the Mesozoic and Cenozoic, the study area was mainly in tropical regions where carbonate deposition prevailed (Murris 1980; Ziegler 2001). Throughout most of the Mesozoic and up to the Lower Miocene, the area was part of a broad, shallow carbonate platform.

The Mesozoic carbonate systems of the Persian Gulf contain most of the extensive reservoir rocks in this area and form one of the richest hydrocarbon provinces in the world. This was mostly due to their vast-scale deposition and presence of source rocks, reservoir, and seal cap facies within the depositional system (Murris 1980). Within this time interval, in the Jurassic and Early Cretaceous periods, maximum marine transgression led to high production of organic matter, and its deposition under anoxic conditions forming organic rich deposits that were transformed over geologic time into petroleum source rocks (Alsharhan and Nairn 1997). Thick evaporites of the Gachsaran Formation, limestones and marls of the Mishan Formation followed by the sandstone, red marls, and siltstones of the Agha Jari Formation characterize the Mio-Pliocene of the region.

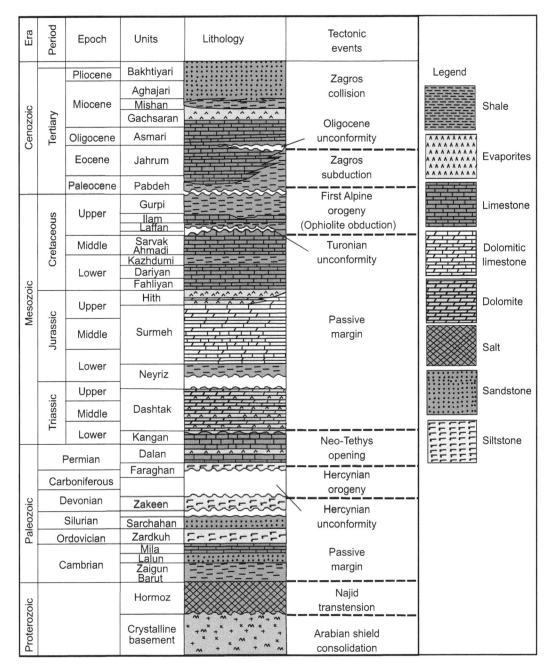

Fig. 2 Generalized stratigraphic column of Iranian sector of the Persian Gulf (modified from Al-Husseini 2008)

Folding accompanied by syntectonic and post-tectonic molasses took place in Plio–Pleistocene (Rabbani 2013).

The morphology of the Persian Gulf is highly affected by the Qatar Arch (Aali et al. 2006). The Qatar Arch is a first-order structure that was created in the central Persian Gulf following the tectonic movements during the Late Precambrian to Early Cambrian in the region (Fig. 3). It is a very large (over 100 km wide and 300 km long) regional gentle anticline (Ziegler 2001). According to offshore seismic data in the study area, this structure has a northeast–southwest direction in the Iranian sector of the Persian Gulf and continues southwards to the Qatar peninsula (Perotti et al. 2011). As Fig. 4 demonstrates, the thicknesses of the Pabdeh, Gurpi and Kazhdumi formations, and Ahmadi Member significantly decrease toward the central parts of the study area with a noticeable thinning which can be due to the effect of the Qatar Arch Paleohigh during depositional time (Alsharhan and Nairn 1997).

Salt diapirism is another significant structural element in the Persian Gulf (<5–20 km in size) piercing the stratigraphic sequences at different levels. Anticlines and domes have been induced by deep-seated salt pillows and salt

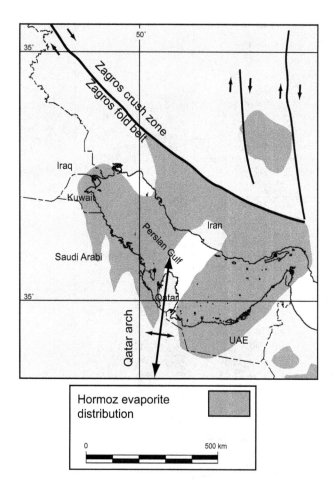

Fig. 3 Location of the Qatar Arch and distribution of Hormoz Salt in the study area (modified after Ghazban and Al-Aasm 2010)

ridges (Kent 1979). Almost all of this salt diapirism originates from the extrusion and remobilization of the Infra-Cambrian Hormoz Salt Series (Edgell 1991). Figure 3 shows the distribution of the Hormoz Salt in the Persian Gulf area.

2.1 Kazhdumi Formation

A transgression and sea-level rise in the Middle Cretaceous resulted in the deposition of the Kazhdumi Formation throughout the Albian (Alsharhan and Kendall 1991). In the Iranian offshore fields, the Kazhdumi Formation consists of calcareous shale and dark bituminous limestone with subordinate argillaceous limestone formed mostly in a neritic environment (Ghazban 2009). In addition, some thin sandstone beds may be present (Ghasemi-Nejad et al. 2009). The Burgan and Nahr Umr formations are the regional equivalents of the Kazhdumi Formation in the Arabian parts of the Persian Gulf (Rahmani et al. 2010). The underlying and overlying formations of the Kazhdumi

Formation are the Sarvak and Dariyan formations, respectively (Rahmani et al. 2013).

2.2 Sarvak Formation

The Sarvak Formation is part of the Bangestan Group and deposited as a result of a significant transgressive phase in the Middle Cretaceous, after regional emergence and periods of clastic and deltaic sedimentation (Alsharhan and Nairn 1997). The carbonates in the Sarvak Formation blanket most of the Persian Gulf area. The bituminous shaly limestone of the Mauddud and Khatiyah members (in the central and western parts of the Persian Gulf), the Ahmadi Member with shaly facies in the northern Persian Gulf, and the Mishrif reefal limestone member in the southern Persian Gulf (Ghazban 2009) are four members of the Sarvak Formation. The Laffan shales overlay the Sarvak Formation with an unconformity surface and act as an efficient regional seal for the Sarvak Reservoir. The Kazhdumi Formation underlies the Sarvak Formation with a transitional contact.

2.3 Gurpi Formation

The Gurpi Formation consists of thin bedded, deep-marine marl, and marly limestone deposited when local dysoxic conditions occurred in the northern Persian Gulf region. The Gurpi Formation can act as a seal for the Ilam reservoir which underlies the Gurpi Formation with an erosional disconformity (Homke et al. 2009). The Upper Aruma and Bahrah–Tayarat are the Gurpi equivalents in the coastal Arabia and Kuwait areas (Rabbani 2013).

2.4 Pabdeh Formation

Neritic to basinal marls and argillaceous limestones of the Pabdeh Formation deposited in a Paleocene–Eocene transgression which resulted from the Late Cretaceous tectonic activities. This formation consists of shale, marl, and argillaceous limestones (Soleimani et al. 2013). A monotonous deep-water shale facies with a limestone unit in its middle part is the main lithology of the Pabdeh Formation. Based on lithological characteristics, the Pabdeh represents deposition in a deep-water, anoxic environment in an overall transgressive sequence. In the northern Persian Gulf, the Asmari and Gurpi formations overlay and underlie the Pabdeh Formation, respectively. There appears to be a transition to clean limestones of the Jahrum Formation toward the southwest (Sharland 2001). Regional equivalents of the Pabdeh Formation are the Umm er Radhuma, Rus, and Dammam formations.

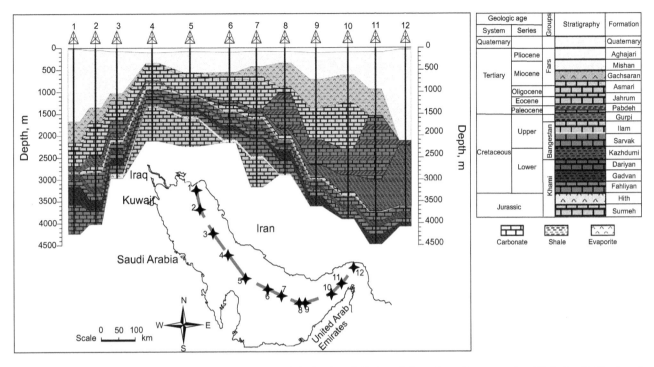

Fig. 4 Lithostratigraphic cross section in the Iranian sector of the Persian Gulf through Jurassic to Quaternary (modified after Rabbani et al. 2014)

3 Materials and methods

A total of 262 samples (including shale, marl, and argillaceous limestone) from the Kazhdumi, Gurpi, and Pabdeh formations and Ahmadi Member were taken from 16 fields within the Iranian sector of the Persian Gulf (Tables 1, 2, 3, 4, 5; Fig. 1). The selected samples were washed with water and detergent several times to remove contaminants from drilling mud additives. Then, the samples were crushed, pulverized, and homogenized. Rock-Eval pyrolysis was performed by a Vinci Rock-Eval 6 instrument in the AGH University of Poland on a 50 mg rock sample. See Espitalié et al. (1977), Lafargue et al. (1998) and Peters et al. (2005) for the details of this method. After completing the pyrolysis, the samples were heated to 850 °C at a rate of 25 °C/min in an oxidation oven and in the presence of air to oxidize (burn) all of the residual carbon. This process generates CO and CO_2 which are measured quantitatively. The parameters measured by this analysis included the total organic carbon (TOC) (wt%), S_1 (mg HC/g rock), S_2 (mg HC/g rock), T_{max} (the temperature at which the S_2 peak is the highest, °C), and S_3 (mg CO_2/g rock). Moreover, the hydrogen index (HI), oxygen index (OI), production index (PI), and migration index (S_1/TOC) were calculated.

Bitumen extractions were performed on approximately 10–15 g of 18 powdered samples from the Kazhdumi,

Gurpi, and Pabdeh formations and also Ahmadi Member by using a Soxhlet apparatus for 72 h with an azeotropic mixture of dichloromethane (DCM) and methanol (CH_3OH) (93:7). The analyzed samples were selected based on higher organic matter content. The extracted bitumen in the rock samples was deasphalted by precipitation with n-hexane. Aliphatic, aromatic, and polar fractions were separated from the deasphalted samples by liquid column chromatography. The saturated fractions in the extracted bitumens were analyzed by gas chromatography (GC) in the AGH University of Poland. A Hewlett Packard 5890 series II GC held at a temperature of 300 °C equipped with a 50 m × 0.2 mm Agilent DB1 column (0.5 μm film thickness) with a constant flow of 0.3 mL/min of nitrogen as a carrier gas was used for this analysis. The column oven was programmed to hold at 30 °C for 5 min and then increase to 320 °C at a rate of 3 °C/min. The oven stays at 320 °C for 20 min. The components eluting the column were detected by a flame ionization detector (FID) held at 325 °C.

The vitrinite reflectance was measured in a random mode according to Taylor et al. (1998) and reported in R_o%. Based on the amount of organic carbon present within the samples, 50 samples were selected for this analysis. The samples were mounted in resin, and then ground and polished using an alumina–ethanol slurry. The analysis was carried out with a Leitz-MPV-SP microscope

Table 1 Rock-Eval and vitrinite reflectance data for the Pabdeh samples

Field	Well	Lith.[a]	MD, m	TOC	T_{max}	S_1	S_1/TOC	S_2	$S_1 + S_2$	S_3	PI	HI	R_o, %
M	M-1	ShMl	1341	0.37	436	0.52	1.41	0.65	1.17	0.87	0.44	176	
		ShMl	1320	0.43	432	0.40	0.93	0.52	0.92	0.97	0.43	121	0.35
	M-2	ShMl	1159	1.59	430	0.11	0.07	2.92	3.03	1.24	0.04	184	0.35
		ShMl	1195	1.36	429	0.12	0.09	2.61	2.73	1.13	0.04	192	
		ShMl	1242	1.25	430	0.17	0.14	1.11	1.28	1.26	0.13		
		ShMl	1280	1.33	431	0.10	0.08	1.46	1.56	1.37	0.06	110	
I	I-1	Ml	2365	0.50	435	0.12	0.24	1.36	1.48	1.65	0.08	272	
		Ml	2355	0.43	433	0.15	0.35	1.37	1.52	1.71	0.10	319	
		Ml	2345	0.31	435	0.10	0.32	0.54	0.64	1.33	0.16	174	
		ShMl	2195	0.33	435	0.10	0.30	0.75	0.85	1.50	0.12	227	
		Ml	2165	0.37	434	0.12	0.32	1.08	1.20	1.54	0.10	292	
		Ml	2135	0.33	434	0.19	0.58	1.39	1.58	1.15	0.12	421	
		Ml	2100	0.39	433	0.12	0.31	1.47	1.59	1.34	0.08	377	
		LimMl	2080	0.34	434	0.10	0.29	0.97	1.07	1.36	0.09	285	
		LimMl	1955	0.55	432	0.12	0.22	2.89	3.01	1.09	0.04	525	
		Ml	1945	0.59	435	0.17	0.29	2.77	2.94	1.08	0.06	469	
		Ml	1895	0.52	433	0.12	0.23	2.24	2.36	0.83	0.05	431	
		Ml	1835	0.58	433	0.16	0.28	2.71	2.87	1.02	0.06	467	
		Ml	1825	0.62	438	0.19	0.31	2.61	2.80	0.96	0.07	421	
		Ml	1805	0.52	431	0.10	0.19	1.87	1.97	1.10	0.05	360	
		Ml	1795	0.46	434	0.13	0.28	1.60	1.73	1.17	0.08	348	
		Ml	1775	0.33	432	0.11	0.33	1.19	1.30	0.92	0.08	361	
		Ml	1765	0.51	434	0.17	0.33	1.57	1.74	0.67	0.10	308	
		Ml	1755	0.63	431	0.18	0.29	1.20	1.38	1.15	0.13	190	
		Ml	1725	0.31	434	0.11	0.35	0.97	1.08	0.86	0.10	313	
		Ml	1705	0.33	430	0.10	0.30	0.96	1.06	0.89	0.09	291	
		Ml	1695	0.45	433	0.08	0.18	0.81	0.89	1.17	0.09	180	
		LimMl	1625	0.33	436	0.14	0.42	0.93	1.07	1.82	0.14	282	
		Ml	1305	0.31	428	0.16	0.52	1.18	1.34	0.78	0.12	381	
	I-2	Ml	1763	0.38	439	0.14	0.37	0.62	0.76	0.70	0.18	163	
K	K-1	Ml	2875	0.35	430	0.13	0.37	0.68	0.81	1.79	0.16	194	
		Ml	2865	0.40	433	0.11	0.28	0.86	0.97	1.25	0.12	215	
		Ml	2715	0.96	425	0.21	0.22	5.20	5.41	1.43	0.04	542	
		Ml	2485	0.53	427	0.11	0.21	1.34	1.45	1.71	0.07	253	
		Ml	2485	0.84	433	0.14	0.17	2.38	2.52	1.91	0.06	283	
		Ml	2455	1.31	433	0.23	0.18	4.54	4.77	1.77	0.05	347	
		Ml	2405	1.44	431	0.26	0.18	5.86	6.12	1.57	0.04	407	
		Ml	2375	1.50	430	0.27	0.18	8.91	9.18	1.66	0.03	594	0.45
		Ml	2298	1.04	434	0.26	0.25	4.69	4.95	1.40	0.05	451	
		Ml	2260	0.74	428	0.20	0.27	2.96	3.16	1.27	0.06	400	
J	J-1	MlSh	2046	0.39	432	0.14	0.36	0.50	0.64	1.06	0.22		
		MlSh	2040	0.71	434	0.29	0.41	1.89	2.18	1.36	0.13		
P	P-3	Ml	2344	0.43	440	0.16	0.37	0.57	0.73	1.46	0.22	132	
		Ml	2329	0.34		0.15	0.44	0.51	0.66	1.45	0.23	149	
		Ml	2316	0.45	433	0.17	0.38	0.55	0.72	1.47	0.24	122	

Table 1 continued

Field	Well	Lith.[a]	MD, m	TOC	T_{max}	S_1	S_1/TOC	S_2	$S_1 + S_2$	S_3	PI	HI	R_o, %
		Ml	2310	0.33	431	0.16	0.49	0.50	0.66	0.96	0.24	152	0.75
		Ml	2272	0.37	431	0.17	0.46	1.33	1.50	1.75	0.11	363	
		Ml	2268	0.33	433	0.18	0.55	0.51	0.69	1.16	0.26	155	
		Ml	2264	0.36	434	0.19	0.52	0.87	1.06	1.49	0.18	240	0.68
		Ml	2252	0.36	430	0.17	0.47	0.62	0.79	0.96	0.22	172	
		Ml	2242	0.43	431	0.19	0.44	1.10	1.29	1.18	0.15	257	
		Ml	2212	0.36	436	0.23	0.64	0.78	1.01	0.94	0.23	216	
		Ml	2204	0.37	431	0.18	0.49	0.61	0.79	1.50	0.23	164	
		Ml	2192	0.44	431	0.21	0.48	1.07	1.28	1.13	0.16	242	
		Ml	2182	0.42	432	0.24	0.57	0.72	0.96	1.35	0.25	172	0.70
		Ml	2172	0.48	436	0.21	0.44	0.93	1.14	1.21	0.18	193	
		Ml	2162	0.41	431	0.20	0.49	0.60	0.80	1.06	0.25	147	
		Ml	2152	0.61	430	0.29	0.47	1.45	1.74	1.36	0.17	237	
		Ml	2143	0.50	429	0.21	0.42	0.70	0.91	1.21	0.23	141	
		Ml	2132	0.45	418	0.30	0.67	1.29	1.59	1.40	0.19	287	0.63
		Ml	2113	0.43	429	0.23	0.53	0.56	0.79	0.94	0.29	130	
		Ml	2106	0.35	434	0.21	0.60	0.62	0.83	1.02	0.25	178	
		Ml	2090	0.30	430	0.18	0.60	0.73	0.91	0.62	0.20	242	0.60
		Ml	2070	0.35		0.16	0.46	0.52	0.68	0.65	0.24	149	0.62
		Ml	2054	0.47	438	0.15	0.32	0.57	0.72	0.49	0.21	122	
		Ml	2046	0.65	437	0.23	0.35	0.90	1.13	2.29	0.20	138	
		Ml	2034	0.45	433	0.22	0.49	0.82	1.04	1.88	0.21	184	
		Ml	2026	0.51	428	0.23	0.45	0.72	0.95	1.89	0.24	140	
		Ml	2014	0.71	420	0.30	0.42	1.19	1.49	1.46	0.20	168	
		Ml	2009	0.66	428	0.19	0.29	0.81	1.00	1.40	0.19	123	
		Ml	2005	0.62	426	0.21	0.34	0.90	1.11	1.17	0.19	145	
		Ml	1996	0.38	428	0.20	0.53	0.50	0.70	1.01	0.29	132	
		Ml	1994	0.70	433	0.29	0.41	1.12	1.41	3.35	0.21	160	0.63
		Ml	1986	0.81	428	0.27	0.33	1.32	1.59	1.83	0.17	162	
		Ml	1978	0.87	423	0.30	0.35	1.55	1.85	2.11	0.16	179	
		Ml	1970	0.75	426	0.29	0.39	1.30	1.59	1.77	0.18	174	
		Ml	1960	0.71	421	0.35	0.49	1.17	1.52	2.05	0.23	164	
		Ml	1946	0.56	426	0.27	0.49	0.74	1.01	1.89	0.27	133	
		Ml	1938	0.63	427	0.25	0.40	1.13	1.38	2.07	0.18	180	
		Ml	1930	0.59	428	0.25	0.42	0.91	1.16	2.15	0.22	153	0.58
		Ml	1922	0.75	426	0.22	0.29	1.16	1.38	1.52	0.16	155	
		Ml	1914	0.70	426	0.22	0.31	0.95	1.17	1.46	0.19	135	
		Ml	1898	0.46	433	0.19	0.41	0.73	0.92	1.38	0.21	158	
		Ml	1887	0.74	426	0.27	0.36	1.34	1.61	1.56	0.17	181	
		Ml	1878	0.76	426	0.28	0.37	1.18	1.46	1.61	0.19	156	
		Ml	1872	1.19	425	0.38	0.32	1.75	2.13	1.53	0.18	147	0.56
		Ml	1856	0.99	423	0.31	0.31	2.04	2.35	1.56	0.13	206	
		Ml	1846	0.98	426	0.31	0.32	1.52	1.83	1.97	0.17	155	
		Ml	1838	0.96	426	0.36	0.38	1.64	2.00	1.95	0.18	172	
		Ml	1819	0.93	425	0.34	0.37	1.45	1.79	1.93	0.19	156	0.54
		Ml	1808	0.83	427	0.41	0.50	1.48	1.89	1.63	0.22	179	
		Ml	1798	0.87	428	0.27	0.31	1.31	1.58	1.65	0.17	150	
		Ml	1788	0.87	436	0.26	0.30	1.18	1.44	1.50	0.18	135	

Table 1 continued

Field	Well	Lith.[a]	MD, m	TOC	T_{max}	S_1	S_1/TOC	S_2	$S_1 + S_2$	S_3	PI	HI	R_o, %
		Ml	1773	1.27	428	0.32	0.25	2.11	2.43	1.86	0.13	166	0.48
		Ml	1736	1.49	431	0.40	0.27	2.66	3.06	1.69	0.13	179	0.46
		Ml	1728	1.47	430	0.38	0.26	2.71	3.09	1.79	0.12	184	
		Ml	1716	1.06	431	0.26	0.24	1.73	1.99	1.51	0.13	163	
		Ml	1702	1.25	429	0.22	0.18	2.16	2.38	1.60	0.09	173	
		Ml	1688	1.37	427	0.41	0.30	4.15	4.56	2.07	0.09	303	
		Ml	1676	0.99	430	0.26	0.26	1.91	2.17	1.80	0.12	194	
		Ml	1668	0.59	429	0.20	0.34	1.25	1.45	1.64	0.14	214	0.36
		Ml	1660	1.01	428	0.27	0.27	2.23	2.50	1.67	0.11	221	
		Ml	1652	0.67	428	0.31	0.46	1.60	1.91	1.71	0.16	239	
		Ml	1642	0.69	428	0.25	0.36	1.21	1.46	1.78	0.17	176	
		Ml	1632	0.83	428	0.32	0.39	1.88	2.20	1.67	0.15	227	
		Ml	1622	1.03	430	0.25	0.24	1.57	1.82	1.86	0.14	152	0.40
		Ml	1608	1.45	428	0.28	0.19	3.09	3.37	1.46	0.08	213	
		Ml	1596	1.56	427	0.27	0.17	3.12	3.39	1.95	0.08	200	0.48
		Ml	1588	1.58	425	0.43	0.27	4.57	5.00	1.65	0.09	290	
		Ml	1580	1.26	425	0.31	0.25	3.15	3.46	1.62	0.09	250	
		Ml	1572	1.66	427	0.68	0.41	5.15	5.83	1.76	0.12	310	
		Ml	1564	0.96	421	0.31	0.32	2.35	2.66	1.77	0.12	245	0.40
		Ml	1556	1.11	426	0.32	0.29	2.59	2.91	1.53	0.11	233	
		Ml	1544	1.07	416	0.40	0.38	2.59	2.99	1.78	0.13	243	
		Ml	1536	1.17	413	0.50	0.43	3.19	3.69	1.47	0.14	273	
		Ml	1528	0.98	404	0.36	0.37	2.33	2.69	1.35	0.13	237	
		Ml	1520	1.25	402	0.67	0.53	3.02	3.69	1.12	0.18	241	
		Ml	1512	1.50	410	0.66	0.44	3.56	4.22	1.59	0.16	238	0.55
O	O-2	Ml	2372	1.11	408	0.80	0.72	3.60	4.40	1.78	0.18	324	
		Ml	2397	1.35	413	0.56	0.41	5.43	5.99	2.05	0.09	402	
		Ml	2423	1.91	423	0.59	0.31	7.66	8.25	2.10	0.07	401	
		Ml	2443	1.98	422	0.46	0.23	7.51	7.97	1.64	0.06	379	
		Ml	2463	1.43	421	0.57	0.40	6.09	6.66	1.92	0.09	426	
		Ml	2488	1.30	427	0.37	0.28	4.00	4.37	3.26	0.08	308	
		Ml	2507	1.99	421	0.40	0.20	8.20	8.60	2.19	0.05	412	
		Ml	2529	2.64	419	0.51	0.19	10.62	11.13	2.69	0.05	402	
		Ml	2550	3.36	419	0.63	0.19	15.55	16.18	1.68	0.04	463	
		Ml	2567	3.00	424	0.49	0.16	13.51	14.00	2.07	0.04	450	
		Ml	2591	2.95	424	0.51	0.17	13.37	13.88	1.56	0.04	453	
		Ml	2608	2.67	425	0.49	0.18	12.12	12.61	1.47	0.04	454	0.56
		Ml	2658	0.86	428	0.19	0.22	2.43	2.62	1.91	0.07	283	
		Ml	2688	1.01	429	0.18	0.18	2.06	2.24	2.40	0.08	204	
		Ml	2702	0.49	429	0.22	0.45	1.22	1.44	1.90	0.15	249	
		Ml	2722	0.55	427	0.15	0.27	0.80	0.95	2.27	0.16	145	0.58
		Ml	2748	3.77	424	0.54	0.14	16.39	16.93	2.00	0.03	435	

The units of the Rock-Eval pyrolysis parameters and indices: TOC wt%, S_1 mg HC/g rock, S_2 mg HC/g rock, $S_1 + S_2$ mg HC/g rock, S_3 mg CO_2/g rock, T_{max} °C, HI mg HC/g TOC, OI mg CO_2/g TOC

[a] ShMl is shaly marl, Ml is marl, Sh is shale, MlSh is marly shale, and LiMl is limy marl

Table 2 Rock-Eval and vitrinite reflectance data for the Gurpi samples

Field	Well	Lith.[a]	MD, m	TOC	T_{max}	S_1	S_1/TOC	S_2	$S_1 + S_2$	S_3	PI	HI	R_o, %
M	M-1	ShMl	1412	0.39	427	0.32	0.82	1.02	1.34	1.07	0.24	262	0.50
		ShMl	1403	0.53	435	0.57	1.08	1.23	1.80	1.10	0.32	233	
		ShMl	1393	0.47	430	0.54	1.15	2.21	2.75	0.88	0.20	470	
		ShMl	1387	0.39		0.49	1.26	0.58	1.07	1.16	0.46	149	0.47
		ShMl	1381	0.66	435	0.43	0.65	3.73	4.16	2.00	0.10	565	
	M-2	ShMl	1396	0.93	429	0.16	0.17	1.27	1.43	0.72	0.11	137	0.48
		ShMl	1426	0.84	426	0.11	0.13	0.96	1.07	0.66	0.10	114	
K	K-1	ShMl	3523	1.91	434	2.07	1.08	7.72	9.79	1.58	0.21	404	
		ShMl	3411	1.95	434	0.58	0.30	8.39	8.97	10.11	0.06	430	
		Ml	3353	1.92	431	0.48	0.25	8.05	8.53	11.56	0.06	419	
		Ml	3250	0.34	428	0.12	0.35	0.70	0.82	1.18	0.15	206	
		ShMl	3238	0.35	429	0.15	0.43	0.89	1.04	1.21	0.14	254	
		Sh	3230	0.74	432	0.12	0.16	1.46	1.58	1.76	0.08	197	
		Sh	3220	0.68	432	0.12	0.18	1.46	1.58	1.50	0.07	215	
		Sh	3210	0.75	431	0.12	0.16	1.48	1.60	1.78	0.08	197	
		Sh	3200	0.76	430	0.11	0.14	1.14	1.25	1.65	0.09	150	
		Sh	3190	0.71	430	0.15	0.21	1.19	1.34	2.04	0.12	168	
		Ml	3175	0.68	431	0.16	0.24	1.35	1.51	2.46	0.10	199	
		Ml	3165	0.67	433	0.13	0.19	1.58	1.71	1.59	0.07	236	
		Ml	3155	0.83	435	0.26	0.31	2.34	2.60	1.71	0.10	282	
		Ml	3145	0.48	434	0.15	0.31	1.39	1.54	1.47	0.10	290	
		Ml	3135	0.38	432	0.12	0.32	1.00	1.12	1.58	0.11	263	
		Ml	3115	0.44	433	0.12	0.27	1.33	1.45	1.53	0.08	302	
		Ml	3105	0.35	435	0.14	0.40	1.27	1.41	1.48	0.10	363	
		Ml	3085	0.60	433	0.16	0.27	1.76	1.92	1.87	0.09	293	
		Ml	3045	0.55	435	0.17	0.31	1.46	1.63	1.50	0.10	265	
J	J-1	MlSh	2348	0.88	435	0.23	0.26	1.07	1.30	1.32	0.18	992	
		MlSh	2345	0.90	433	0.31	0.35	1.14	1.45	1.43	0.21	947	
		MlSh	2340	0.78	430	0.18	0.23	0.75	0.93	1.29	0.19	972	
		MlSh	2335	0.96	433	0.29	0.30	1.31	1.60	1.53	0.18	986	
		MlSh	2140	0.39	426	0.32	0.82	0.60	0.92	1.24	0.35	786	0.55
I	I-2	ShMl	1891	1.62	430.0	0.41	0.25	3.46	3.87		0.11	214	
		ShMl	1942	0.62	434.0	0.21	0.34	0.58	0.79		0.27	94	
		Ml	1980	0.82	432.0	0.21	0.26	1.31	1.52		0.14	160	
	I-1	Sh	2696	0.80	426	0.12	0.15	0.70	0.82	1.50	0.14	88	0.56
		ShMl	2660	1.83	424	0.77	0.42	6.92	7.69	3.67	0.10	378	
		ShMl	2650	1.81	403	0.54	0.30	3.55	4.09	8.21	0.13	196	
		Ml	2640	0.68	413	0.35	0.51	1.43	1.78	3.51	0.20	210	
		ShMl	2625	0.50	432	0.12	0.24	1.38	1.50	2.13	0.08	276	
		ShMl	2605	0.62	433	0.12	0.19	1.97	2.09	2.16	0.06	318	
		ShMl	2595	0.47	433	0.11	0.23	1.33	1.44	1.94	0.08	283	
		LimMl	2585	0.63	430	0.10	0.16	1.06	1.16	1.85	0.08	168	
		Ml	2575	0.50	432	0.09	0.18	1.06	1.15	2.00	0.08	212	

Table 2 continued

Field	Well	Lith.[a]	MD, m	TOC	T_{max}	S_1	S_1/TOC	S_2	$S_1 + S_2$	S_3	PI	HI	R_o, %
		Ml	2565	0.51	434	0.10	0.20	0.96	1.06	1.86	0.10	188	
		Ml	2555	0.50	433	0.24	0.48	1.22	1.46	1.93	0.17	244	
		Ml	2535	0.77	431	0.24	0.31	2.43	2.67	2.10	0.09	316	
		Ml	2525	0.70	432	0.20	0.29	1.98	2.18	1.98	0.09	283	
		Ml	2515	0.69	431	0.20	0.29	1.88	2.08	1.98	0.09	272	
		Ml	2503	0.54	434	0.15	0.28	1.56	1.71	1.96	0.08	289	
		Ml	2495	0.81	432	0.18	0.22	1.74	1.92	2.58	0.09	215	
		Ml	2475	0.98	431	0.36	0.37	2.25	2.61	2.86	0.14	230	
		Ml	2445	1.17	430	0.50	0.43	3.32	3.82	3.81	0.13	284	
		Ml	2413	0.80	430	0.26	0.33	2.80	3.06	1.75	0.08	350	
		Ml	2405	0.84	432	0.22	0.26	2.71	2.93	1.82	0.08	323	
		Ml	2395	0.90	431	0.14	0.16	3.00	3.14	1.77	0.04	333	
		Ml	2385	0.99	431	0.17	0.17	3.13	3.30	1.79	0.05	316	
		Ml	2375	1.05	430	0.22	0.21	3.67	3.89	1.96	0.06	350	
O	O-2	Ml	2985	0.34		0.23	0.68	0.62	0.85		0.27	182	0.65
		Ml	3165	0.63	433	0.48	0.76	0.92	1.40		0.34	146	0.70
		Ml	3362	1.28	439	0.75	0.59	2.06	2.81		0.27	161	0.75

The units of the Rock-Eval pyrolysis parameters and indices: TOC wt%, S_1 mg HC/g rock, S_2 mg HC/g rock, $S_1 + S_2$ mg HC/g rock, S_3 mg CO_2/g rock, T_{max} °C, HI mg HC/g TOC, OI mg CO_2/g TOC

[a] ShMl is shaly marl, Ml is marl, Sh is shale, and MlSh is marly shale

Table 3 Rock-Eval and vitrinite reflectance data for the Ahmadi samples

Field	Well	Lith.[a]	MD, m	TOC	T_{max}	S_1	S_1/TOC	S_2	$S_1 + S_2$	S_3	PI	HI	R_o, %
J	J-1	Ml	2561	1.60	440	1.83	1.14	4.25	6.08	1.38	0.30	842	
		Ml	2561	0.52	432	0.37	0.71	0.61	0.98	1.24	0.38	750	0.67
I	I-2	Ml	2123	1.37	422	0.41	0.30	7.98	8.39	0.53	0.05	582	
		Ml	2138	2.68	424	1.19	0.44	16.94	18.13	0.70	0.07	632	
D	D-1	Ml	1924	0.42		0.08	0.19	0.36	0.44	2.01	0.19	86	0.39
		Ml	1846	0.30		0.11	0.37	0.21	0.32	1.58	0.34	70	0.36
C	C-1	MlSh	1930	1.25	428	0.78	0.62	1.66	2.44	2.15	0.32	133	0.50
		MlSh	1910	1.25	431	0.62	0.50	1.67	2.29	1.93	0.27	134	
		MlSh	1886	1.08	432	0.89	0.82	1.75	2.64	1.48	0.34	162	
		MlSh	1902	1.02	426	0.90	0.88	1.73	2.63	2.23	0.34	170	
B	B-1	ShMl	2520	0.45	432	0.21	0.47	0.64	0.85	0.85	0.25	143	
		ShMl	2510	0.51	433	0.18	0.35	0.70	0.88	0.87	0.20	138	
		MlSh	2490	0.98	435	0.16	0.16	1.11	1.27	1.21	0.13	113	
		MlSh	2450	1.34	425	0.35	0.26	5.42	5.77	1.16	0.06	404	
		Ml	2420	3.86	421	0.96	0.25	14.16	15.12	2.73	0.06	367	
		Ml	2380	0.38	435	0.16	0.42	0.52	0.68	0.73	0.24	138	0.57
O	O-2	Ml	3453	0.44	442	0.46	1.05	1.94	2.40	0.82	0.19	441	0.80
M	M-2	MlSh	1555	0.33	419	0.05	0.15	0.11	0.16	1.09	0.33	33	

The units of the Rock-Eval pyrolysis parameters and indices: TOC wt%, S_1 mg HC/g rock, S_2 mg HC/g rock, $S_1 + S_2$ mg HC/g rock, S_3 mg CO_2/g rock, T_{max} °C, HI mg HC/g TOC, OI mg CO_2/g TOC

[a] Ml is marl, MlSh is marly shale, and ShMl is shaly marl

Table 4 Rock-Eval and vitrinite reflectance data for the Kazhdumi samples

Field	Well	Lith.[a]	MD, m	TOC	T_{max}	S_1	S_1/TOC	S_2	$S_1 + S_2$	S_3	PI	HI	R_o, %
M	M-1	ShMl	1555	0.34	431	0.46	1.35	0.51	0.97	1.11	0.48	150	
	M-2	ShMl	1588	0.30	410	0.05	0.17	0.10	0.15	1.14	0.35	33	
F	F-1	Sh	1468	0.19	416	0.24	1.26	0.16	0.40	0.74	0.60	84	0.50
D	D-1	Ml	2026	0.39	425	0.19	0.49	0.71	0.90	1.77	0.21	182	0.41
C	C-1	LimMl	2108	0.72	422	0.75	1.04	1.00	1.75	2.33	0.43	139	0.47
		Sh	2091	0.86	423	0.85	0.99	1.23	2.08	2.89	0.41	143	
		Sh	2086	1.13	428	0.99	0.88	2.04	3.03	2.94	0.33	181	
		Sh	2070	1.68	428	1.62	0.96	3.47	5.09	1.61	0.32	207	
		Sh	2060	2.83	429	2.24	0.79	6.62	8.86	1.41	0.25	234	0.54
		MlSh	2033	2.79	423	1.48	0.53	6.45	7.93	2.47	0.19	231	0.47
		Sh	2029	2.64	426	1.27	0.48	4.44	5.71	2.84	0.22	168	
		Sh	2022	2.47	426	1.29	0.52	5.08	6.37	2.70	0.20	206	0.51
		Sh	2014	2.19	426	1.26	0.58	4.86	6.12	2.69	0.21	222	
		Sh	2006	1.99	430	1.19	0.60	3.84	5.03	2.34	0.24	193	0.56
		Sh	1998	1.94	430	1.00	0.52	3.49	4.49	2.94	0.22	180	
		Sh	1991	2.13	429	1.20	0.56	3.79	4.99	2.77	0.24	178	0.53
		MlSh	1987	1.33	430	0.86	0.65	2.24	3.10	2.14	0.28	168	0.61
		MlSh	1979	1.01	431	0.79	0.78	1.71	2.50	1.51	0.32	169	0.64
		MlSh	1970	0.70	430	0.75	1.07	1.05	1.80	1.61	0.42	150	
		Ml	1967	0.51	429	0.54	1.06	0.72	1.26	1.78	0.43	141	
		Ml	1952	0.34	425	0.57	1.68	0.62	1.19	1.65	0.48	182	
		Ml	1940	1.16	428	0.72	0.62	1.54	2.26	2.00	0.32	133	
B	B-1	Sh	2970	1.02	437	0.14	0.14	1.08	1.22	0.96	0.11	106	
		Sh	2960	1.17	435	0.20	0.17	1.76	1.96	1.26	0.10	151	0.69
		Sh	2950	1.34	437	0.29	0.22	2.09	2.38	1.32	0.12	157	
		Sh	2930	1.66	438	0.31	0.19	2.21	2.52	1.95	0.12	133	0.68
		Sh	2920	1.38	440	0.18	0.13	1.89	2.07	1.71	0.09	137	
		Sh	2910	0.95	440	0.18	0.19	1.32	1.50	0.92	0.12	138	0.66
		Sh	2900	0.80	461	0.10	0.13	0.50	0.60	1.44	0.17	63	
		MlSh	2890	1.22	434	0.38	0.31	2.25	2.63	1.00	0.14	184	0.66
		MlSh	2880	1.43	427	0.94	0.66	3.67	4.61	0.83	0.20	257	
		MlSh	2870	1.39	425	1.32	0.95	4.62	5.94	0.83	0.22	332	0.65
		MlSh	2860	0.83	435	0.21	0.25	1.34	1.55	1.06	0.14	161	
		MlSh	2850	0.78	434	0.28	0.36	1.27	1.55	1.01	0.18	162	
		MlSh	2840	0.96	433	0.24	0.25	1.74	1.98	1.15	0.12	181	0.64
		MlSh	2830	0.83	437	0.25	0.30	1.12	1.37	1.02	0.18	135	
		MlSh	2820	0.93	427	0.71	0.76	3.10	3.81	0.90	0.19	333	0.62
		MlSh	2800	0.74	436	0.15	0.20	0.44	0.59	1.48	0.25	59	
		MlSh	2790	0.82	436	0.16	0.19	0.90	1.06	1.27	0.15	109	
		Sh	2780	1.03	436	0.20	0.19	1.30	1.50	1.14	0.13	126	0.62
		Sh	2770	1.02	437	0.18	0.18	1.10	1.28	1.16	0.14	107	
		MlSh	2760	1.31	434	0.40	0.31	2.43	2.83	1.59	0.14	185	0.61
		MlSh	2750	0.78	436	0.21	0.27	0.96	1.17	1.07	0.18	124	
		ShMl	2740	0.72	429	0.18	0.25	1.28	1.46	0.85	0.12	178	0.60
O	O-2	ShMl	3548	0.57	436	0.32	0.56	1.89	2.21	2.11	0.14	332	0.80
		ShMl	3595	0.70	464	0.18	0.26	3.81	3.99	1.43	0.05	544	0.81
		ShMl	3603	0.54	463	0.17	0.31	3.21	3.38	0.91	0.05	594	0.82

Table 4 continued

Field	Well	Lith.[a]	MD, m	TOC	T_{max}	S_1	S_1/TOC	S_2	$S_1 + S_2$	S_3	PI	HI	R_o, %
J	J-1	Ml	2612	0.77	423	0.41	0.53	3.97	4.38	0.30	0.09	516	
		Ml	2600	0.60	416	0.60	1.00	0.55	1.15		0.52		0.38
		Ml	2610	0.57	420	0.63	1.11	0.69	1.32		0.48	121	
		Ml	2620	0.71	422	1.06	1.49	1.24	2.30		0.46	175	
		Ml	2670	0.67	412	0.33	0.49	0.56	0.89	2.60	0.37		
		Ml	2676	0.92	429	0.61	0.66	1.31	1.92	2.20	0.32	142	
		Ml	2692	2.41	422	3.39	1.41	8.70	12.09	0.90	0.28	361	
		Ml	2720	1.44	429	0.99	0.69	4.35	5.34	1.20	0.19	302	
I	I-2	Ml	2211	0.77	423	0.41	0.53	3.97	4.38	0.31	0.09	516	
		Ml	2242	0.41	429	0.17	0.41	0.54	0.71	0.34	0.24	132	

The units of the Rock-Eval pyrolysis parameters and indices: TOC wt%, S_1 mg HC/g rock, S_2 mg HC/g rock, $S_1 + S_2$ mg HC/g rock, S_3 mg CO_2/g rock, T_{max} °C, HI mg HC/g TOC, OI mg CO_2/g TOC

[a] ShMl is shaly marl, Ml is marl, Sh is shale, MlSh is marly shale, and LiMl is limy marl

in the organic petrography laboratory of Research Institute of Petroleum Industry (RIPI) in Iran. A sapphire glass standard with a 0.589 % reflectance value was used for calibration. The measurements were performed under reflected light at a wavelength of 546 nm with an oil immersion objective with ×125 magnifications. At least, 50 readings were performed for each sample.

Basin modeling is a useful method for investigating the burial and thermal evolutions of sedimentary basins. In this study, six selected wells were modeled using PetroMod-1D modeling software (version 11, Schlumberger). The selected wells include B-1 (within the Field B, in the western part of the study area), C-1 (within the Field C, in the western part of the study area), D-1 (within the Field D, in the Central Persian Gulf), F-1 (within the Field F, in the Central Persian Gulf), J-1 (within the Field J, in the eastern part of the study area), and P-1 (within the Field P, in the eastern part of the study area). The location of the fields is shown in Fig. 1. Important 1D model input parameters involve the burial depths, thickness of the strata, erosion thickness and time, lithologies, kerogen types and kinetics, and further geochemical parameters such as the initial %TOC and HI (Table 6). The lithological information was inferred from unpublished well log data from National Iranian Oil Company (NIOC). The absolute ages were obtained from the timescale and regional chronostratigraphic subdivisions of Gradstein et al. (2004). The thermal evolution is modeled based on boundary conditions including the sediment water interface temperature (SWIT, in °C), paleo-water depth (PWD, in meter), and heat flow (in mW/m²).

The upper boundary condition for calculating the temperature development in a sedimentary basin is the SWIT (Yalçin et al. 1997). The PetroMod-1D software estimates the SWIT values through time based on the approach developed by Wygrala (1989). This estimation is based on the paleogeographical position of the area through geological time, variations in the mean surface paleo temperatures versus latitude and geological time; and water depth during the time of deposition (Yalçin et al. 1997). The SWIT calculations of this study were based on the paleo-latitude of the Northern Arabian plate.

The PWD values are required to calculate the SWIT. The PWD is dependent on combination of tectonic subsidence and changes in global sea levels. The depositional environment of each formation gives information about the PWD. A PWD of 0 m was considered for erosional events or phases of non-deposition and values of 20 m were applied for times of carbonate deposition. Negative PWD values were not used in this study.

The heat flow is the lower boundary condition of heat transfer into a sedimentary basin (Yalçin et al. 1997). It is an important input parameter in basin modeling and usually difficult to define for the geological past. Therefore, thermal history models are commonly calibrated against maturity and temperature profiles. In this study, bottom-hole temperature and vitrinite reflectance data were used for the temperature and maturity calibrations, respectively. Recently, heat flow values in the range of 60–68 mW/m² were shown to be in accordance with the vitrinite reflectance measurements in the central Persian Gulf (Mohsenian et al. 2014). The easy %R_o kinetic model of Sweeney and Burnham (1990) was applied to calculate the thermal maturity levels of the studied formations. Petroleum generation stages were calculated assuming mainly Type II kerogen and using a reaction kinetic dataset based on Burnham (1989).

Table 5 Isoprenoids ratios measured for the studied samples

Field	Well	Formation	Lith.	Depth, m	Pr/Ph	Pr/n-C_{17}	Ph/n-C_{18}
I	I-1	Gurpi	Ml	2445	0.42	0.41	0.51
	I-1	Gurpi	Ml	2395	0.28	0.51	0.60
L	L-1	Gurpi	MlSh	2857	0.36	0.55	0.77
	L-1	Gurpi	ShMl	2767	0.31	0.51	0.73
H	H-1	Gurpi	Sh	2199	0.12	0.44	0.69
E	E-1	Kazhdumi	Sh	1631	0.20	0.56	0.94
G	G-1	Kazhdumi	ShMl	1504	0.30	0.36	0.63
N	N-1	Kazhdumi	Sh	2000	0.56	0.84	1.42
A	A-1	Kazhdumi	Sh	3070	0.41	0.44	0.79
C	C-1	Kazhdumi	Sh-MlSh	2048	0.42	0.62	0.94
	C-1	Kazhdumi	Sh	1999	0.59	0.53	0.94
	C-1	Kazhdumi	MlSh	1982	0.34	0.58	0.97
L	L-1	Pabdeh	Ml	2238	0.13	0.23	0.61
	L-1	Pabdeh	Ml	2137	0.39	0.55	0.81
	L-1	Pabdeh	Ml	1968	0.28	0.55	0.82
	L-1	Ahmadi Mbr.	MlSh	3148	0.27	0.38	0.59
H	H-1	Ahmadi Mbr.	Sh	2574	0.12	0.37	0.80
G	G-1	Ahmadi Mbr.	Ml	1412	0.11	0.54	0.60

Table 6 Examples of input data for burial, thermal maturity, and hydrocarbon generation modeling in well P-1

Formation	Top, m	Base, m	Thickness, m	Eroded, m	Deposition from, Ma	Deposition to, Ma	Eroded from, Ma	Eroded to, Ma	Lithology
Bakhtiyari	70	100	30	0	10.87	1.5	1.5	0	Sandstone and marl
Upper Fars	100	914	814		13.3	10.87			Marl and Lime-Marly
Lower Fars	914	1200	286		20	13.3			Evaporite
Asmari	1200	1514	314		23	20			Limestone
Pabdeh	1514	2972	1458	200	59.33	35	35	23	Limestone and marl
Gurpi	2972	3695	723	200	81.01	64.02	64.02	59.33	Limestone and marl
Ilam	3695	3715	20		83.43	81.01			Limestone
Laffan	3715	3716	2		88.29	83.43			Shale
Sarvak-Mishrif	3716	3799	83	50	92.97	90.71	90.71	88.29	Limestone
Sarvak-Ahmadi	3799	3827	28		98	95			Limestone and marl
Sarvak-Mauddud	3827	3912	85		102.77	98			Limestone
Kazhdumi	3912	4045	133		115	102.77			Limestone and marl

4 Results and discussions

4.1 Rock-Eval data

Tables 1, 2, 3 and 4 and Fig. 5 show the results of the Rock-Eval pyrolysis in the studied wells. The cross-plot of S_1 versus TOC discriminates the nonindigenous and indigenous nature of the hydrocarbons present in the source rock samples (Hunt 1996). A migration index (S_1/TOC) greater than 1.5 reveals that migrated hydrocarbons affected the samples, whereas an index less than 1.5 points to an indigenous nature for the hydrocarbons. All of the studied samples have migration indices lower than 1.5, indicating that the analyzed samples were not polluted by migrated hydrocarbons (Fig. 6; Tables 1, 2, 3, 4).

The TOC content and genetic potential (summation of S_1 and S_2 peaks of Rock-Eval pyrolysis) provide important information about hydrocarbon generation potential of the source rocks. The TOC contents of the studied samples are in the range of 0.2–3.86 wt% with values generally lower

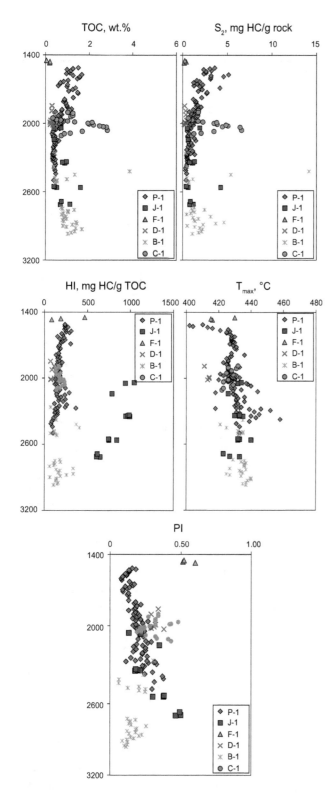

Fig. 5 Geochemical logs showing Rock-Eval data in the studied wells

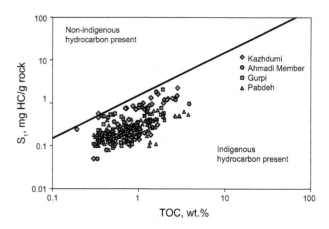

Fig. 6 Plot of S_1 versus TOC for distinguishing between indigenous and nonindigenous hydrocarbons present in the samples in the Kazhdumi (*right*), Gurpi (*middle*), and Pabdeh (*left*) formations (after Hunt 1996). The inclined line represents S_1/TOC = 1.5

values measured for the studied samples are in the range of 0.5–17 mg HC/g rock with average of 2.5 mg HC/g rock. The cross-plots of genetic potential versus TOC (Fig. 7a) and S_2 versus TOC (Fig. 7b) indicate that the studied samples can be generally regarded as having a fair generative potential of hydrocarbon. HI values are in the range of 33–991 mg HC/g TOC (Tables 1, 2, 3, 4), and fair petroleum generation potential of the studied samples is also evident by cross-plot of HI versus TOC (Fig. 7c). Generally, the studied samples show lower TOC, genetic potential, and HI values in the central wells of the Persian Gulf (Tables 1, 2, 3, 4). This is in agreement with increasing the thickness of the studied source rocks from the Central Persian Gulf toward the eastern and western parts (Fig. 4). The occurrence of uplift in the Central Persian Gulf due to the presence of the Qatar Arch resulted in relatively poor preservation of organic matter in this part of the Persian Gulf compared to the adjacent areas.

The type of organic matter present in the source rocks can be evaluated based on the modified Van Krevelen diagram of HI versus T_{max}. The analyzed samples mainly plotted in the zone of mixed Type II–III kerogens and Type III kerogen of this diagram (Fig. 8a). Moreover, in the S_2–TOC plot (Fig. 8b), most of the samples fall in the zone of mixed Types II–III kerogens grading to Type III. The local source rock evaluation study of Ghasemi-Nejad et al. (2009) in the South Pars Field also reveals this type of kerogen for the Kazhdumi Formation. This type of kerogen may originate from mixtures of terrigenous and marine organic matter with varying oil and gas generation potential or may also originally be a marine Type II organic matter which has partially been oxidized during deposition. As shown in the following section, the latter interpretation is the more likely.

than 2 wt% (Tables 1, 2, 3, 4). The genetic potential varies between 0.2 and 18.13 mg HC/g rock with values mostly lower than 6 mg HC/g rock (Tables 1, 2, 3, 4). The S_2

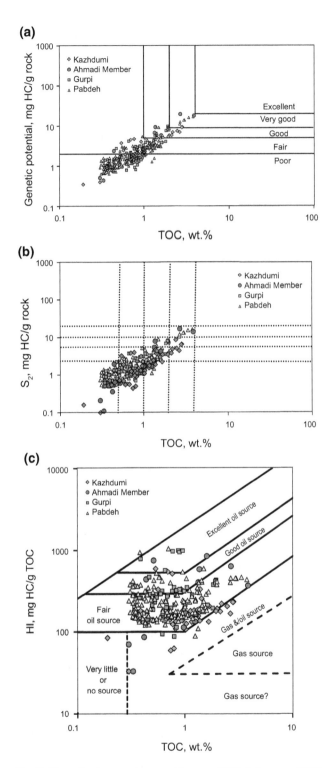

Fig. 7 Plots of **a** the genetic potential versus TOC, **b** S_2 versus TOC, and **c** HI versus TOC diagram showing source rock quality of the studied samples

4.2 Normal alkane and isoprenoids

All of the studied rock samples illustrate unimodal normal alkane distribution patterns, typically ranging from n-C_{15}

to n-C_{34} (Fig. 9). Normal alkanes less than C_{14} are absent, probably because of evaporative loss during sample preparation. The maximum peak is generally detected in the range from n-C_{17} to n-C_{27}. This normal alkane distribution pattern is characteristic of source rocks with strong input from marine organic matter. The pristane to phytane ratio (Pr/Ph) is considered as an indicator of the redox condition of the depositional environment. Low Pr/Ph ratios (<1) reflect an anoxic depositional environment, while greater values reveal more oxic conditions (Peters et al. 2005). In the studied samples, phytane is dominant over pristane and the Pr/Ph ratio displays values lower than 0.6 (Table 5). These values suggest marine reducing depositional conditions which is verified by Pr/n-C_{17} lower than 1 and Ph/n-C_{18} greater than 0.5 (Table 5). In the cross-plot of Pr/n-C_{17} versus Ph/n-C_{18}, the studied samples fall in the zone of marine Type II kerogen deposited under reducing conditions (Fig. 10).

4.3 Thermal maturity

The evaluation of thermal maturity of organic matter in the studied samples was carried out using vitrinite reflectance ($\%R_o$), pyrolysis T_{max}, and production index (PI) values. All of the Rock-Eval S_2 values are greater than 0.5 mg HC/g rock (Tables 1, 2, 3, 4), so T_{max} is reliable for thermal maturity evaluation (Tissot and Welte 1984). The measured vitrinite reflectance values are in good agreement with pyrolysis T_{max} data (Fig. 11). The mean vitrinite reflectance values of the studied samples are in the range of 0.35–0.82 $\%R_o$ (Tables 1, 2, 3, 4) showing thermally immature to mature stage of hydrocarbon generation in the analyzed samples. This is supported by T_{max} and PI values in the range of 402–464 °C and 0.03–0.35, respectively (Tables 1, 2, 3, 4). Thermal maturity has also been estimated by HI–T_{max} plot which indicates that the studied samples contain immature to late mature organic matter (Fig. 8a). The lowest vitrinite reflectance, T_{max}, and PI values were recorded in the central wells, while the highest values were measured for the samples from the eastern wells (Tables 1, 2, 3, 4). Generally, the Kazhdumi Formation and Ahmadi Member with older age and deeper burial are more thermally mature than the Pabdeh and Gurpi formations.

4.4 Burial and thermal history modeling

To generate reliable burial and thermal history models, the accurate timing and duration of erosional events should be fully constrained. As a consequence of eustatic sea-level changes and epeirogenic movements, several regional unconformities, erosion, and hiatuses occurred through the sedimentary succession of the study area (Sharland 2001).

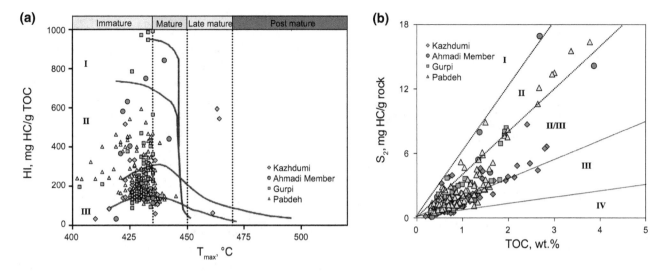

Fig. 8 a Modified Van Krevelen diagram of HI versus T_{max} and **b** S_2 versus TOC for the analyzed samples

In the Mid-Late Cretaceous and then in the Cenozoic, hiatus and erosion affected the studied area as a consequence of collision of the Arabian and Eurasian plates forming the Zagros Mountains (Zagros orogeny). The widespread Turonian unconformity occurred in the study area as a result of localized uplift (following the initiation of ophiolite obduction along the northeastern plate margin) and a global eustatic fall in sea level. The upper part of the Sarvak Formation is sometimes absent due to erosion during the Turonian unconformity. The pre-Neogene sediments are one of the most important erosional surfaces in the Tertiary sequences in the study area. Tectonics and eustasy integrated to cause a major relative sea-level fall through the Oligocene time resulting in widespread erosion and non-deposition across the entire region (Ghazban 2009). In general, total erosions between 50 and 200 m were considered in our models (Table 6). However, after carrying out a comprehensive sensitivity analysis and trying out several scenarios, the amount of erosions was found to have negligible impact on the present-day maturity and temperature trends. The results of the best fit models are presented here.

4.4.1 Well P-1, within the Field P

In well P-1, a constant heat flow value of 63 mW/m^2 (from the Late Cretaceous onward) gives the best fit between the measured and calculated vitrinite reflectance and bottom-hole temperatures (Fig. 12a). Compared to other wells, the studied source rocks have attained higher levels of maturity in the well P-1 possibly because of the deeper burial. The Ahmadi Member and Kazhdumi Formation, with maximum burial depths greater than 4000 m, are at the late oil window in this well with maximum burial temperatures of 152 and 160 °C and calculated vitrinite reflectance values of 1.2 %R_o and 1.25 %R_o, respectively (Figs. 13, 14). The onset of the oil window (0.55–0.7 %R_o) was in the Early Eocene (49 Ma) at a depth of approximately 1800 m (Fig. 12a). Within the Middle-Late Eocene (38 Ma) and at a depth greater than 2500 m, these source rocks entered the main oil window (0.7–1 %R_o), and during the Late Miocene time (8 Ma), they reached the late oil window (1–1.3 %R_o). The transformation ratio (TR), which is defined as the ratio of generated hydrocarbons to the total generation potential of a source rock (Shalaby et al. 2011), reached 96 % for the Kazhdumi Formation and Ahmadi Member (Figs. 13, 14). The Gurpi Formation has just reached the late oil window with a maximum burial temperature of 143 °C and calculated vitrinite reflectance of 1 %R_o (Figs. 12a, 15). Hydrocarbon generation begun from the late Middle Eocene (40 Ma) at a burial depth of approximately 1800 m. Main oil generation occurred during the Late Oligocene (24 Ma) at a depth of approximately 2400 m. The Pabdeh Formation with a maximum burial temperature of about 124 °C and calculated vitrinite reflectance of 0.8 %R_o is interpreted to be within the main oil generation window (Figs. 12a, 16). Oil generation began in the Oligocene (30 Ma) at a burial depth of

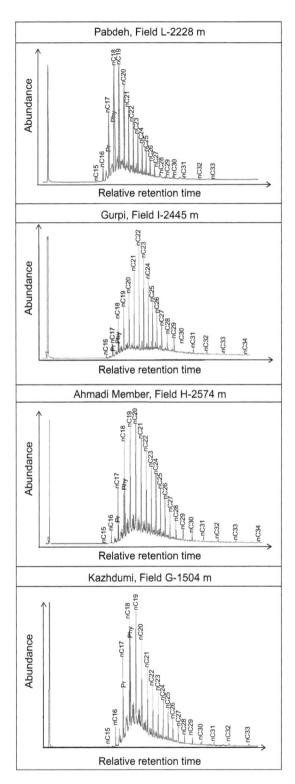

Fig. 9 Examples of gas chromatograms for the selected samples

approximately 1800 m and the main oil window occurred in the Late Miocene (7 Ma). The TR reached 84 % and 40 % for the Gurpi and Pabdeh formations, respectively.

4.4.2 Well J-1, within Field J

In well J-1, a constant heat flow value of 72 mW/m^2 (from the Late Cretaceous onward) gives the best fit between the measured and calculated vitrinite reflectance and bottom-hole temperatures (Fig. 12b). Based on the burial/thermal history model, the Kazhdumi Formation and Ahmadi Member are in the main oil window in this well with maximum burial temperatures and calculated vitrinite reflectance values of approximately 115 °C and 0.75 %R_o, respectively (Figs. 12b, 13, 14). These source rocks reached the required levels of thermal maturity for the onset of the oil window from the late Middle Eocene at burial depths greater than 1500 m. At a burial depth greater than 2600 m, the Kazhdumi Formation and Ahmadi Member reached the main oil generation window in the Late Miocene and Pliocene, respectively. The TR of these source rocks reached approximately 20 % in this well. The Gurpi and Pabdeh formations are early mature with estimated maximum burial temperature and calculated vitrinite reflectance values of approximately 90 °C and 0.6 %R_o, respectively (Figs. 12b, 15). The oil generation in the Gurpi Formation started from the Late Miocene (9 Ma) at a depth of 2100 m. The Pabdeh Formation has just reached the required thermal maturity for the hydrocarbon generation (Figs. 12b, 16). The generated hydrocarbons by these formations are not significant with TR values lower than 7 %.

4.4.3 Well F-1, within Field F

Well F-1 is modeled with a constant heat flow value of 71 mW/m^2 from the Late Cretaceous onward. This value led to a good match between the measured and calculated vitrinite reflectance and bottom-hole temperatures (Fig. 17a). All of the studied source rocks are thermally immature in this well, with estimated maximum burial temperatures lower than 80 °C and calculated vitrinite reflectance values lower than 0.5 %R_o (Figs. 13, 14, 15, 16).

4.4.4 Well D-1, within Field D

In well D-1, the best fit between the calculated and measured vitrinite reflectance and bottom-hole temperatures was obtained assuming a constant heat flow of 60 mW/m^2 for the Lower Cretaceous onward (Fig. 17b). In this well, the Pabdeh Formation is not present and the facies change to the Jahrum Formation. Other studied rock units are immature in this well with estimated maximum burial temperatures lower than 80 °C and calculated vitrinite reflectance lower than 0.5 %R_o (Figs. 13, 14, 15, 16).

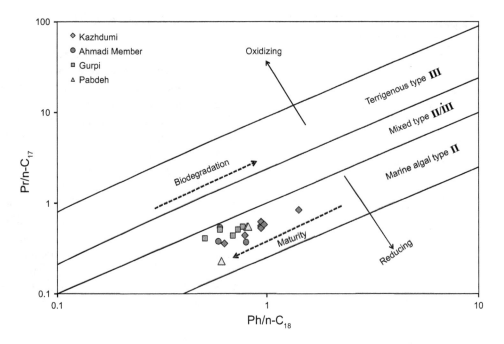

Fig. 10 Cross-plot of Pr/n-C$_{17}$ versus Ph/n-C$_{18}$ for the studied crude samples

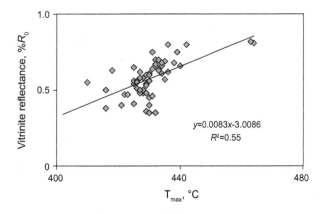

Fig. 11 Plot of T_{max} versus vitrinite reflectance for the studied samples

4.4.5 Well C-1, within Field C

In well C-1, a constant heat flow value of 73 mW/m^2 (from the Late Cretaceous onward) gives the best fit between the measured and calculated vitrinite reflectance and bottom-hole temperatures (Fig. 18a). The Pabdeh Formation changes to the Jahrum Formation and is not present in this well. The Kazhdumi Formation is at an early mature stage, with a maximum burial temperature of approximately 100 °C (occurred at Late Eocene) and calculated vitrinite reflectance of 0.65 %R_o (Fig. 13). The onset of the oil window was in the Middle Eocene (47 Ma) at a depth greater than 1400 m for this

formation (Fig. 18a). The generated hydrocarbon of the Kazhdumi Formation is not significant in this well location with TR lower than 10 % (Fig. 13). The Ahmadi Member and Gurpi Formation are thermally immature with maximum burial temperatures lower than 85 °C and calculated vitrinite reflectance lower than 0.5 %R_o (Figs. 14, 15).

4.4.6 Well B-1, within Field B

Well B-1 was modeled with a constant heat flow value of 72 mW/m^2 from the Late Cretaceous onward. With this value both the temperature and maturity trends have a reasonably good fit with the observed data (Fig. 18b). The Kazhdumi Formation and Ahmadi Member are interpreted to be early mature, with maximum burial temperatures of 106 and 97 °C and calculated vitrinite reflectance values of 0.67 %R_o and 0.62 %R_o, respectively (Figs. 13, 14). The onset of oil generation from the Kazhdumi Formation and Ahmadi Member occurred in the Oligocene (at a burial depth of 1500 m) and Late Miocene (at a depth of 1788 m), respectively. TR values lower than 10 % indicate that the generated hydrocarbons by these source rocks are not significant. The Gurpi and Pabdeh formations, with maximum burial temperatures lower than 90 °C and calculated vitrinite reflectance lower than 0.5 %R_o, are thermally immature (Figs. 15, 16).

As discussed previously, the studied source rocks have fair hydrocarbon generation potential in the study area. In source rocks with fair generation potential, the necessary

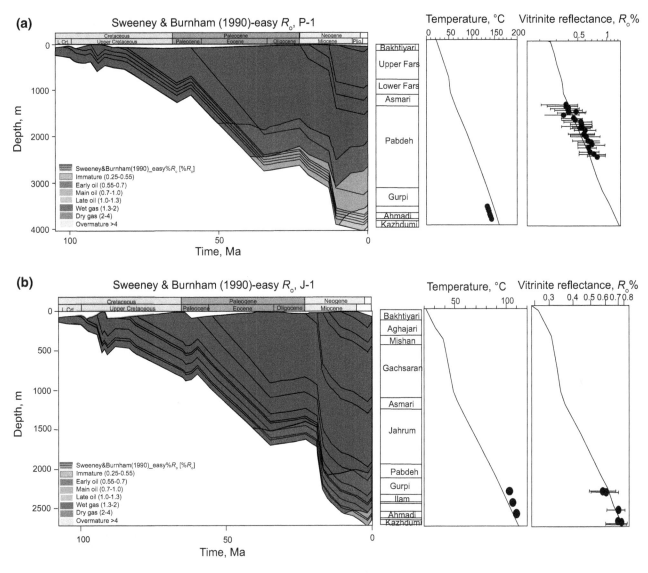

Fig. 12 Burial and thermal modeling of the eastern wells P-1 (*top*) and J-1 (*bottom*). The calibrations of the measured and calculated bottom-hole temperature and vitrinite reflectance data are also shown

TR required to reach the onset of expulsion was estimated to be in the range of 45 %–55 % (Bordenave and Hegre 2010). Only in the well P-1, are the TR values of the Kazhdumi Formation, Ahmadi Member, and Gurpi Formation in the required range for hydrocarbon expulsion when start of the expulsion is in the Early to Late Miocene.

Due to shallower depth of burial and younger age, the Gurpi and Pabdeh formations generally have lower thermal maturity than the Kazhdumi Formation and Ahmadi Member in the study area. All of the studied source rocks are immature in the central wells. The presence of the Qatar Arch and distribution of Hormoz Salt in the Persian Gulf region can be possible reasons for lower thermal maturity in the central parts. The Qatar Arch has deformed the sedimentary cover by an order of magnitude more than

the diapiric structures. A basement high is thus inferred in the core of the Qatar Arch and it has separated the Persian Gulf into northwest and southeast parts (Konert et al. 2001). The presence of this paleohigh at the central part of the Persian Gulf caused different burial depths for the studied rock units in the region such that they have shallower burial depth around the Qatar Arch, while being more deeply buried in the surrounding areas (Alsharhan and Nairn 1997). Lower thermal maturity of the studied formations in the central part can be a result of this lower burial depth in this part of the study area (Fig. 4).

Salt has a thermal conductivity two to four times greater than that of other sedimentary rocks (Bjørlykke 2010). It can have a great impact on the maturity of the organic matter and timing of hydrocarbon generation. Because the

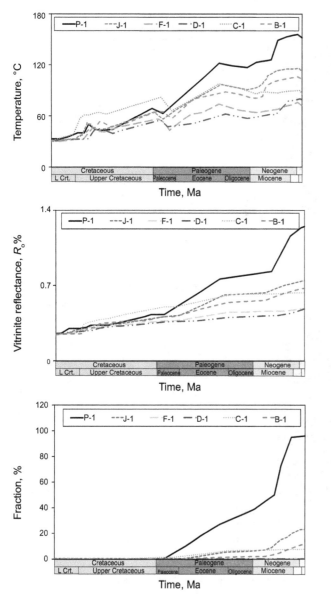

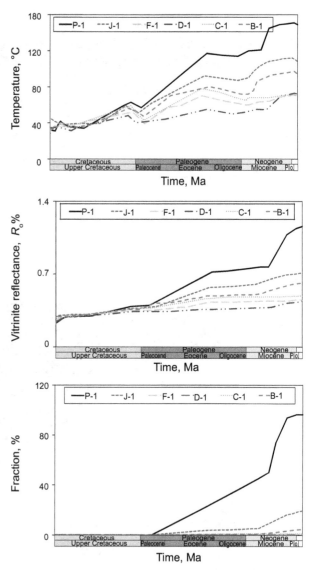

Fig. 13 Evolution of temperature, maturity, and transformation ratio (TR) for the Kazhdumi Formation in the investigated wells of the study area

Fig. 14 Evolution of temperature, maturity, and transformation ratio (TR) for the Ahmadi Member in the investigated wells of the study area

heat is transferred more easily to the source rock units situated above the salt layers, they become more mature compared to adjacent source rocks not affected by salt. The salt diapirs of the Late Proterozoic Hormoz Formation in the area are present only in southeast and northwest of the Qatar Arch, while they are absent around the crest of this arch (Husseini 2000) (Fig. 3). The absence of salt-related phenomena around the crest of the arch is possibly due to the lack or reduced thickness of the Hormoz Formation in this region (Konert et al. 2001). So the presence of the Hormoz Formation in the southeast and northwest of the Qatar Arch can be considered as another possible reason

for higher thermal maturity in these areas relative to the central part influenced by the Qatar Arch.

5 Summary and conclusions

The Kazhdumi Formation, the Ahmadi Member of the Sarvak Formation, and the Gurpi and Pabdeh formations were introduced as source rock candidates of the Middle Cretaceous–Early Miocene petroleum system in the Persian Gulf. In this study, hydrocarbon generation potential, depositional environment, and thermal maturity of 262

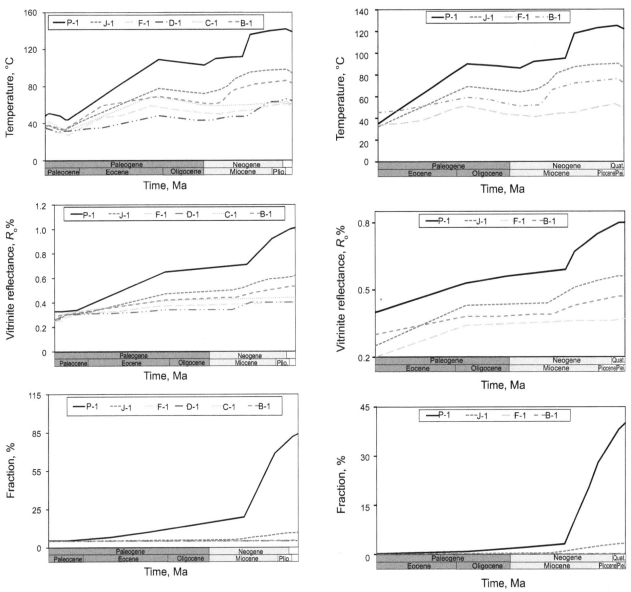

Fig. 15 Evolution of temperature, maturity, and transformation ratio (TR) for the Gurpi Formation in the investigated wells of the study area

Fig. 16 Evolution of temperature, maturity, and transformation ratio (TR) for the Pabdeh Formation in the investigated wells of the study area

cutting samples of these rock units were analyzed in 16 fields located in the Iranian sector of the Persian Gulf. Also, by using PetroMod 1D software, burial and thermal histories were modeled for six selected wells in the study area to analyze the thermal maturity evolution and hydrocarbon generation histories of the Kazhdumi Formation, Ahmadi Member, and the Gurpi and Pabdeh formations. Bottom-hole temperatures and measured vitrinite reflectance values were used for calibration of models.

- Based on Rock-Eval pyrolysis data and normal alkane distribution patterns, the studied source rock candidates

have fair hydrocarbon generation potential and deposited under marine reducing conditions with marine organic matter as the main input.

- Vitrinite reflectance, Rock-Eval pyrolysis T_{max}, and PI values indicate a wide range of maturities between thermally immature to mature for the studied rock units. The Kazhdumi Formation is more thermally mature than the other source rocks. The highest level of maturity is observed in the eastern parts of the study area.

- The constant heat flow values in the range of 63–73 (from the Late Cretaceous onward) give the best fit

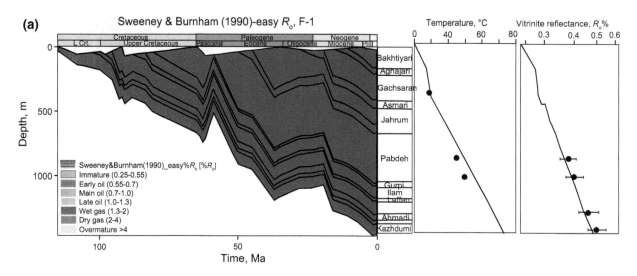

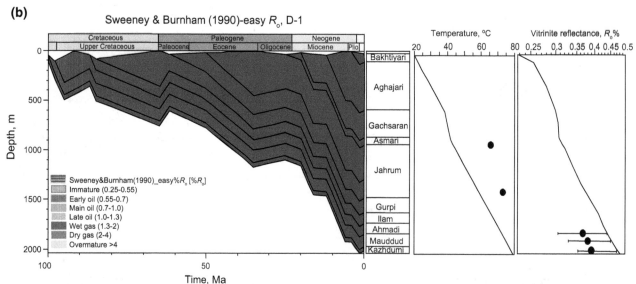

Fig. 17 Burial and thermal modeling of the central wells F-1 (*top*) and D-1 (*bottom*). The calibrations of the measured and calculated bottom-hole temperature and vitrinite reflectance data are also shown

between the measured and calculated bottom-hole temperatures and vitrinite reflectance values in the studied wells.

- The studied source rock candidates are not sufficiently mature for hydrocarbon generation in the central wells (D-1 and F-1).
- The Kazhdumi Formation is early mature in the western wells (B-1 and C-1) and is in the main oil window in the eastern wells (J-1 and P-1). The hydrocarbon generation from the Kazhdumi Formation started from the Early Eocene, whereas the main phase of generation begun during Late Miocene.
- The results of the burial and thermal modeling indicate that the Ahmadi Member is immature in well C-1 and it is early mature in the B-1 well. In the eastern wells (P-1 and J-1), the Ahmadi Member is in the main oil

window. The oil generation from the Ahmadi Member may have begun from Early Eocene and the main oil window occurred in the Late Miocene.

- The Pabdeh and Gurpi formations are thermally immature for hydrocarbon generation in the western wells. They are early mature in well J-1 and are in the main oil window in well P-1. The hydrocarbon generation from the Gurpi Formation started in the Middle Eocene and the main phase of oil generation was in the Late Oligocene. The onset of oil generation from the Pabdeh Formation was in the Oligocene and the main oil window was within the Late Miocene.
- Due to the higher thermal maturity, the Kazhdumi Formation and Ahmadi Member probably have a more significant role in charging the reservoirs of the study area than the Gurpi and Pabdeh formations.

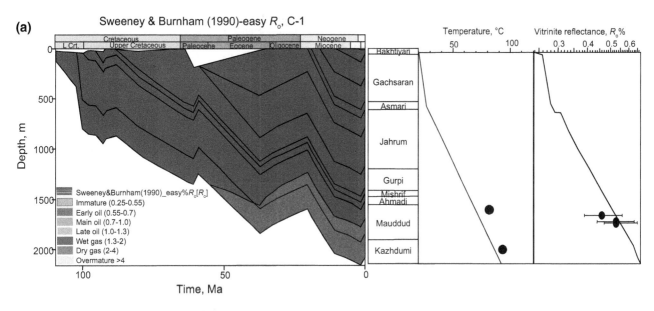

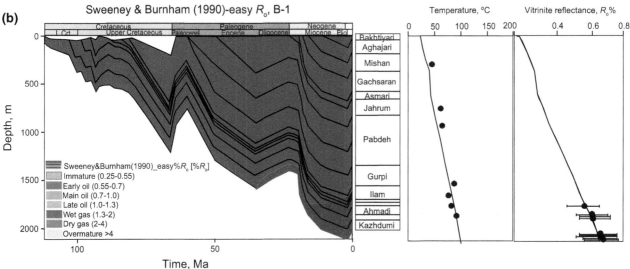

Fig. 18 Burial and thermal modeling of the central wells C-1 (*top*) and B-1 (*bottom*). The calibrations of the measured and calculated bottom-hole temperature and vitrinite reflectance data are also shown

- The onset of oil expulsion in the studied source rocks was after deposition of related cap rocks, which allows the accumulation of generated hydrocarbons in the available reservoirs.

Acknowledgments The authors thank the Research Institute of Petroleum Industry (RIPI) and especially Dr. Mohammad Reza Kamali for providing facilities to complete this study.

References

Aali J, Rahimpour-Bonab H, Kamali MR. Geochemistry and origin of the world's largest gas field from Persian Gulf, Iran. J Pet Sci Eng. 2006;50(3–4):161–75.

Alavi M. Regional stratigraphy of the Zagros fold-thrust belt of Iran and its proforeland evolution. Am J Sci. 2004;304(1):1–20.

Al-Husseini MI. Launch of the Middle East geological time scale. GeoArabia. 2008;13(4):185–8.

Alizadeh B, Sarafdokht H, Rajabi M, Opera A, Janbaz M. Organic geochemistry and petrography of Kazhdumi (Albian–Cenomanian) and Pabdeh (Paleogene) potential source rocks in southern part of the Dezful Embayment, Iran. Org Geochem. 2012;49:36–46.

Alsharhan AS, Kendall CGSC. Cretaceous chronostratigraphy, unconformities and eustatic sealevel changes in the sediments of Abu Dhabi, United Arab Emirates. Cretac Res. 1991;12(4):379–401.

Alsharhan AS, Nairn AEM. Sedimentary basins and petroleum geology of the Middle East. Amsterdam: Elsevier; 1997.

Bjørlykke K. Petroleum geoscience from sedimentary environments to rock physics. Berlin: Springer; 2010.

Bordenave ML. The Middle Cretaceous to Early Miocene petroleum system in the Zagros Domain of Iran, and its Prospect Evaluation. Presented at the Annual Meeting of American Association of Petroleum Geologists, Houston, Texas; 10–13 March 2002.

Bordenave ML, Burwood R. Source rock distribution and maturation in the Zagros Orogenic Belt: provenance of the Asmari and Bangestan Reservoir oil accumulations. Org Geochem. 1990;16(1–3):369–87.

Bordenave ML, Hegre JA. Current distribution of oil and gas fields in the Zagros Fold Belt of Iran and contiguous offshore as the result of the petroleum systems. Geol Soc Lond Spec Publ. 2010;330(1):291–353.

Bordenave ML, Huc AY. The Cretaceous source rocks in the Zagros foothills of Iran. Oil Gas Sci Technol—Rev IFP. 1995;50(6):727–52.

Burnham AK. A simple kinetic model of petroleum formation and cracking. Lawrence Livermore National Laboratory Report UCID-21665 1989.

Edgell HS. Proterozoic salt basins of the Persian Gulf area and their role in hydrocarbon generation. Precambrian Res. 1991;54(1):1–14.

Espitalié J, Laporte JL, Madec M, Marquis F, Leplat P, Paulet J, et al. Méthode rapide de caractérisation des roches mètres, de leur potentiel pétrolier et de leur degré d'évolution. Oil Gas Sci Technol—Rev IFP. 1977;32(1):23–42.

Ghasemi-Nejad E, Head MJ, Naderi M. Palynology and petroleum potential of the Kazhdumi Formation (Cretaceous: Albian–Cenomanian) in the South Pars field, northern Persian Gulf. Mar Pet Geol. 2009;26(6):805–16.

Ghazban F. Petroleum geology of the Persian Gulf. Tehran: Tehran University Press; 2009.

Ghazban F, Al-Aasm IS. Hydrocarbon induced diagenetic dolomite and pyrite formation associated with the Hormoz Island Dome, Offshore Iran. J Pet Geol. 2010;33:183–96.

Gradstein FM, Ogg JG, Smith A, et al. Geologic timescale 2004. Cambridge: Cambridge University Press; 2004.

Haghi AH, Kharrat R, Asef MR, Rezazadegan H. Present-day stress of the central Persian Gulf: implications for drilling and well performance. Tectonophysics. 2013;608:1429–41.

Homke S, Vergés J, Serra-Kiel J, Bernaola G, Sharp I, Garcés M, et al. Late Cretaceous-Paleocene formation of the proto–Zagros foreland basin, Lurestan Province, SW Iran. Geol Soc Am Bull. 2009;121(7–8):963–78.

Hunt JM. Petroleum geochemistry and geology. New York: W.H. Freeman; 1996.

Husseini MI. Origin of the Arabian plate structures: amar collision and Najd rift. GeoArabia. 2000;5(4):527–42.

Kent PE. The emergent Hormuz salt plugs of southern Iran. J Pet Geol. 1979;2(2):117–44.

Konert G, Afifi AM, Al-Hajri SA, Droste HJ. Paleozoic stratigraphy and hydrocarbon habitat of the Arabian Plate. GeoArabia. 2001;6(3):407–42.

Lafargue E, Marquis F, Pillot D. Rock-Eval 6 applications in hydrocarbon exploration, production, and soil contamination studies. Oil Gas Sci Technol—Rev IFP. 1998;53(4):421–37.

Mohsenian E, Fathi-Mobarakabad A, Sachsenhofer RF, Asadi-Eskandar A. 3D basin modelling in the Central Persian Gulf, offshore Iran. J Pet Geol. 2014;37(1):55–70.

Murris RJ. Middle East; stratigraphic evolution and oil habitat. AAPG Bull. 1980;64(5):597–618.

Opera A, Alizadeh B, Sarafdokht H, Janbaz M, Fouladvand R, Heidarifard MH. Burial history reconstruction and thermal maturity modeling for the Middle Cretaceous-Early Miocene petroleum system, southern Dezful Embayment, SW Iran. Int J Coal Geol. 2013;120:1–14.

Perotti CR, Carruba S, Rinaldi M, Bertozzi G, Feltre L, Rahimi M. The qatar—south fars arch development (Arabian Platform, Persian Gulf): insights from seismic interpretation and analogue modelling. In: Schattner U, editor. New frontiers in tectonic research—at the Midst of Plate convergence. Croatia: InTech; 2011. p. 325–52.

Peters KE, Walters CC, Moldowan JM. The biomarker guide. Cambridge: Cambridge University Press; 2005.

Rabbani AR. Petroleum geochemistry, offshore SE Iran. Geochem Int. 2007;45(11):1164–72.

Rabbani AR. Geochemistry of crude oil samples from the Iranian sector of the Persian Gulf. J Pet Geol. 2008;31(3):303–16.

Rabbani AR. Petroleum geology and geochemistry of the Persian Gulf. Tafresh: Tafresh University; 2013.

Rabbani AR, Kotarba MJ, Baniasad AR, Hosseiny E, Wieclaw D. Geochemical characteristics and genetic types of the crude oils from the Iranian sector of the Persian Gulf. Org Geochem. 2014;70:29–43.

Rabbani A, Tirtashi RB. Hydrocarbon source rock evaluation of the super giant Ahwaz oil field, SW Iran. Aust J Basic Appl Sci. 2010;4(5):1–14.

Rahmani O, Aali J, Junin R, Mohseni H, Padmanabhan E, Azdarpour A, Zarza S, Moayyed M, Ghazanfari P. The origin of oil in the Cretaceous succession from the South Pars oil layer of the Persian Gulf. Int J Earth Sci. 2013;102:1337–55.

Rahmani O, Aali J, Mohseni H, Rahimpour-Bonab H, Zalaghaie S. Organic geochemistry of Gadvan and Kazhdumi formations (Cretaceous) in South Pars field, Persian Gulf, Iran. J Pet Sci Eng. 2010;70(1–2):57–66.

Shalaby MR, Hakimi MH, Abdullah WH. Geochemical characteristics and hydrocarbon generation modeling of the Jurassic source rocks in the Shoushan Basin, north Western Desert, Egypt. Mar Pet Geol. 2011;28(9):1611–24.

Sharland PR. Arabian plate sequence stratigraphy. Manama: Gulf PetroLink; 2001.

Soleimani B, Bahadori AR, Meng F. Microbiostratigraphy, microfacies and sequence stratigraphy of upper cretaceous and paleogene sediments, Hendijan oilfield, Northwest of Persian Gulf, Iran. Nat Sci. 2013;5(11):1165–82.

Sweeney J, Burnham AK. Evaluation of a simple model of vitrinite reflectance based on chemical kinetics. AAPG Bull. 1990;74(10):1559–70.

Taylor GH, Teichmüller M, Davis A, Diessel CFK, Littke R, Robert P. Organic petrology. Berlin: Gebrüder Borntraeger; 1998.

Tissot BP, Welte DH. Petroleum formation and occurrence. New York: Springer; 1984.

Wygrala BP. Integrated study of an oil field in the Southern Po Basin, Northern Italy. Zentralbibliothek d Kernforschungsanlage. 1989;2313:1–217.

Yalçin MN, Littke R, Sachsenhofer RF. Thermal history of sedimentary basins. In: Welte D, Horsfield B, Baker D, editors. Petroleum and basin evolution. Berlin: Springer; 1997. p. 71–167.

Ziegler MA. Late Permian to Holocene paleofacies evolution of the Arabian Plate and its hydrocarbon occurrences. GeoArabia. 2001;6(3):445–504.

4

An economic evaluation method of coalbed methane resources during the target selection phase of exploration

Dong-Kun Luo[1] · Liang-Yu Xia[1]

Abstract Because forecasting a development program during the target selection phase of exploration for coalbed methane (CBM) is impossible, the conventional method that relies on a conceptual (or detailed) development program cannot be used during the economic evaluation of CBM resources. Hence, this study focuses on establishing an economic evaluation model based on the characteristics of the target selection phase. The discounted cashflow method is applied to the construction of the model with the assumption that there is a uniform distribution of production wells. The computational error generated by the assumption is corrected by introducing a correction factor based on the production profile of single CBM wells. The case study demonstrates that the blocks lacking economic value can be screened out, and the most advantageous targets can be found by computing the resource values in the best- and worst-case scenarios. This technique can help to reduce wasted investments and improve the quality of decision-making in selecting targets for exploration.

Keywords Coalbed methane · Exploration target selection · Economic evaluation · Scenario analysis

✉ Liang-Yu Xia
    xialiangyu_paper@sina.com

[1] School of Business Administration, China University of
    Petroleum (Beijing), Beijing 102249, China

Edited by Xiu-Qin Zhu

1 Introduction

According to the standard "Specifications for Coalbed Methane Resources/Reserves"(DZ/T 0216-2010) issued by the Ministry of Land and Resources of the People's Republic of China, coalbed methane (CBM) exploration is divided into a target selection phase and an exploration phase (Ministry of Land and Resources 2011). In the target selection phase, a comprehensive study of data, obtained by exploration and analogy, geological surveys, and coal mine production, is conducted to locate CBM exploration targets for the resource evaluation phase. The CBM resources selected in the target selection phase are classified as prospective resources.

Economic evaluations must be performed for CBM resources (reserves) at various phases of the exploration to satisfy the economic efficiency principle (Ministry of Land and Resources 2011; Attanasi 1998; Moore 2012). However, most economic evaluations currently target CBM reserves at or above a proven level (Kirchgessner et al. 2002; Robertson 2009; Zhang et al. 2004; Wang et al. 2004). A few scholars (Mu and Zhao 1996), who studied economic evaluation methods for the exploration phase, have recommended the use of adjusted conventional natural gas evaluation parameters to perform economic evaluations in resource-rich areas that have been explored only at a low level. However, studies on CBM's economic evaluation methods in the target selection phase are rare.

The discounted cashflow method is the most widely applicable economic evaluation methods. Applying the discounted cashflow method to the evaluation of CBM resources usually relies on a conceptual or detailed development program (Shimada and Yamaguchi 2009; Wong et al. 2010; Sander et al. 2011; Robertson 2009; Sander and Connell 2014; Chen et al. 2012a; Zhou et al. 2013; Yang

2008; Cao and Wang 2011).However, during the target selection phase, drilling and exploration work have yet to begin; therefore, no conditions exist for forecasting the development program, which makes it difficult for the conventional methods and procedures to be used to perform an evaluation. Due to the present difficulties with economic evaluation, geological evaluation remains the primary method used during the target selection phase (Zhao and Zhang 1999; Wang et al. 2009; Liu et al. 2001; Chen et al. 2012b; Hou et al. 2014).

The economic value of resources is jointly determined by many factors, including geological issues (Li et al. 2000). If the exploration targets are screened only using geological parameters, some resources that have superior geological conditions but little economic value may enter the exploration sequence and remain until their lack of economic value is shown. Because this situation can cause investments to be wasted unnecessarily, building an economic evaluation method that can promote decision-making during the target selection phase is necessary.

The discounted cashflow method is still employed to build the economic evaluation model, but it is used in an approach that is based on the characteristics of the target selection phase, and it is different from the traditional method, which depends on the use of development programs.

2 Economic evaluation model and target selection

2.1 NPV method

Commonly used evaluation indices in financial evaluations based on the discounted cashflow method include the financial net present value (NPV), the internal rate of return, and the payback period (National Development and Reform Commission 2006; Ministry of Construction of the People's Republic of China 2010).The financial NPV is the best indicator for economic evaluations of oil and gas resources (Luo 2002). Therefore, the financial NPV was selected as the basis for a CBM economic evaluation model. The formula for calculating the financial NPV is as follows:

$$\mathrm{NPV} = \sum_{t=0}^{T} (CI - CO)_t (1+i)^{-t} \qquad (1)$$

where, CI is the cash inflow, CO s the cash outflow, t is the number of the evaluation period (t takes value between 0 and T), I is the benchmark discount rate, and T is the number of evaluation periods.

According to formula (1), the primary task of calculating the NPV index is to forecast the amount of cashflow

generated by follow-up exploration and development activities, including cash inflow and cash outflow. The cash outflow includes exploration investments, development investments, liquidity, operating costs, business taxes and surcharges, and adjusted income taxes. The cash inflows include sales income, subsidy income, asset residual value recovery, and liquidity recovery (Ministry of Construction 2010). Asset residual value recovery is not considered because it is offset by the cost of land restoration cost when the well site is abandoned.

The procedure for applying the discounted cashflow method to evaluate oil and gas projects involves several steps. First, a conceptual (or detailed) development program is constructed that includes drilling and recovery projects, ground engineering projects, and the site's capacity for construction and annual gas production. Next, the essential constituents of the cashflow are estimated based on the development program. Finally, the financial evaluation indices are calculated based on the cashflow (Ministry of Construction 2010). Evaluation is difficult to perform using this procedure because it is hard to obtain the required geological, technical, and economic information during the target selection phase. Therefore, building an economic evaluation model that targets the characteristics of the target selection phase is necessary.

2.2 Characteristics of the target selection phase

Although drilling and exploration are not performed during the CBM target selection phase, the amount of CBM resources can still be inferred from geological parameters obtained from coal mine exploration data, such as the coal's rank, thickness, depth, pressure, and gas content (Zhang et al. 2002; Liu et al. 2001), and the amount of recoverable resources can be determined using geological analogy forecasts (Wang et al. 2003).

The number of development wells can be inferred from the recoverable area and the control area of a single CBM well. Although the total number of wells can be estimated, simulating and forecasting the production profile is impossible because gas testing cannot be performed by drilling exploration wells, and production data from test wells (a well group) cannot be obtained (Kang et al. 2012; Shao et al. 2013; Yang et al. 2008).Therefore, a well-drilling plan cannot be formulated based on the production profile of a single well or well group. Moreover, the well-drilling plan significantly affects economic evaluation results because it is also the basis for estimating the annual investment into drilling and recovery engineering projects. This problem must be resolved appropriately.

2.3 Economic evaluation model

2.3.1 Modeling approach and procedures

The modeling approach and procedures (Fig. 1) were established according to the characteristics of the target selection phase. The steps are as follows:

(1) Use geological information on the coalfield to infer the availability of CBM resources through a comprehensive study. Determine the conversion rate for turning resources into reserves and the recovery ratio based on analogous geological conditions; calculate the amount of recoverable reserves; and estimate the amount of exploration work required to verify the amount of reserves simultaneously.

(2) Comprehensively consider the recoverable reserves, market scale, and demand for gas (Luo and Xia 2009). Formulate a productivity plan that includes the amount of resources produced, the construction capacity, the annual gas supply capacity, and the number of years that this amount of gas can be supplied. If the amount of recoverable reserves is large but the market demand is small, determine the effective resource capacity and formulate a plan based on the market demand. If the amount of recoverable reserves is small but the market demand is large, formulate a plan according to the recoverable resource capacity that ensures a stable gas supply over a certain period of time. Determine the number of production wells through the gas-bearing area and the designated single well control area (the well distance).Assuming the wells drilled are uniformly distributed across the production period, the number of wells drilled annually is the total number of wells divided by the length of the production period.

(3) Cashflow estimate: Estimate the exploration investment based on the exploration workload. Estimate the annual drilling and recovery engineering project investment based on the number of production wells per year. Estimate the ground engineering project investment based on the construction capacity. Estimate the sales income based on the annual gas production capacity. Estimate the operating costs based on the gas production capacity and the number of wells. Calculate other cash inflows and outflows based on the relevant provisions.

(4) Based on the cashflow estimate, establish an economic evaluation model that uses the financial NPV formula. Because investments of the same amount that occur at different times have different time values, the discounted present values are also

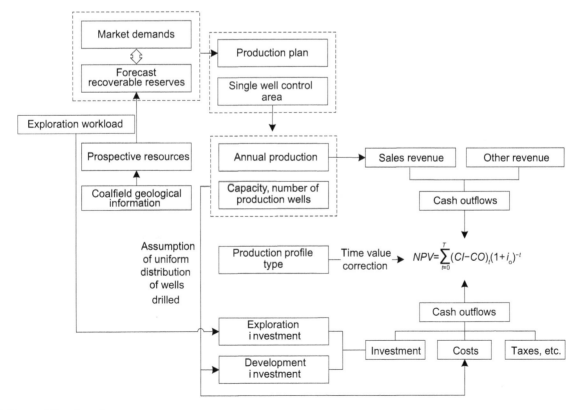

Fig. 1 The modeling flow chart

different. The assumption that drilled wells exhibit a uniform distribution changes the times of drilling and recovery of the investment in the engineering project, generating computational errors. To correct the errors, a time value correction factor is introduced into the model. The following method is used to resolve the inability to forecast and determine the designated yield fraction in the distribution of wells drilled: First, differentiate between the typical CBM single well (well group) production profile types and compute the time value correction factor for each type. If the profile type corresponding to the object to be evaluated can be determined, make the correction using the corresponding correction factor. If the affiliated profile type cannot be determined, a scenario analysis method can be used to estimate the resource values for each of the possible profile types to provide reference information for decision-making.

2.3.2 Economic evaluation model

The time value correction factor is defined as r_{COV}. The CMB economic evaluation model can be obtained using formula (1) (assuming the exploration period is one year and that the investment into the ground engineering project occurs early in the production period):

$$
\begin{aligned}
NPV = & \sum_{t=2}^{1+T_d} R(t) \cdot (1+i)^{-t} - I_E \cdot (1+i)^{-1} \\
& - r_{COV} \cdot \sum_{t=2}^{1+T_d} I_D \cdot (1+i)^{-t} \\
& - I_S \cdot (1+i)^{-2} - \sum_{t=2}^{1+T_d} C_L(t) \cdot (1+i)^{-t} \\
& - \sum_{t=2}^{1+T_d} T_X(t) \cdot (1+i)^{-t}
\end{aligned}
\tag{2}
$$

where T_d is the production period, $R(t)$ is the total income during year t, I_E is the exploration investment, I_D is the annual average drilling and recovery engineering project investment calculated with the assumption that the wells drilled are uniformly distributed, I_S is the ground engineering project investment, $C_L(t)$ is the operating cost in year t, and $T_X(t)$ is the amount of taxes in year t.

The time value correction factor r_{COV} is calculated as follows:

The average annual investment into the drilling and recovery engineering project is

$$
I_D = I_d \times N
\tag{3}
$$

where I_d is the single well-drilling and recovery engineering project investment, and N is the annual average number of wells drilled. N can be calculated using the following formula:

$$
N = \frac{A}{a \times T_p}
\tag{4}
$$

where A is the area of the region that produces resources, a is the single well control area, and T_p is the production period.

In formula (2), $\sum_{t=2}^{1+T_d} I_D \times (1+i)^{-t}$ is the sum of the drilling and recovery engineering project investments' discounted present value for each year, assuming that the wells drilled are uniformly distributed. Substituting formulas (3) and (4) into this expression yields the following:

$$
\begin{aligned}
\sum_{t=2}^{1+T_d} I_D \times (1+i)^{-t} &= \frac{I_d A}{a T_P} \sum_{t=2}^{1+T_d} (1+i)^{-t} \\
&= \frac{I_d A}{a T_P} \times \frac{(1+i)^{T_p} - 1}{i(1+i)^{T_p}}
\end{aligned}
\tag{5}
$$

where $\frac{I_d A}{a T_P}$ is the average annual investment into the drilling and recovery engineering project, and $\frac{(1+i)^{T_p}-1}{i(1+i)^{T_p}}$ is the discount factor when the uniform distribution of wells drilled is expressed by r_{even}. In addition, the discount factor for the real distribution of the wells drilled is r_{act}; then,

$$
r_{COV} \times \frac{I_d A}{a T_P} \times r_{even} = \frac{I_d A}{a T_P} \times r_{act}.
\tag{6}
$$

The formula for computing the time value correction factor obtained from Eq. (6) is as follows:

$$
r_{COV} = \frac{r_{act}}{r_{even}}
\tag{7}
$$

Shao et al. (2013) have summed up four gas production modes (production profile types) for CBM wells. Mode I is used as an example in this work to explain the method for determining the time value correction factor for the distribution of wells drilled.

In Fig. 2, No. 1 is the production profile curve drawn for Mode I, and No. 3 is the annual gas supply capacity curve for the target. To meet the stable gas supply requirement indicated in No. 3, it is necessary to set up a reasonable annual well-drilling plan. The corresponding well-drilling distribution curve (No. 3) is obtained by simulating the schedule of the production plan.

Assuming there is a stable gas supply for 20 years with a benchmark discount rate of 12 %, r_{even} is 7.5. For the well-drilling distribution displayed in No. 2, r_{act} is 10.5.

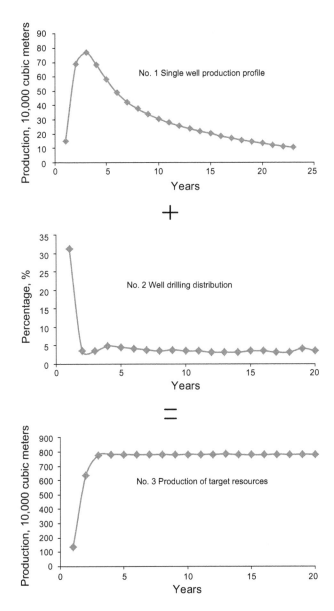

Fig. 2 Relationship between the well-drilling distribution and a single well's production profile

Therefore, the calculated time value correction factor is 1.4.

Parameter values, such as a stable production duration or gradual reduction rate, under the same production profile type, also affect the time value correction factor; however, these effects can be ignored during the selection phase under the premise of reasonable type differentiation; therefore, an average correction factor is used in the calculations.

2.4 A target selection method based on the economic evaluation

The primary factors are selected to function as scenario parameters in the economic evaluation, such as the recovery ratio, size of the single well control area, production profile type (or time value correction factor), and amount invested. Best- and worst-case scenarios are constructed based on the values of the scenario's parameter. The CBM resource NPVs for these two scenarios are calculated separately, to screen the exploration targets. If the NPV in the best-case scenario is less than 0, the target is not worth further exploration and development, and should be abandoned. If the NPV in the worst-case scenario is greater than 0, the target should have priority in exploration and development. If the NPV is greater than 0 in the best-case scenario and less than 0 in the worst-case scenario, a prudent decision should be made after undertaking further study of the evaluation target or re-evaluating it after the completion of appropriate exploratory work.

3 Case study

The resource forecast data for a specific CBM target are as follows (Tables 1 and 2):

Scenarios are constructed based on the value interval of the recovery ratio, the size of the single well control area, the value of the time correction factor, and the single well-drilling and recovery engineering project investment in Tables 1 and 2. In the best-case scenario, the recovery ratio is 40 %, the size of a single well control area is 0.5 km, the time correction factor is 1.1, the single well-drilling and recovery engineering project investment is 500,000 CNY/well, and the calculated NPV is −25 million CNY. In the worst-case scenario, the recovery ratio is 20 %, the size of the single well control area is 0.2 km, the time correction factor is 1.4, the single well-drilling and recovery engineering project investment is 1.2 million CNY/well, and the calculated NPV is −423 million CNY. This target should be eliminated based on the calculated results.

4 Conclusions

Full consideration of the data acquired while designing the method of estimating the essential cashflow constituents can ensure the operability of the established model, whereas it can lower the accuracy of its evaluation. Moreover, the implied assumption in the model affects the results of the calculations. For example, when selecting the size of the single well control area and the recovery ratio as scenario parameters during a scenario analysis, the implied assumption is that the well distance and the recovery ratio are independent, essential factors. The relationship between the two (i.e., the well distance and the target's recovery ratio are inversely related) is not considered. This assumption leads to an overly optimistic result for the best-

Table 1 Geological and technical data

Item	Unit	Value
Potential resource capacity	100 million cubic meters	2000
Conversion rate for turning resources into reserves	%	20
Recovery ratio	%	20–40
Size of the area producing resources	km^2	200
Size of the single well control area	km^2	0.2–0.5
Time correction factor	–	1.1, 1.4

Table 2 Economic data

Item	Unit	Value
Market scale	100 million cubic meters	Sufficiently large
Number of years of stable gas supply	Year	20
Exploration investment	100 million	1
Single well-drilling and recovery engineering project investment	10,000 CNY/Well	50–120
Known unit productivity ground engineering project investment	10,000 CNY/100 million cubic meters	1000
Topography and geomorphology correction coefficient	–	1.2
Unit production cost	CNY/cubic meter	0.35
Price	CNY/cubic meter	1.5
Subsidy	CNY/cubic meter	0.4
Commodity rate	%	95
Benchmark discount rate	%	20

case scenario and an overly pessimistic result for the worst-case scenario. Therefore, an actual example should be integrated to examine the model's computation error. If the error is too large, it will be necessary to develop a method for controlling it.

In addition, the differentiation between the typical CBM production profile types is a key part of evaluating the value of a resource. Although scholars (Kang et al. 2012; Shao et al. 2013) have already sorted out the types of CBM production modes from the viewpoint of an economic evaluation, the suitability of these divisions requires further study.

Acknowledgments This study was funded by the Beijing Natural Science Foundation (9144033) and the National Social Science Fund for Major Projects (11&ZD164).

References

Attanasi ED. Relative importance of physical and economic factors in Appalachian coalbed gas assessment. Int J Coal Geol. 1998;38:47–59.

Cao Y, Wang XZ. Economic evaluation of CBM gas development projects. Nat Gas Ind. 2011;31(11):103–6 (in Chinese).

Chen XZ, Tang DZ, Xu H, Qu YG, et al. Geological evaluation system of potential coalbed methane exploration and development blocks with low and medium coal ranks. J Jilin Univ (Earth Science Edition). 2012a;42(2):115–9 (in Chinese).

Chen YH, Yang YG, Luo JH. Uncertainty analysis of coalbed methane economic. assessment with Monte Carlo method. Proc Environ Sci. 2012b;12:640–5.

Hou HH, Shao LY, Tang Y, Luo XL, et al. Criteria for selected areas evaluation of low rank CBM based on multi-layered fuzzy mathematics: a case study of Turpan-Hami Basin. Geol China. 2014;41(3):1002–7 (in Chinese).

Kang YY, Shao XJ, Wang CF. Production characteristics and affecting factors of high-mid rank coalbed methane wells: taking Fanzhuang and Hancheng mining areas as examples. Pet Explor Dev. 2012;39(6):728–32 (in Chinese).

Kirchgessner DA, Masemoreb SS, Piccot SD. Engineering and economic evaluation of gas recovery and utilization technologies at selected US mines. Environ Sci Policy. 2002;5:397–409.

Liu HL, Wang HY, Zhang JB. Coal bed methane resource and its exploration direction in China. Petr Explor Dev. 2001;28(1):9–11 (in Chinese).

Li YH, Zhang SA, Wang H. Analysis of the effect of geological conditions in the economic evaluation of coal bed methane exploitation projects. Coal Geol China. 2000;12(2):26–8 (in Chinese).

Luo DK. Oil and gas exploration investment economic evaluation methods. Pet Geol Recov Effic. 2002;9(1):21–3 (in Chinese).

Luo DK, Xia LY. Economic evaluation method for CBM prospect resources. J Daqing Pet Inst. 2009;33(4):115–9 (in Chinese).

Ministry of Construction, People's Republic of China. Petroleum construction project economic evaluation methods and parameters. Beijing: China Planning Press. 2010. 22-23 (in Chinese).

Ministry of Land and Resources, People's Republic of China. Specifications for coalbed methane resources/reserves. China Standard Press, Beijing 2011, pp. 5–7 (in Chinese).

Moore T. Coalbed methane: a review. Int J Coal Geol. 2012; 101:36–81.

Mu XZ, Zhao QB. Discussion on the economic analysis methods for regions with lower degrees of coalbed methane resource exploration in China. China Coalbed Methane. 1996;1:21–2 (in Chinese).

National Development and Reform Commission, Ministry of Construction, People's Republic of China. Construction Project Economic Evaluation Methods and Parameters. China Planning Press, Beijing, 2006, pp. 2–13 (in Chinese).

Robertson EP. Economic analysis of carbon dioxide sequestration in Powder River basin coal. Int J Coal Geol. 2009;77:234–41.

Sander R, Allinson WG, Connell LD, Neal PR. Methodology to determine the economics of CO_2 storage in coal seams with enhanced coalbed methane recovery. Energy Proc. 2011;4:2129–36.

Sander R, Connell LD. A probabilistic assessment of enhanced coal mine methane drainage (ECMM) as a fugitive emission reduction strategy for open cut coal mines. Int J Coal Geol. 2014;131:288–303.

Shao XJ, Wang CF, Tang DZ, et al. Productivity mode and control factors of coalbed methane wells: a case from Hancheng region. J China Coal Soc. 2013;38(2):271–6 (in Chinese).

Shimada S, Yamaguchi K. Economic assessment of enhanced coalbed methane recovery for low rank coal seam. Energy Proc. 2009;1: 1699–704.

Wang B, Li JM, Zhang Y, Wang HY, et al. Geological characteristics of low rank coalbed methane. China. Pet Explor Dev. 2009; 36(1):30–3 (in Chinese).

Wang HY, Zhang JB, Liu HL. The prediction of proved economic reserves and development prospect of coalbed methane in China. Pet Explor Dev. 2003;30(1):15–7 (in Chinese).

Wang XH, Lu X, Jiang WD, Xian BA, et al. Economic evaluation of coalbed development in Fanzhuang block, Qinshui coalbed gas field. Nat Gas Ind. 2004;24(5):137–9 (in Chinese).

Wong S, Macdonald D, Andrei S, et al. Conceptual economics of full scale enhanced coalbed methane production and CO_2 storage in anthracitic coals at South Qinshui basin, Shanxi, China. Int J Coal Geol. 2010;82:280–6.

Yang WJ. Study of the economic evaluation of CBM development project. China Coalbed Methane. 2008;5(1):38–40 (in Chinese).

Yang S, Kang YS, Zhao Q, et al. Method for predicting economic peak yield for a single well of coalbed methane. J China Univ Mining Technol. 2008;18:521–6.

Zhang XM, Zhuang J, Zhang SA. China coalbed methane geology and resource evaluation. Beijing: Science Press; 2002. p. 211–24 (in Chinese).

Zhang SA, Wang ZP, Li YH. Economic evaluation methods and prediction model for coalbed methane development. J China Univ Mining Technol. 2004;33(3):314–7 (in Chinese).

Zhao QB, Zhang GM. Important parameters for coalbed methane evaluation and block selection principles. Pet Explor Dev. 1999;26(2):23–7 (in Chinese).

Zhou FD, Hou WW, Allinson G, et al. A feasibility study of ECBM recovery and CO_2 storage for a producing CBM field in Southeast Qinshui Basin, China. Int J Greenhouse Gas Control. 2013;19:26–40.

5

Supply-based optimal scheduling of oil product pipelines

Hao-Ran Zhang[1] · Yong-Tu Liang[1] · Qiao Xiao[2] · Meng-Yu Wu[1] · Qi Shao[1]

Abstract Oil product pipelines have features such as transporting multiple materials, ever-changing operating conditions, and synchronism between the oil input plan and the oil offloading plan. In this paper, an optimal model was established for a single-source multi-distribution oil product pipeline, and scheduling plans were made based on supply. In the model, time node constraints, oil offloading plan constraints, and migration of batch constraints were taken into consideration. The minimum deviation between the demanded oil volumes and the actual offloading volumes was chosen as the objective function, and a linear programming model was established on the basis of known time nodes' sequence. The ant colony optimization algorithm and simplex method were used to solve the model. The model was applied to a real pipeline and it performed well.

Keywords Oil products pipeline · Scheduling optimization · Linear programming (LP) model · Ant colony optimization algorithm (ACO) · Simplex method (SM)

Abbreviations

Subscripts and sets

$i, i' \in I$	The set of offloading station numberings
$j \in J$	The set of old batch numberings
	$J = \{J_{old} \cup J_{new}\}$
J_{old}	The set of old batch numberings
J_{new}	The set of new batch numberings
$\tau \in T_C$	The set of time node numberings
$\tau_r \in T$	The set of all time nodes
	$T = \{T_{vc} \cup T_{ac} \cup T_{ab} \cup T_b \cup T_o\}$
T_{vc}	The set of all time nodes when the head of the batch demanded by the station reaches there
T_{ac}	The set of batch start-offloading time nodes of each station
T_{ab}	The set of batch end-offloading time nodes of each station
T_b	The set of input plan flow rate-changing time nodes
T_o	The set of plan-start time and end times

Parameters

V_i	Volume coordinates of station i, equaling to the filled volume of the pipe segment from the initial station to the station i, whose volume coordinate equals to 0
V_{oi}	Volume coordinates of batch j. For the old batch, the volume coordinate equals to the filled volume of the pipe segment from the initial station to the position of its head at plan-start time. For the new batch, the volume coordinate equals to minus sum of volume of earlier injected new batches
$V_{xi,j}$	Volume of the batch j needed by station i

✉ Yong-Tu Liang
  liangyt21st@163.com

[1] Beijing Key Laboratory of Urban Oil and Gas Distribution Technology, China University of Petroleum, Beijing 102249, China

[2] CNPC Trans-Asia Gas Pipeline Company Ltd., Beijing 100007, China

Edited by Xiu-Qin Zhu

$Q_{\text{xmax } i}$ Maximum offloading flow rate at station i

$Q_{\text{xmin } i}$ Minimum offloading flow rate at station i

$Q_{\text{cmax } i}^{\tau}$ Maximum flow rate of the pipeline segment between station i and $i + 1$ from time node τ to $\tau + 1$

$Q_{\text{cmin } i}^{\tau}$ Minimum flow rate of the pipeline segment between station i and $i + 1$ from time node τ to $\tau + 1$

$Q_{k\tau}$ Input flow rate at initial station from time node τ to $\tau + 1$

Continuous variables

t_{τ} Time corresponding to the time node τ

$V_{si,j}$ Actual offloading volume of batch j at station i

$V_{pi,j}$ Offloading volume at station i in the time-window from time node τ to $\tau + 1$

$M_{1i,j}$ Relaxation artificial variables of objective function

$M_{2i,j}$ Tightening artificial variables of objective function

Discrete variables

$\tau_{ai,j}$ Time node number when the batch j oil head reaches station i

$\tau_{ci,j}$ Time node number when the batch j is being offloaded at station i

$\tau_{bi,j}$ Time node number when offloading of batch j is finished at station i

τ_{tc} Time node number of plan-start time

τ_{tb} Time node number of plan-end time

1 Introduction

Approximately, 17.95 million barrels of oil products are imported and exported everyday around the world (BP 2014), most of which are transported to different cities by pipelines. As oil product pipelines are developing at an incredible pace, the topological structure and operation of oil product pipelines are becoming more complex than ever, adding difficulty in making schedules. The main issues concerned are how to make a more rational batch-scheduling plan and how to meet the consumption demand of each region along the pipeline in a safe and economic way. The oil products pipeline has features such as multiple oil products, ever-changing working conditions, and synchronism between the injection plan of the pipeline's initial station and the offloading plan of the offloading stations along the pipeline. Milidiú and dos Santos Liporace (2003) proved that the scheduling plan of oil batches is a non-deterministic polynomial complete (NPC) issue if the batch sequence constraint is considered. Presently, batch-dispatchers use manual or semi-automatic methods to create the batch-scheduling plan for most of the supply-based pipelines. In other words, there exists no mature

algorithm that can automatically make the scheduling plan that meets with the demands of actual operation.

Much research focuses on these complex scheduling issues. Determination of time expression is a fundamental step of building a scheduling model, and can directly affect the size of the model and the selection of algorithm. Currently, there are two major time expressions for scheduling models available, namely discrete-time and continuous-time expression.

Discrete-time expression, dividing the period studied into several isometric- or length-specified time-windows, takes the time node of time-windows as the scheduling plan's event nodes and analyzes the logical relationship between variables. Using the method of discrete-time expression can simplify the non-linear coupling relationship between variables as well as reduce the difficulty of building and solving a model. The research of Rejowski and Pinto (2003) embodied the advantage of a discrete-time expression when dealing with time-related electrical price issues. Magatão et al. (2004), using a discrete-time expression, solved the issue of pipeline network scheduling plans, which also reflects its preponderance in simplifying large and complex models. Zyngier and Kelly (2009) proved that the introduction of a stock constraint will add to the model's complexity and improve the discrete-time expression. Herrán et al. (2010) resolved the issue of multi-injection and tracking batch interfaces through discrete-time expression. de Souza Filho et al. (2013) combined a discrete-time expression with a heuristic algorithm and set up a mixed integer linear programming (MILP) model to resolve the issue of scheduling aiming at minimizing power costs. Despite the fact that the research above has demonstrated that discrete-time expression works well when dealing with sub-problems like tracking batch interfaces, the final solution given by a discrete-time expression may not possess practical applicability in that the practical planning cycle is more than a week and the long-time step length may lead to poor optimality, while the short one may result in excessive model and dimension disasters.

Continuous-time expressions divide the time-window according to the happening and ending of an event, the beginning and ending time of which are known or unknown. In other words, there exists an uncertainty for time-window's length and time nodes. Analyzing the occurring and ending condition of events and inter-connection between events is essential for continuous-time expressions. Although the adoption of continuous-time expressions may lead to a more complex model structure and stronger coupling link between variables, it can minimize the size of a model and improve the solving efficiency. Based on the previous research of the MILP discrete method for dendritic pipeline network scheduling,

MirHassani and Ghorbanalizadeh (2008) have proposed a continuous-time MILP method, the result of which demonstrates that the introduction of continuous-time can apparently enhance the calculating efficiency. During the past several years, optimizing of schedule issues on the basis of continuous-time models for different pipelines has become an issue of interest in academia. For instance, some researchers aim at single-source pipelines (Cafaro and Cerdá 2004, 2008; Relvas et al. 2006), some focus on tree-structure pipelines (Mirhassani and Ghorbanalizadeh 2008; Castro 2010; Cafaro and Cerdá 2011), or mesh-structure pipelines (Cafaro and Cerdá 2012). However, at present, the scheduling plan given by a continuous-time MILP model is just an approximate scheduling which contains only a general time zone and approximate injection as well as offtake volume for each station instead of a detailed operating time.

In the subsequent research, many researchers began to study the algorithm for a detailed scheduling plan on the basis of an approximate scheduling plan. Cafaro et al. (2011) chose the simplest monophyletic transfer pipe as the research object, and obtained an approximate scheduling plan and then developed a step-by-step algorithm for detailed planning. Whereafter, a detailed scheduling plan that can achieve simultaneously offloading operations was developed (Cafaro et al. 2012). Recently, on the basis of the previous research, Cafaro et al. (2015) established a mixed integer non-linear programming (MINLP) model and made a detailed scheduling plan for a real monophyletic pipeline, considering the hydraulic coupling non-linear constraint.

Nevertheless, the current continuous-time expression MILP model ignores the time nodes such as batches' arrival and batch delivery operation's starting and ending moment, which will inevitably bring about uneconomical operating period distribution and excessive time-window offset. Those will decrease the model's practicability. Moreover, a large number of models take the limit of download and injection size as known parameters which are time-related. It would be more reasonable if it is replaced by a limit of operating flow rate. This paper uses the time-continuous expression method to establish a LP model on the basis of known time nodes' sequence. The objective function is the minimum deviation between the demand batch volume and the actual offloading volume at each station. To accelerate solving speed, the hybrid algorithm of ACO (ant colony optimization) and the SM (simplex method) is used to solve the model.

Section 2 of this paper describes the scheduling issue and gives the model's assumption conditions. Section 3 establishes the objective function of the model and describes the constraints of the model. In Sect. 4, the model-solving process is discussed. Section 5 verifies the

correctness and applicability of the model with two examples. We end with our conclusions in Sect. 6.

2 Problem description

2.1 Supply-based scheduling of single-source products pipeline

Some products pipelines serve the refinery, with the responsibility of transporting the refined oil to the downstream market. Firstly, the refinery's production plan is made on the basis of the downstream market's demand. Next, the injecting plan at the initial station is made according to the production plan. The batch-dispatchers can work out the offloading plan based on the supply of the initial station, called the supply-based schedule. Thus, the demand of the downstream market can be satisfied in time, and at the same time, human resources are saved and inventory cost is sharply reduced.

There are three kinds of stations in the single-source pipeline system: initial station (input station), offloading stations (intermediate stations), and terminal station, as shown in Fig. 1. The initial station is linked with a refinery and the terminal station is an oil depot with a large storage capacity. The oil products are sequentially transported in the pipeline and the batch sequence is known. The optimal research about the pipeline scheduling plan based on supply is to determine the offloading station's actual offloading volume on the basis of known conditions, including the input schedule at the initial station and demanded volumes at offloading stations.

Parameters are taken into account, including pipeline information, oil input sequence, volume and flow rate of the initial station in a certain period of time, demand volumes of each offloading station, upper and lower limits of flow rate, and each station's offloading flow rate limits. The decision variables are actual offloading volumes and the starting and ending time of offloading operation.

2.2 Modeling hypotheses

The scheduling of an oil products pipeline system is subject to several constraints. In order to improve solving efficiency, some assumptions are made:

(1) Inventory constraints of the terminal station are neglected and it can receive any type of oil at any time.
(2) When batches move in products pipeline in order, contamination will occur inevitably between the two adjacent oils. The mixed oil section is considered as an interface.

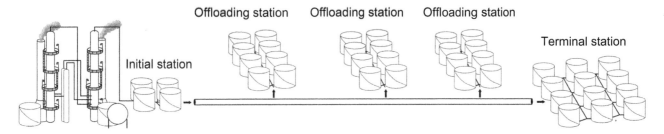

Fig. 1 Oil products pipeline system

(3) The offloading stations along the pipeline own a certain number of tanks, which can store the oil temporarily. The paper assumes that the demands given by each offloading station have considered stock volumes, irrespective of the tank capacity constraints, and oil storage conditions of each intermediate station.

(4) The oil is incompressible.

(5) Any offloading flow rate of each offloading station is constant within a time-window.

3 Mathematical formula

3.1 Objective function

The minimum deviation between each batch demanded volume of each station and the actual offloading volume is defined as the objective function.

$$\min f = \sum_i \sum_j \left| V_{xi,j} - V_{si,j} \right| \quad i \in I, j \in J \tag{1}$$

Since this objective function is non-continuous, it is difficult to solve. Artificial relaxation variables and artificial tightening variables are introduced to linearize the objective function.

$$\min f = \sum_i \sum_j \left(M_{1i,j} + M_{2i,j} \right) \quad i \in I, j \in J \tag{2}$$

$M_{1i,j}$, $M_{2i,j}$ mentioned above should meet the following constraints:

$$V_{xi,j} - V_{si,j} + M_{1i,j} \geq 0 \tag{3}$$

$$V_{si,j} - V_{xi,j} + M_{2i,j} \geq 0 \tag{4}$$

$$M_{1i,j} \geq 0, M_{2i,j} \geq 0 \quad i \in I, j \in J \tag{5}$$

If $V_{xi,j} - V_{si,j} \geq 0$, according to Eq. (3), the minimum value of $M_{1i,j}$ is 0. The minimum value of $M_{2i,j}$ is equal to $V_{xi,j} - V_{si,j}$ according to Eq. (4). If $V_{xi,j} - V_{si,j} \leq 0$, according to Eq. (3), the minimum value of $M_{1i,j}$ is equal to $V_{si,j} - V_{xi,j}$. The minimum value of $M_{2i,j}$ is 0. Therefore, Eq. (2) is equivalent to Eq. (1).

3.2 Model constraints

3.2.1 Time node constraints

If the order of all the time nodes is known—Sect. 4 will describe how to determine the order—all the time nodes are numbered. The following is the corresponding expression:

$$t_\tau \leq t_{\tau+1} \quad \tau \in T_C, t_\tau \in T. \tag{6}$$

For any given station, the arrival time of any batch's oil head cannot be later than that of the next batch oil head.

$$\tau_{ai,j} < \tau_{ai,j+1} \quad i \in I, j \in J, \tau_{ai,j} \in T_C \tag{7}$$

For the same batch, the arrival time of the oil head at any station cannot be later than that of the next station.

$$\tau_{ai,j} < \tau_{ai+1,j} \quad i \in I, j \in J, \tau_{ai,j} \in T_C \tag{8}$$

The time when a station starts to offload the demanded batch cannot be earlier than oil head's arriving time. The time when a station finishes offloading the demanded batch cannot be later than the arrival time of the next batch.

$$\tau_{ai,j} < \tau_{ci,j} \tag{9}$$

$$\tau_{ai,j+1} > \tau_{bi,j} \quad i \in I, j \in J, \tau_{ai,j}, \tau_{ci,j}, \tau_{bi,j} \in T_C \tag{10}$$

The time that a station starts to offload the batch cannot be later than the ending time.

$$\tau_{ci,j} < \tau_{bi,j} \quad i \in I, j \in J, \tau_{ci,j}, \tau_{bi,j} \in T_C \tag{11}$$

The time when a station starts and ends to offload the batch cannot be earlier than the scheduled starting time and cannot be later than the scheduled ending time.

$$\tau_{tc} < \tau_{ci,j} < \tau_{tb} \tag{12}$$

$$\tau_{tc} < \tau_{bi,j} < \tau_{tb} \quad i \in I, j \in J, \tau_{ci,j}, \tau_{bi,j} \in T_C \tag{13}$$

3.2.2 Offloading plan constraints

The actual offloading volume of any batch at any station is the sum of offloading volumes during all the time-windows from offloading staring time node to the ending node.

$$V_{si,j} = \sum_{\tau=\tau_{ci,j}}^{\tau_{bi,j}-1} V_{pi,\tau} \quad i \in I, j \in J, \tau_{ci,j}, \tau_{bi,j} \in T_C \tag{14}$$

Due to the limit of the offloading flow rate, the offloading volume within any time-window should not be larger than that of the maximum offloading flow rate multiplied by the length of the time-window or less than that of the minimum offloading flow rate multiplied by the length of the time-window.

$$V_{pi,\tau} \leq t_{\tau+1}Q_{\mathrm{xmaxi}} - t_\tau Q_{\mathrm{xmaxi}} \tag{15}$$

$$V_{pi,\tau} \geq t_{\tau+1}Q_{\mathrm{xmini}} - t_\tau Q_{\mathrm{xmini}} \tag{16}$$
$$i \in I, \tau \in \{\tau_{ci,j}, \ldots, \tau_{bi,j} - 1\}, j \in J$$

3.2.3 Batch transportation constraints

Considering the hydraulic constraints, the flow rate of the pipeline should be within a certain range. When there exists a gasoline and diesel mixed interface, in order to reduce the amount of mixed oil, the Reynolds number of the fluid in the pipeline must be larger than the critical Reynolds number. The minimum flow rate of the pipeline between station i and station $i + 1$ multiplied by the length of the time-window should not be larger than the difference between the volume input by the initial station within this time-window and the offloading volumes of station i and all stations before station i within the time-window. The allowable maximum flow rate of the pipeline between station i and station $i + 1$ multiplied by the length of time-window should not be less than the difference between the volume input by the first station within this time-window and the offloading volumes of station i and all stations before station i within the time-window.

$$t_{\tau+1}Q^\tau_{\mathrm{cmini}} - t_\tau Q^\tau_{\mathrm{cmini}} \leq t_{\tau+1}Q_{k\tau} - t_\tau Q_{k\tau} - \sum_{i'=1}^{i} V_{pi',\tau} \tag{17}$$

$$t_{\tau+1}Q^\tau_{\mathrm{cmaxi}} - t_\tau Q^\tau_{\mathrm{cmaxi}} \geq t_{\tau+1}Q_{k\tau} - t_\tau Q_{k\tau} - \sum_{i'=1}^{i} V_{pi',\tau} \tag{18}$$
$$i \in I, \tau \in T_C, t_\tau \in T$$

According to the conservation of volume, the volume coordinate of batch j plus the total input volume of the initial station before time $t_{\tau_{ai,j}}$ minus the total offloading volumes at stations before station i during the time-windows from $t_{\tau_{ai,j}}$ to $t_{\tau_{ci,j}}$ is equal to the volume coordinate of station i.

$$V_{oj} + \sum_{\tau=1}^{\tau_{ai,j}-1} (t_{\tau+1}Q_{k\tau} - t_\tau Q_{k\tau}) - \sum_{i'=1}^{i-1} \sum_{\tau=\tau_{ci',j}}^{\tau_{ai,j}-1} V_{pi',\tau} = V_i \tag{19}$$
$$i \in I, j \in J, \tau_{ci,j}, \tau_{ai,j} \in T_C, t_\tau \in T$$

4 Model solving

According to the objective function and the constraints if the sequence of all the time nodes is determined (all of the discrete variables are determined), an LP model can be established as shown in Sect. 3 and solved by SM. Therefore, finding the optimal sequence of time nodes is significant to solve this issue. While, as coupling with a large-scale LP model, this issue is more complex than traditional sequencing issues, such as the traveling salesman problem (TSP). Dynamic programming is one of the widely used algorithms for such kinds of issue. However, when dealing with a large quantity of time nodes, due to the curse of dimensionality, the practicability will be limited when solving the model through this method. On the other hand, intelligence algorithms have been utilized to solve complex programming issues, for instance, genetic algorithm (GA), particle swarm optimization algorithm (PSO), ant colony optimization algorithm (ACO), etc. Considering the constraints of the model and the fact that the optimal sequence does not have much difference with each feasible sequence, the ACO algorithm is more suitable to solve the model as it has better convergence in terms of optimizing the sequence.

In the ACO algorithm, all artificial ants are placed at an initial position of a multi-dimension space at the beginning. The objective for these ants is to find the food's position (the optimal solution). The objective function can be regarded as the food concentration to evaluate each position. During each iteration operation, each ant will select an orientation randomly and move in a specific step length to explore a new position, and then ants will be reallocated to a few best explored positions. In this way, as the explored region expands, the result will converge to a better solution. Finally, the optimal solution can be found.

As the target is to find the optimal sequence of time nodes, the positions in the ACO algorithm can be represented by sequences. A possible time node sequence is necessary since the initial position is very important for the ACO algorithm. Given that all the offloading stations do not offload batches, because the injecting plan is known, the batch interface can be traced and batch's arrival at stations can be simulated accordingly. Thus, the sequence of time nodes when batches arrive at stations, the injecting flow rate changes, and study horizon's beginning as well as end can be further calculated. The batch's offloading starting moment is close to the one when the batch's head reaches the station, and the finish time is close to one when the next batch's head reaches the station, providing that the batch's arrival is within the study horizon. The start of the offloading operation should be close to the head of the study horizon and the ending moment should be next to the end of the study horizon if the batch's arriving time is beyond the study horizon. In this way, the initial time node sequence is generated. During each iteration operation, two time nodes will be selected randomly and the firstly chosen one is plugged after the second one, and then we judge whether this changed sequence can meet the formulas

(7)–(13). If not, two new time nodes' orders will be exchanged randomly until they meet those constraints. Then an LP model can be established and solved by the simplex method. Therefore, a detailed scheduling plan and the value of its objective function can be obtained. Sorting all the explored positions on the basis of the value of their objective function, ants are reallocated to a few of the best positions, awaiting the next round of relocation.

The structure of the algorithm is as follows:

(1) To calculate the initial sequence and take it as the initial position of ants.
(2) To make ants move randomly to generate new sequences.
(3) According to those new sequences, establish the LP models and solve them by SM to obtain their objective function values.
(4) Sorting all the explored positions on basis of their objective function values and allocating ants to a few of the best positions.
(5) Repeat step 2 until the value of the objective function is less than the allowable maximum error or the iteration number is the maximum.
(6) To output the optimizing result.

The flowchart of the algorithm is shown in Fig. 2.

5 Example

In this section, two examples aiming at a certain real pipeline are given through the proposed model, using an Intel Core i7-4770k (3.50 GHz) computer with 8 parallel threads and MATLAB calculating software. In the first example, an operation case in summer is presented, in which all the demand of offloading stations is rational. In other words, there exists a promising solution that is capable of satisfying all stations' offloading demands. In the second example, another example in wintertime is presented, in which an irrational demand has arisen. By virtue of it, the convergence of the model is demonstrated.

5.1 Basic data

Taking an oil products pipeline as the research object, if the batch input plan of the initial station and the oil-filled state in the pipeline at the initial moment and the demanded volume by each offloading station are known, the scheduling plan for the pipeline can be made in the studied horizon. The length of the pipeline is 112 km. This pipeline transports different types of gasoline and diesel. There are six stations: the initial station (IS), 1# offloading station (1#OS), 2# offloading station (2#OS), 3# offloading station (3#OS), 4# offloading station (4#OS), and the terminal station (TS). Table 1 shows the basic data of the pipeline.

Considering the hydraulic requirements, the flow rate of the pipeline between stations should be controlled within a certain range, as shown in Table 2.

According to the design pressure constraint, restrictions on equipment such as pumps, and the application range of flow rate meters in the offloading stations, there exists flow rate range constraints of the stations as shown in Table 3.

The ant colony algorithm parameters are assigned as follows: The number of ants is 50 and the maximum number of iterations is 100. Considering the slight computational and round-off errors, the maximum error is set at 5.

5.2 Example one

The starting time of the plan is 0 and the end time of the plan is set at 71.8 h. In summer, each offloading station gives the demanded volumes based on the market requirements. Combined with the oil-filled state in the pipeline at the initial moment, the offloading plan of each offloading station within the study horizon can be made. The sequence of the batch is 95# gasoline–92# gasoline–0# diesel–92# gasoline–95# gasoline–92# gasoline. There is a mixed oil interface between 95# and 92# gasoline in the pipeline at the initial moment, at a location of 18.5 km away from the initial station. Table 4 shows the volume of the new batch input at the initial station:

Table 5 shows the volume coordinate of each batch calculated. The volume coordinates of old batches are positive values, and the volume coordinates of new batches are negative values.

Table 6 shows the inputting flow rate of the initial station.

Table 7 shows the volume of each batch that offloading stations demand.

With known conditions and model above, the offloading plan can be made as shown in Table 8.

Table 9 shows the total offloading amount of each batch and the deviation between each batch's offloading volume and the demanded volume.

As Table 9 shows, the deviation between the demanded and offloading volume is small enough to be accepted. The offloading volumes normally satisfy the volumes demanded of each station.

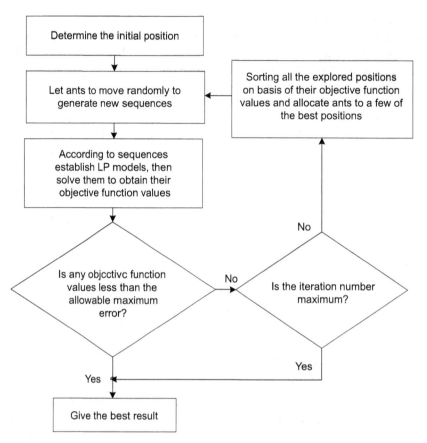

Fig. 2 Flowchart of the algorithm

Table 1 Basic pipeline data

Pipeline segment	Outer diameter, mm	Wall thickness, mm	Distance, km
IS—1#OS	323.9	7.1	18.5
1#OS—2#OS	323.9	7.1	32.7
2#OS —3#OS	323.9	7.1	18.5
3#OS—4#OS	273.1	6.4	27.4
4#OS—TS	273.1	6.4	14.9

Table 2 Allowable flow rate range of the pipeline segment between stations

Pipeline segment	Maximum flow rate, m^3/h	Minimum flow rate, m^3/h (no gasoline and diesel mixed interface)	Minimum flow rate, m^3/h (a gasoline and diesel mixed interface)
IS—1#OS	500	30	50
1#OS—2#OS	500	30	50
2#OS —3#OS	500	30	50
3#OS—4#OS	400	30	50
4#OS—TS	400	30	50

The batch transportation diagram, which is shown in Fig. 3, is based on the initial state as well as the inputting and offloading plans. In the diagram, the 95# gasoline is in blue, while the 92# gasoline is red, and 0# diesel is green. The rectangles on the left side of vertical axis denote the initial state of the pipeline. The rectangles on the horizontal

Table 3 Flow rate range of all stations

Station	Maximum flow rate, m³/h	Minimum flow rate, m³/h
1#OS	300	30
2#OS	300	30
3#OS	300	30
4#OS	300	30

Table 4 Example one: first station input volumes of the new batches

Batch number	Oil type	First station input volume, m³
2	92# gasoline	3515.7
3	0# diesel	13,791.0
4	92# gasoline	1747.2
5	95# gasoline	2288.7
6	92# gasoline	5998.5

Table 5 Example one: volume coordinate of each batch

Batch number	Oil type	Volume coordinate, m³
1	95# gasoline	7497.8
2	92# gasoline	1393.6
3	0# diesel	−3515.7
4	92# gasoline	−17306.7
5	95# gasoline	−19053.9
6	92# gasoline	−21342.6

Table 6 Example one: input flow rate of the first station

Time, h	Flow rate, m³/h
0.00–13.27	265
13.27–52.45	455
52.45–71.80	310

axis represent the injection plan at the initial station. As for other rectangles in the diagram, they define the offloading plans. The quantity of offloading and injection flow is in accordance with the width of these rectangles. The black line in the diagram represents the batch interface's migration process. As Fig. 3 shows, each offloading operation is conducted within an allowable time range.

Figure 4 shows the flow rate in each pipeline segment. All the intermediate stations are only allowed to offtake instead of injecting, since the studied pipeline is one with single-source and multiple distributions. Thus, the flow rate between 4#OS and TS is the minimum one along the pipeline. As illustrated in formulas 17 and 18, the minimum flow rate should be adjusted if there is an interface between gasoline and diesel. During the periods from 13.27 to 41.27 and 44.07 to 71.80 h, there exists a diesel oil–gasoline interface in the pipeline. Therefore, during these periods the lower limit of flow rate is 50 m³/h. From Fig. 4, the flow rate of the terminal station at any time is no less than the lower limit.

Because the demands of each offloading station are satisfied, the value of the objective function is less than the maximum error in the iteration process, and hence the computing program stops. The computational results of this example are shown in Table 10.

5.3 Example two

For the same pipeline, a scheduling plan lasting 71.8 h in winter season is made. Each offloading station gives the demanded volume of specific batch based on the market. The sequence is 95# gasoline–92# gasoline––10# diesel–92# gasoline–95# gasoline–92# gasoline. There is a mixed oil interface between 95# and 92# gasoline in the pipeline at the initial moment, at a location of 65 km away from the initial station. Table 11 shows the volume of the new batches input at the initial station:

Table 12 shows the volume coordinate of each calculated batch.

Table 7 Example one: each offloading station's demanded volume of each batch

Station number	Batch number	Oil type	Demanded volume, m³
1	3	0# diesel	1974
1	6	92# gasoline	1596
2	2	92# gasoline	2482
2	3	0# diesel	4170
2	4	92# gasoline	154
2	5	95# gasoline	1527
2	6	92# gasoline	40
3	3	0# diesel	1727
3	4	92# gasoline	71
4	1	95# gasoline	2000

Table 8 Example one: offloading plan of each station

Station number	Batch number	Offloading volume, m^3	Offloading flow rate, m^3/h	Offloading start time, h	Offloading end time, h	Time period, h
1	3	377.219	115.357	18.35	21.62	3.27
1	3	253.180	167.669	21.62	23.13	1.51
1	3	254.278	122.840	23.13	25.20	2.07
1	3	349.533	81.098	25.20	29.51	4.31
1	3	739.207	63.890	29.51	41.08	11.57
1	6	210.954	219.516	57.44	58.36	0.92
1	6	1393.400	111.472	58.36	70.86	12.50
2	2	610.442	173.915	9.76	13.27	3.51
2	2	1177.532	231.798	13.27	18.35	5.08
2	2	694.070	212.254	18.35	21.62	3.27
2	3	907.698	210.603	25.20	29.51	4.31
2	3	888.335	162.699	29.51	34.97	5.46
2	3	895.973	146.641	34.97	41.08	6.11
2	3	1477.587	162.909	41.08	50.15	9.07
2	4	84.848	30.631	52.45	55.22	2.77
2	4	69.055	31.106	55.22	57.44	2.22
2	5	145.859	108.044	58.36	59.71	1.35
2	5	1296.949	148.528	59.71	68.44	8.73
2	5	84.334	75.299	68.44	69.56	1.12
2	6	38.831	62.631	70.86	71.48	0.62
3	3	420.852	68.879	34.97	41.27	6.30
3	3	414.434	74.539	41.27	46.64	5.37
3	3	332.333	94.682	46.64	50.15	3.51
3	3	291.244	126.628	50.15	52.45	2.30
3	3	268.199	96.823	52.45	55.22	2.77
3	4	70.706	31.994	57.40	59.61	2.21
4	1	2000.100	226.000	0.00	8.85	8.85

Table 9 Example one: offloading volume and deviation

Station number	Batch number	Oil type	Offloading volume, m^3	Deviation, m^3
1	3	0# diesel	1973.417	0.583
1	6	92# gasoline	1595.354	0.645
2	2	92# gasoline	2482.044	0.044
2	3	0# diesel	4169.594	0.406
2	4	92# gasoline	153.903	0.096
2	5	95# gasoline	1526.843	0.157
2	6	92# gasoline	38.831	1.169
3	3	0# diesel	1727.062	0.062
3	4	92# gasoline	70.706	0.294
4	1	95# gasoline	2000.100	0.100

Table 13 shows the inputting flow rate of the initial station.

Table 14 shows the volume of each batch that each offloading station demands.

With known conditions and model above, the offloading plan can be made as shown in Table 15.

Table 16 shows the total offloading amount of each batch and the deviation between each batch's offloading volume and the demanded volume.

As the sum of the fifth batch's demand volumes is larger than the total input volume of the fifth batch, this means that the demand is not reasonable. The results show that

Table 10 Computational results

Example	Cont. var.	Disc. var.	Non-zero par.	# of con.	# of iter.	CUP time, s	Total deviation
1	232	40	40	345	1	0.176	3.381
2	276	48	41	409	100	281.222	397.770

Cont. var. the number of continuous variables, *Disc. var.* the number of discrete variables, *Non-zero par.* the number of non-zero parameters, *# of con.* the number of constraints, *# of iter.* the number of iterations, *CUP time* the calculation time

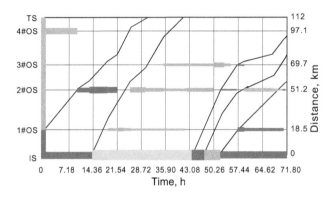

Fig. 3 Example one: batch transportation diagram (*blue* the 95# gasoline; *red* the 92# gasoline; *green* the 0# diesel)

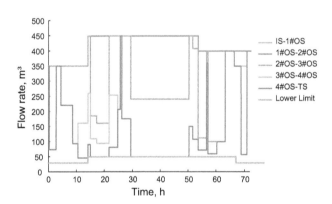

Fig. 4 Example one: flow rate in each pipeline segment and its lower limit

Table 11 Example two: first station input volumes of the new batches

Batch number	Oil type	First station input volumes, m³
2	92# gasoline	4962.0
3	−10# diesel	12,391.5
4	92# gasoline	2861.8
5	95# gasoline	2469.7
6	92# gasoline	7296.0

Table 12 Example two: volume coordinate of each batch

Batch number	Oil type	Volume coordinate, m³
1	95# gasoline	7497.8
2	92# gasoline	4896.5
3	−10# diesel	−4962.0
4	92# gasoline	−17353.5
5	95# gasoline	−20215.3
6	92# gasoline	−22685.0

Table 13 Example two: First station input flow rate

Time, h	Flow rate, m³/h
0.00–14.18	350
14.18–53.56	450
53.56–71.80	400

there exists a big deviation between the actual offloading volume and the demanded volume of the fifth batch at 4#OS, in accordance with the real situation. The offloading volumes of the rest of the batches all meet the demanded volumes.

The batch transportation diagram is shown in Fig. 5. In the diagram, the −10# diesel is in yellow. All the offloading operations are conducted within a reasonable time range.

The diesel oil-gasoline interfaces exist in the pipeline during the period from 14.18 h to 69.98 h. Therefore, the lower limit is 50 m³/h. For other periods, the lower limit is 30 m³/h. Figure 6 shows the flow rate in each pipeline segment. As shown in Fig. 6, the flow rate along the pipeline is always within the reasonable range.

The computational results of example two are shown in Table 10. Since the objective function has no zero-solution in this example, the calculation runs until it reaches the

Table 14 Example two: each offloading station's demanded volume of each batch

Station number	Batch number	Oil type	Demanded volumes, m^3
1	2	92# gasoline	676
2	2	92# gasoline	1867
2	3	−10# diesel	4349
2	4	92# gasoline	945
2	5	95# gasoline	1851
2	6	92# gasoline	227
3	2	92# gasoline	3161
3	3	−10# diesel	5848
3	4	92# gasoline	265
4	2	92# gasoline	1424
4	5	95# gasoline	1000

Table 15 Example two: offloading plan of each station

Station number	Batch number	Offloading volume, m^3	Offloading flow rate, m^3/h	Offloading start time, h	Offloading end time, h	Time period, h
1	2	676.329	276.053	0.30	2.75	2.45
2	2	594.769	264.342	15.03	17.28	2.25
2	2	1271.924	288.418	17.28	21.69	4.41
2	3	4348.498	209.062	29.48	50.28	20.80
2	4	81.584	42.054	51.62	53.56	1.94
2	4	863.342	287.781	53.56	56.56	3.00
2	5	1851.000	300.000	57.03	63.20	6.17
2	6	226.890	50.420	66.36	70.86	4.50
3	2	545.329	129.225	4.44	8.66	4.22
3	2	495.307	256.636	8.66	10.59	1.93
3	2	836.752	188.458	10.59	15.03	3.59
3	2	466.504	70.046	15.03	21.69	2.25
3	2	605.966	196.106	21.69	24.78	3.09
3	2	211.501	243.105	24.78	25.65	0.87
3	3	892.336	273.723	26.22	29.48	3.26
3	3	3971.502	190.938	29.48	50.28	20.8
3	3	983.877	299.963	50.28	53.56	3.28
3	4	264.833	39.765	53.56	60.22	6.66
4	2	415.262	115.672	10.59	14.18	3.59
4	2	144.047	169.467	14.18	15.03	0.85
4	2	135.250	60.111	15.03	17.28	2.25
4	2	195.554	44.343	17.28	21.69	4.41
4	2	533.769	172.741	21.69	24.78	3.09
4	5	604.482	292.020	68.79	70.86	2.07

maximum allowable iteration number. Thus, the convergence and stability need to be further discussed. Using same data in example 2, the calculation is repeated four times, and the iterating processes are shown in Fig. 7. The calculations, converging until they have iterated for respectively 61, 66, 69, 72, and 77 times, all converge to the same value in the end. The stability and convergence are demonstrated.

6 Conclusions

The proposed model considers the batches' arriving time as time nodes and takes the influence of mixed oil interface on minimum flow rate into account, which increases the difficulty of calculation. Thus, a continuous-time expression scheduling model is built and then a hybrid algorithm consisting of ACO and SM is applied to solve the model.

Table 16 Example two: offloading volume and deviation

Station number	Batch number	Oil type	Offloading volume, m³	Deviation, m³
1	2	92# gasoline	676.329	0.329
2	2	92# gasoline	1866.693	0.307
2	3	−10# diesel	4348.498	0.502
2	4	92# gasoline	944.926	0.074
2	5	95# gasoline	1851.000	0.000
2	6	92# gasoline	226.890	0.110
3	2	92# gasoline	3161.359	0.359
3	3	−10# diesel	5847.715	0.285
3	4	92# gasoline	264.833	0.167
4	2	92# gasoline	1423.881	0.119
4	5	95# gasoline	604.482	395.518

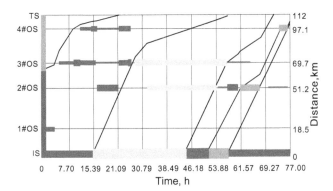

Fig. 5 Example two: batch transportation diagram (−10# diesel is in *yellow*)

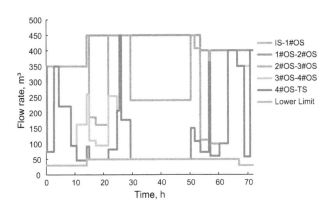

Fig. 6 Example two: flow rate in each pipeline segment and its lower limit

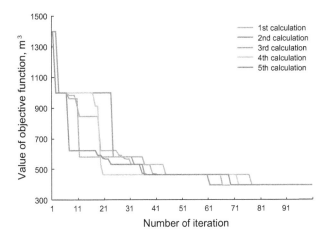

Fig. 7 Example two: process of iterations

can provide guidance to the scheduling plan for the actual operation. In further research, pipeline's hydraulic calculation will be taken into consideration in order to enhance the model's practicability and accuracy.

Acknowledgments The work is part of the Program of "Study on the mechanism of complex heat and mass transfer during batch transport process in products pipelines" funded under the National Natural Science Foundation of China (grant number 51474228). The authors are grateful to all study participants.

As shown in the examples, the calculation speed is relatively fast although the model is large. The scheduling plan obtained in this model has minimum deviation from the demand, which is also in accordance with the actual field situation, At the same time, the model's convergence and stability are verified. Therefore, the accuracy, efficiency, and practicability of this model are evident, and the results

References

BP. BP Statistical Review of World Energy. June 2014. www.bp.com/statisticalreview.

Cafaro DC, Cerdá J. Optimal scheduling of multiproduct pipeline systems using a non-discrete MILP formulation. Comput Chem Eng. 2004;28:2053–68.

Cafaro DC, Cerdá J. Efficient tool for the scheduling of multiproduct pipelines and terminal operations. Ind Eng Chem Res. 2008; 47(24):9941–56.

Cafaro DC, Cerdá J. A rigorous mathematical formulation for the scheduling of tree-structure pipeline networks. Ind Eng Chem Res. 2011;50:5064–85.

Cafaro DC, Cerdá J. Rigorous scheduling of mesh-structure refined petroleum pipeline networks. Comput Chem Eng. 2012;38: 185–203.

Cafaro VG, Cafaro DC, Méndez CA, et al. Detailed scheduling of operations in single-source refined products pipelines. Ind Eng Chem Res. 2011;50:6240–59.

Cafaro VG, Cafaro DC, Méndez CA, et al. Detailed scheduling of single-source pipelines with simultaneous deliveries to multiple offtake stations. Ind Eng Chem Res. 2012;51:6145–65.

Cafaro VG, Cafaro DC, Méndez CA, et al. MINLP model for the detailed scheduling of refined products pipelines with flow rate dependent pumping costs. Comput Chem Eng. 2015;72:210–21.

Castro PM. Optimal scheduling of pipeline systems with a resource-task network continuous-time formulation. Ind Eng Chem Res. 2010;49:11491–505.

de Souza Filho EM, Bahiense L, Ferreira Filho VJM. Scheduling a multi-product pipeline network. Comput Chem Eng. 2013; 53:55–69.

Herrán A, de la Cruz JM, de Andrés B. A mathematical model for planning transportation of multiple petroleum products in a multi-pipeline system. Comput Chem Eng. 2010;34:401–13.

Magatão L, Arruda LVR, Neves F Jr. A mixed integer programming approach for scheduling commodities in a pipeline. Comput Chem Eng. 2004;28:171–85.

Milidiú R L, dos Santos Liporace F. Planning of pipeline oil transportation with interface restrictions is a difficult problem. PUC. 2003.

MirHassani SA, Ghorbanalizadeh M. The multi-product pipeline scheduling system. Comput Math Appl. 2008;56(4):891–7.

Rejowski R, Pinto JM. Scheduling of a multiproduct pipeline system. Comput Chem Eng. 2003;27:1229–46.

Relvas S, Matos HA, Barbosa-Póvoa APFD, et al. Pipeline scheduling and inventory management of a multiproduct distribution oil system. Ind Eng Chem Res. 2006;45:7841–55.

Zyngier D, Kelly JD. Multi-product inventory logistics modeling in the process industries. Optimization and Logistics Challenges in the Enterprise. 2009;30:61–95.

6

Experimental analysis of pressure characteristics of catalyst powder flowing down a cyclone dipleg

Zhi-Gang Wei[1] · Chao-Yu Yan[1] · Meng-Da Jia[1] · Jian-Fei Song[1] · Yao-Dong Wei[1]

Abstract An experiment was carried out for investigating pressure behavior of catalyst powders, with a Sauter mean diameter of 63.6 μm, flowing downward in a cyclone dipleg with 150 mm inner diameter and 9000 mm high. Time mean pressure and time series of pressure fluctuations were measured at different axial positions in th dipleg with particle mass fluxes ranging from 50.0 to 385.0 kg m^{-2} s^{-1}. The experimental results showed that the time mean pressure in the dipleg increased progressively from the top section to the bottom section. The experimental phenomena displayed that the fluidization patterns in the dipleg can be divided into two types on the whole, namely the dilute–dense coexisting falling flow and the dense conveying flow along the dipleg. In the dilute–dense coexisting falling flow, the dilute phase region was composed of a length of swirling flow below the inlet of dipleg and a dilute falling flow above the dense bed level. With increasing particle mass flux, the dilute–dense coexisting falling flow was gradually transformed to be the dense conveying flow, and the exit pressure of the dipleg increased considerably. The pressure fluctuations were closely related to the fluidization patterns inside the dipleg. In the dilute–dense coexisting falling flow, the pressure fluctuations in the dilute flow region originated from particle clusters, propagating downward as a pressure wave; however, the pressure fluctuations in the dense flow region originated from rising gas bubbles, propagating upward. When the dense conveying flow was formed in the dipleg, the pressure fluctuations originated mainly from instability of the feed and the compressed gas, propagating downward. The standard deviation of the pressure fluctuations indicated that the intensity of pressure fluctuations first increased and then decreased with increasing particle flux.

Keywords Cyclone · Dipleg · Particle flow · Fluidization pattern · Pressure fluctuations

List of Symbols

Variables

g Gravity acceleration, m/s^2
G_s Particle mass flux, kg/m^2 s
P Pressure, Pa
t Time, s
Z Axial height or axial coordinates, m
V_s Particle velocity, m/s

Greek Letters

ρ_s Particle density, kg/m^3
ε Voidage (−)
ΔP Pressure difference, Pa
ΔZ Axial coordinate difference, m

1 Introduction

In gas–solid circulating fluidized (CFB) beds, the particles are collected by cyclones and then returned into a fluidized bed along a cyclone dipleg (dip-leg) conveying system. In the cyclone dipleg, the particles are conveyed mainly under gravity action from a low-pressure top section to a high-pressure bottom region in which the flow behavior is

✉ Chao-Yu Yan
   yanchaoyu@cup.edu.cn

[1] State Key Laboratory of Heavy Oil Processing, China
     University of Petroleum, Beijing 102249, China

Edited by Xiu-Qin Zhu

different from that in the riser or in the downcomer, although they have the same tubular structure. One of the flow characteristics in the dipleg is that the particle velocity is greater than gas velocity, so the direction of slip velocity between gas and particles is upward. Another one is the particle flow is usually against a negative pressure gradient. These two characteristics lead to the complexity of fluidization pattern and flow instability in the dipleg, which demonstrates that there are more than one different fluidization pattern coexisting and distinctive pressure fluctuations in the dipleg (Kunni and Levenspiel 1991; Hoffmann and Stein 2002; Cortés and Gil 2007).

Many researchers have studied diplegs, including Geldart et al. (1993), Li et al. (1997), Wang et al. (2000a,b), and Gil et al. (2002). Their emphases were focused on measuring the axial pressure distribution along the dipleg axial height and the fluidization regime transitions in the dipleg. Wang et al. (2000a,b) and Leung and Wilson (1973) investigated the gas-particle flow instability in the dipleg, and the effect of particle mass flux and negative pressure on the instability. Srivastava et al. (1998) and Zhang et al. (1998) found that under certain operating conditions the particle circuit manifested instability characterized by low frequency oscillations of particle circulation rates, and this instability originated in the standpipe. However, little work has been done on gas-particle flow instability involving the cyclone dipleg; furthermore, the relationship between the pressure fluctuations and the fluidization regime transitions, and the effect particle mass flux on the pressure fluctuations, is rarely reported.

In this work, a series of experiments were conducted with a catalyst powder in a pilot-plant scale CFB unit, to examine the pressure behavior in a 150-mm inner diameter (i.d.) and 9000-mm height cyclone dipleg under a wide range of operating conditions. The time mean pressure and the time series of pressure fluctuations in the cyclone dipleg are discussed based on the measured pressure data. In addition, the pressure fluctuation mechanism was also studied.

2 Experimental

Figure 1 shows the experimental setup employed in this work. It mainly consists of a fluidized bed (600 mm i.d. and 8000 mm height), a riser (200 mm i.d. and 12,500 mm height), and a dipleg (150 mm i.d. and 9000 mm height) connected with a cyclone. The cyclone has a cylinder with 400 mm i.d. The dipleg is made of plexiglas, making it easy to observe the internal flow phenomena. The dipleg exit was dipped into a dense bubbling fluidization bed without any valve. Particles separated from the cyclone flowed into the dipleg and then discharged into the

fluidized bed at the bottom exit of the dipleg. After the particles were conveyed back into the fluidized bed, the particles were pneumatically transported upward by gas through the riser and into the cyclone for separation again. The dipleg was equipped with a set of pressure transducers along its axial height for measuring pressure (as shown in Fig. 1). The measuring range of each pressure transducer was chosen based on the expected maximum pressure at each measure point. The accuracy of the pressure transducer was within ± 0.5 %. All signals from the pressure transducer were recorded simultaneously for a 30-s length of time. The axial base point ($Z = 0$) was set at the inlet of the dipleg, and the positive direction of axis Z is upward as shown in Fig. 1. The particle mass flux was measured by diverting the particles flowing into a collection vessel (not drawn in Fig. 1) for a known period and recording the mass accumulated in that vessel. The particle mass flux was controlled by a valve in the inclined pipe. The superficial gas velocity in the fluidized bed was 0.17–0.18 m s^{-1}. The solid particles employed in the experiments were equilibrium fluid catalytic cracking (FCC) catalyst powders, which had a bulk density of 1030 kg m^{-3}, a particle density of 1560 kg m^{-3}, and a Sauter mean diameter of 63.6 µm. The particle size distribution of FCC catalysts is displayed in Fig. 2.

3 Results and discussion

3.1 Flow phenomena in the dipleg

While the gas-particles flowed steadily in the dipleg under the designed operating conditions, the fluidization regime could be identified by visual observation. Based on the experimental phenomena, the fluidization regimes in the dipleg can be divided into two types, namely the dilute–dense coexisting falling flow and the dense conveying flow. In the dilute–dense coexisting falling flow, the dilute phase region is composed of a length of swirling flow below the inlet of dipleg and a dilute falling flow above the dense bed level. For example, when particle mass flux G_s was small (such as $G_s < 50.0$ kg m^{-2} s^{-1}), from the top section to the bottom section of the dipleg, the swirling flow, the dilute falling flow, and the dense flow appeared in the dipleg, respectively, as shown in Fig. 3a, b. The swirling flow was just located below the cyclone bottom and showed a helical particle flow along the wall of the dipleg. As the particles moved down, the swirling flow gradually straightened with the particles spreading over-section and forming the dilute falling flow mode. It was observed that the particles looked like heavy rain through the dipleg. At the bottom of the dipleg, the dense flow resembled a bubbling fluidized bed. There were some gas bubbles rising in the dense bed, which

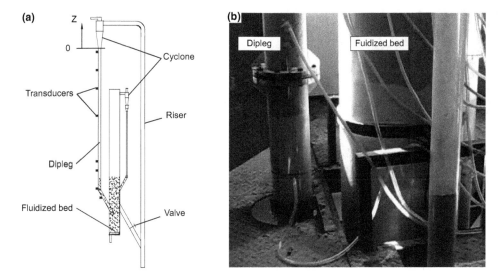

Fig. 1 Schematic diagram of the experimental setup and its photo (non-operating status)

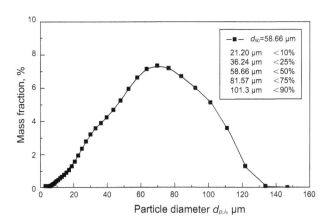

Fig. 2 Particle size distribution of FCC catalysts

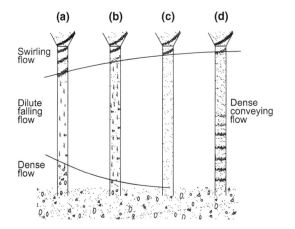

Fig. 3 Diagram of fluidization regimes in the dipleg

increasing G_s, the particle concentration in the dilute falling flow region increased, and the existing regions of swirling flow and dense flow reduced. The interface between the dilute falling flow and the dense flow moved up and down unsteadily caused by the eruption of rising gas bubbles. While $G_s = 200.0$–250.0 kg m^{-2} s^{-1}, the interface disappeared, and there were no gas bubbles in the dipleg. The fluidization regime was transferred from the coexistence of a dilute phase and dense phase flow to the dense conveying flow, as shown in Fig. 3c. Further increasing G_s (such as $G_s > 350.0$ kg m^{-2} s^{-1}), the fluidization regime became the dense conveying flow with a high particle concentration, as shown in Fig. 3d. The gas phase from the cyclone was carried by particles moving downwards. The particle flowing process showed the pressure fluctuation characteristics within the whole dipleg, and the particle concentration distribution looked like a bamboo joint type near the bottom of the dipleg. Figure 4 shows the photos of fluidization regimes in the dipleg during operation. Figure 4a shows the coexistence of dilute falling flow and dense flow, and Fig. 4b shows the dense conveying flow.

3.2 Time mean pressure profiles in the dipleg

For this submerged dipleg, the typical time mean pressure profiles along the axial direction were characterized by a progressively increasing pressure from top to bottom, as shown in Fig. 5. The time mean pressure profiles could be divided into three regions according to the local pressure gradient or the linear pressure distribution. The region I (swirling flow) was just below the inlet of dipleg, where the pressure gradient was small or even zero due to an annular flow. The region II (dilute falling flow) was observed at the

was the fluidized gas bypassing from the exit of the dipleg. So the particle phase moved downward and gas phase moved upward in the dense bed flow regime. With

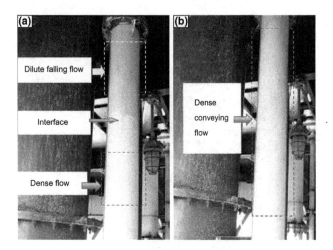

Fig. 4 Photos of fluidization regimes in the dipleg. **a** Dilute falling flow and dense flow. **b** Dense conveying flow

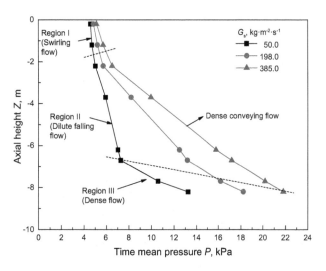

Fig. 5 Time mean pressure profiles along the axial height of the dipleg

the whole shape of pressure profile gradually became smooth, and the pressure difference between the dipleg inlet at its top and dipleg outlet at its bottom increased. Consequently, the pressure buildup range for balancing the negative pressure extended from the just dense flow pattern region in the lower region to the whole dipleg length, forming the dense conveying flow regime as shown in Fig. 5.

The relationship between pressure and axial position can be used to calculate the apparent voidage ε in the dipleg if the wall friction and acceleration action of particles are neglected, as shown in the following equation:

$$\Delta P = g(1 - \varepsilon)\rho_s \Delta Z \text{ or } \varepsilon = 1 - \frac{1}{\rho_s g}\frac{\Delta P}{\Delta Z}, \quad (1)$$

where ΔP and ΔZ are the pressure difference and axial coordinate difference between two pressure transducers, respectively. So, the corresponding voidage profiles in the dipleg are obtained according to the interval pressures. Figure 6 shows that the apparent voidage decreased continuously with particle flowing downwards in the dipleg, in accordance with the description for a downcomer given by Li et al. (1997). Figure 6 also demonstrates that the voidage in most parts of the dipleg decreased with increasing particle mass flux except in the bottom region of the dipleg. As a result, the voidage distribution tended to uniformity with increasing particle mass flux.

On the other hand, the particle velocity can be obtained by means of the following continuity equation with known G_s and ε:

$$V_s = \frac{G_s}{\rho_s(1 - \varepsilon)}. \quad (2)$$

The cross-sectional particle velocity obtained showed that the velocity was gradually decreasing along the particle flow direction in the dipleg, where the particles flow

top section of the dipleg and showed an intermediate pressure gradient, which varied with increasing particle mass flux. The region III (dense flow) appeared at the bottom of dipleg and displayed a large pressure gradient. The pressure gradient in region III decreased with increasing particle mass flux. In addition, the exit pressure of the dipleg increased considerably with increasing particle mass flux.

The pressure profiles are essentially similar to those obtained by Geldart et al. (1993), Li et al. (1997), and Wang et al. (2000a). However, the dipleg employed in this work has a larger diameter and longer length than those used in others' research. Furthermore, the exit of dipleg was submerged deeply into the bubbling fluidized bed in this work. So while the particle mass flux was increasing,

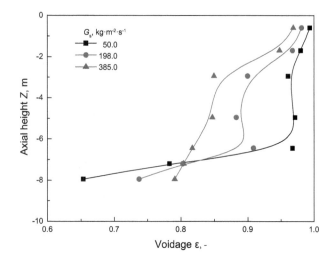

Fig. 6 Voidage profiles in the dipleg

downward from the low-pressure top region to the high-pressure bottom region. Such phenomenon was not observed in the downcomer. Moreover, since the particle velocity was related to the particle mass flux and the voidage, the particle velocity increased with increasing particle mass flux.

3.3 Pressure fluctuation

The experimental results indicated that the gas-particle flow was instable in the dipleg. This instable gas-particle flow can be expressed in the form of pressure fluctuation. Figure 7 shows the pressure profiles with time measured along the dipleg. The time mean values of these dynamic pressures are shown in Fig. 5. The dynamic pressure profiles in Fig. 7 consist of two different types of pressure fluctuations: one is of low frequency and high amplitude and another is of high frequency and low amplitude.

The pressure fluctuations were related to the particle mass flux and fluidization pattern inside the dipleg. When G_s was small (such as $G_s < 50.0$ kg m^{-2} s^{-1}), the pressure fluctuation in Fig. 7a was of high frequency and low amplitude in the dilute falling flow, and low frequency and high amplitude in the dense flow. The former was caused by the falling particle clusters interacting with the up flowing gas, and the latter was caused by the bubbles rising in the dense flow. As G_s increased, the particles had a greater tendency to increase the population of clusters rather than to increase the cluster size, resulting in the pressure fluctuation with high frequency and low amplitude, as shown in Fig. 7b. However, the particle feed from the cyclone into the dipleg showed considerable instability, which caused the pressure fluctuation with low frequency and high amplitude. Meanwhile, the surface of the dense flow moved up and down unsteadily, showing significant pressure fluctuations with low frequency and high amplitude. When G_s increased (such as $G_s > 350.0$ kg m^{-2} s^{-1}) to form dense conveying flow, the pressure fluctuations are shown in Fig. 7c. The down-flowing clusters of particles fall inside the dipleg, and as a result, quantities of gas were carried downwards by the particles, which has been proven by the measured results of Geldart et al. (1993) and Li et al. (1997). Gas-particle flow in the dipleg was flowing against negative pressure, and the slip velocity was upward to balance the negative pressure, resulting in compression of the carried gas. The gas was compressed and expanded alternately to induce the particle concentration and the pressure non-uniformity. Therefore, the pressure fluctuations are attributed to the unstable feed and the compressed gas.

Furthermore, the transfer direction of pressure fluctuations can be identified according to the similarity of pressure profiles, as shown in Fig. 7. The particles in the dipleg

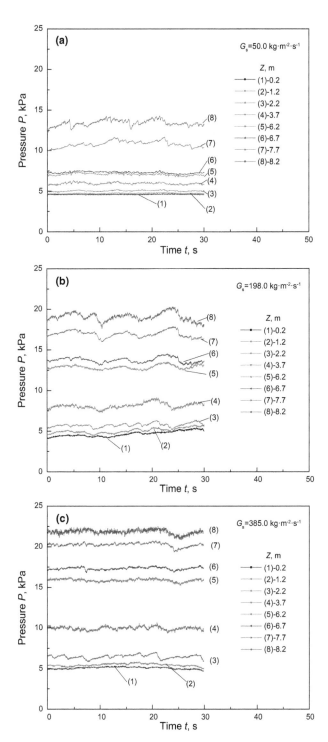

Fig. 7 Pressure fluctuation profiles in the dipleg. **a** $G_s = 50.0$ kg m^{-2} s^{-1}, **b** $G_s = 198.0$ kg m^{-2} s^{-1}, **c** $G_s = 385.0$ kg m^{-2} s^{-1}

were essentially free flowing with the help of gravity, but gas was dependent on the particle flow. The pressure fluctuation propagated downward in the fluidization pattern of dilute falling flow and dense conveying flow, and upward in the fluidization pattern of dense flow.

3.4 Standard deviation of the pressure fluctuations

Because the amplitude of the pressure fluctuation is related to the particle mass flux in the dipleg, the pressure fluctuations can reflect the fluidization regimes of the particle transport types. The pressure fluctuation intensity can be described by the standard deviation of fluctuation pressure. Figure 8 shows the standard deviation profiles of pressure fluctuations along the dipleg under different particle mass fluxes. The pressure fluctuations are attributed to many factors in the operation of diplegs, including particle clusters, gas bubbles, wall friction, and instability of feed and discharge, which enable the pressure fluctuations to be multi-scale and complex in nature (Bi 2007; Jing and Li 1999).

When G_s was small (such as $G_s < 50.0$ kg m^{-2} s^{-1}), the fluidization pattern in the dipleg was featured by the coexistence of dilute falling flow and dense flow. In the upper dilute region, the pressure fluctuations originated from the particle clusters flowing downwards and the unstable feed from the cyclone. While in the bottom dense region, the local pressure fluctuations originated from the local bubbles and unstable discharge. With increasing G_s, the particle concentration increased. The pressure fluctuation mainly resulted from the unstable feed, the friction between particles and wall, the gas compressed by particles, etc. The pressure fluctuation intensity reached a maximum during the fluidization pattern being transformed (while $G_s = 200.0$–250.0 kg m^{-2} s^{-1}). With further increase of G_s (such as $G_s > 350.0$ kg m^{-2} s^{-1}), the gas-particle flow evolved into dense conveying flow. The amplitude of the pressure fluctuations gradually decreased. The reason is that the increasing particle mass flux can reduce the possibilities of forming a non-uniform distribution of particle concentration and causing pressure

fluctuations. It was clear that the influence of gas turbulence was weakened and the interaction between particles or clusters was dominant as particle flux increasing. The continuous agglomeration of particles and the crush of clusters led to the fluctuation of dynamic pressure.

As discussed above, the interaction and the coupling among gas-particle flow and unstable feed created the complexity of local pressure fluctuations. On the other hand, the pressure fluctuations could produce an alternating load to induce the vibration of the cyclone and dipleg. Sometimes the alternating load could induce resonance vibration of the cyclone system and possibly cause a fatigue fracture of the cyclone shell or support rod (Zheng et al. 2011). So, the pressure fluctuation in the dipleg has the potential harm to the operation of the cyclone system.

4 Conclusions

This paper presents an experimental study on the pressure behaviors of a catalyst powder flowing down in a dipleg with 150 mm i.d. and 9000 mm high. The experimental phenomena displayed that the fluidization regimes in the dipleg could be divided into two types on the whole: the dilute–dense coexisting falling flow and the dense conveying flow. In the dilute–dense coexisting falling flow, the dilute phase region was composed of a length of swirling flow below the inlet of dipleg and a dilute falling flow above the dense bed level. As particle mass flux increased, the fluidization regimes gradually developed into dense conveying flow. Meanwhile, the time mean pressure increased progressively. The time series of pressure fluctuations showed that there existed a gas-particle flow instability in the dipleg, which consisted of low frequency with high amplitude and high frequency with low amplitude. The former was caused by the falling particle clusters, the latter was the result of the other factors, such as bubbles rising in the dense flow, instability feed, and compressed gas. The results indicated that the particle mass flux had strongly influence on the time mean pressure and pressure fluctuations.

Acknowledgments The authors acknowledged the support from the National Natural Science Foundation of China (Grant No. 21176250, 21566038), and by the Science Foundation of China University of Petroleum, Beijing (No. 2462015YQ0301).

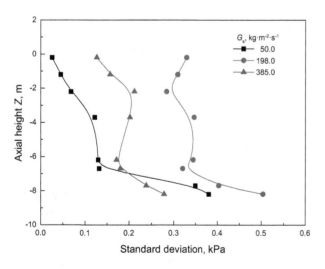

Fig. 8 Standard deviation of pressure fluctuations in the dipleg

References

Bi HT. A critical review of the complex pressure fluctuation phenomenon in gas–solids fluidized beds. Chem Eng Sci. 2007;62(13):3473–93.

Cortés C, Gil A. Modeling the gas and particle flow inside cyclone separators. Prog Energy Combust Sci. 2007;33(5):409–52.

Geldart D, Broodryk N, Kerdoncuff A. Studies on the flow of solid down cyclone diplegs. Powder Technol. 1993;76(2):175–83.

Gil A, Cortés C, Romeo LM, et al. Gas-particle flow inside cyclone diplegs with pneumatic extraction. Powder Technol. 2002;128(1):78–91.

Hoffmann AC, Stein LE. Gas cyclones and swirl tubes principles, design and operation. Berlin: Springer-Verlag; 2002. p. 235–56.

Jing S, Li H. Study on the flow of fine powders from hoppers connected to a moving-bed standpipe with negative pressure gradient. Powder Technol. 1999;101(3):266–78.

Kunni D, Levenspiel O. Fluidization engineering. 2nd ed. Boston: Butterworth-Heinemann; 1991. p. 371–8.

Leung LS, Wilson LA. Downflow of solids in standpipes. Powder Technol. 1973;7(6):343–9.

Li Y, Lu Y, Wang F, et al. Behavior of gas–solid flow in the downcomer of a circulating fluidized bed reactor with a V-valve. Powder Technol. 1997;91(1):11–6.

Srivastava A, Agrawal K, Sundaresan S, et al. Dynamics of gas-particle flow in circulating fluidized beds. Powder Technol. 1998;100(2–3):173–82.

Wang J, Bouma JH, Dries H. An experimental study of cyclone dipleg flow in fluidized catalyst cracking. Powder Technol. 2000a;112(3):221–8.

Wang SJ, Geldart D, Beck MS, et al. A behavior of a catalyst powder flowing down in the dipleg. Chem Eng J. 2000b;77(1–2):51–6.

Zhang JY, Rudolph V, Zhang JY. Flow instability in non-fluidized standpipe flow. Powder Technol. 1998;97(2):109–17.

Zheng M, Chen G, Han J. Failure analysis on two austenitic stainless steels applied in cyclone separators of catalytic cracking unit. Eng Fail Anal. 2011;18(1):88–96.

7

Pre-stack basis pursuit seismic inversion for brittleness of shale

Xing-Yao Yin[1] · Xiao-Jing Liu[1] · Zhao-Yun Zong[1]

Abstract Brittleness of rock plays a significant role in exploration and development of shale gas reservoirs. Young's modulus and Poisson's ratio are the key parameters for evaluating the rock brittleness in shale gas exploration because their combination relationship can quantitatively characterize the rock brittleness. The high-value anomaly of Young's modulus and the low-value anomaly of Poisson's ratio represent high brittleness of shale. The technique of pre-stack amplitude variation with angle inversion allows geoscientists to estimate Young's modulus and Poisson's ratio from seismic data. A model constrained basis pursuit inversion method is proposed for stably estimating Young's modulus and Poisson's ratio. Test results of synthetic gather data show that Young's modulus and Poisson's ratio can be estimated reasonably. With the novel method, the inverted Young's modulus and Poisson's ratio of real field data focus the layer boundaries better, which is helpful for us to evaluate the brittleness of shale gas reservoirs. The results of brittleness evaluation show a good agreement with the results of well interpretation.

Keywords Brittleness · Shale gas · Amplitude variation with angle · Basis pursuit · Bayesian framework

✉ Xing-Yao Yin
xyyin@upc.edu.cn

[1] School of Geosciences, China University of Petroleum
(Huadong), Qingdao 266580, Shandong, China

Edited by Jie Hao

1 Introduction

Shale gas is a very important type of unconventional resource. The term refers to the unconventional gas stored in shale reservoirs. With the development of seismic exploration, a large amount of practice in unconventional shale reservoirs indicated that rock brittleness is one of the critical parameters to be taken into consideration in the evaluation of hydraulic fracturing. The study of shale brittleness is very important for shale gas exploration and development. An empirical brittleness cut-off is defined based on Young's modulus and Poisson's ratio (Grieser and Bray 2007; Rickman et al. 2008) as they control the relationship between stress and strain given by Hooke's Law (Sena et al. 2011). A high-value anomaly of Young's modulus and a low-value anomaly of Poisson's ratio can be used to evaluate the rock brittleness and to infer "sweet spots" of shale gas reservoirs (Harris et al. 2011; Zong et al. 2013). Seismic inversion is the fundamental scientific tool used to obtain parameters concerning lithology and physical properties (Yin et al. 2015). Therefore, the estimation of Young's modulus and Poisson's ratio from pre-stack seismic data is a helpful guide for evaluating the brittleness of shale.

Amplitude variation with angle (AVA) inversion can be used to estimate the subsurface elastic properties from pre-stack seismic reflection data. However, the geophysical inversion problem in nature is an ill-conditioned problem because slight noise contained in the observed data will lead to enormous changes in the estimated parameters. Another problem of AVA inversion is that there are many models adequately fitting the data because the seismic data are band limited. It is common to add additional constraints to stabilize the inversion process and to reduce the number of solutions. This is generally referred to as regularization.

The regularization method was proposed by Tikhonov (1963), and the L-curve (Hansen 1992) was presented for selecting the regularization parameters which balance the data fitting term and trade-off function. A Bayesian approach is another method for stabilizing the inversion performance that treats the model parameter as a random variable with a probability distribution (Duijndam 1987; Buland and Omre 2003; Tarantola 2005; Yuan and Wang 2013; Yin and Zhang 2014). We seek the parameter estimates with a maximum posterior distribution combined with prior information of model parameters and the likelihood function. The prior information can be the probability distribution of the model parameters and the geological information. In special cases, the regularization function is equivalent to the prior information in the Bayesian method.

The sparse solutions are full band; therefore, the sparse estimations are often viewed as high-resolution estimations (Levy and Fullagar 1981; Sacchi 1997; Alemie and Sacchi 2011). The inversion results via l_2 norm regularization (Tikhonov 1963) or assumption of Gaussian probability distribution (Downton 2005) do not lead to high resolution because the estimates lack sparsity. The sparse reflection coefficients generate the blocky layer elastic parameters which suppress the side lobes. The development of the theory of sparse representation promotes the sparse inversion method. Theune et al. (2010) investigated the Cauchy and Laplace statistical distributions for their potential to recover sharp boundaries between adjacent layers. Based on the reflection dipole decomposition described by Chopra et al. (2006), Zhang et al. (2009, 2011) studied the basis pursuit inversion (BPI) of post and pre-stack seismic data, respectively, and got the sparse reflection coefficients and blocky layer elastic parameters, which is a high-resolution inversion method. Pérez et al. (2013) proposed a hybrid Fast Iterative Shrinkage-Thresholding Algorithm (FISTA) least-squares strategy that inverts the location of reflection by the FISTA algorithm (Beck and Teboulle 2009) first and then reevaluates the sparse (high resolution) reflection coefficients. However, this lacks the low-frequency information if we only use the seismic data and sparse regularization. The low-frequency information should be incorporated into the objective function to enhance the meaning of the inversion results and meanwhile promote the stability of the inversion implementation (Yin et al. 2008, 2014; Zong et al. 2012a; Yuan et al. 2015).

The ultimate goal of pre-stack inversion is to obtain the elastic information that can be used for evaluating hydrocarbon potential and the brittleness of the reservoir and to infer ideal drilling locations of "sweet spots" (Sena et al. 2011). Different linear approximations (Aki and Richards 1980; Gray et al. 1999; Russell et al. 2011) of the Zoeppritz equation introduced by Zoeppritz (1919) help us to directly

estimate the elastic parameters (e.g., P-wave velocity, S-wave velocity, Lamé parameters, bulk modulus, density) in which we are interested. The Young's modulus and Poisson's ratio can be calculated from the P-wave velocity, S-wave velocity, and density which can be inverted directly via Aki-Richards approximation. The density is difficult to invert, which will have a deleterious influence on the estimation of Young's modulus. Parameters estimated indirectly will bring in more uncertainty in the inversion results (Zhang et al. 2009). In order to estimate the Young's modulus (Y), Poisson's ratio (P) and density (D) directly, Zong et al. (2012b) derived the linear approximation equation based on Young's modulus, Poisson's ratio, and density and inverted the elastic parameters by Bayesian framework via Cauchy distribution as prior information. The approximation can be named as the YPD approximation. Zong et al. (2013) reformulated the elastic impedance equation in terms of Young's modulus, Poisson's ratio and density based on the YPD approximation, and introduced a stable inversion method named elastic impedance varying with incident angle inversion with damping singular value decomposition (EVA-DSVD) inversion. In this study, we propose a model constrained basis pursuit inversion method to estimate the Young's modulus, Poisson's ratio, and density with the YPD approximation. The model constraint term is added into the objective function through a Bayesian framework. We also take a decorrelation of model parameters before inversion. The introduced model constraint promotes the stability of the inversion. Basis pursuit ensures the sparsity of reflection coefficients and the blocky structure of layer parameters. Model synthetic gather data with different signal-to-noise ratios are studied to test the proposed inversion method. The application on real data from shale reservoirs shows that Young's modulus and Poisson's ratio inverted by the proposed inversion method are reasonable for brittleness evaluation. The result of brittleness evaluation fits well with well interpretation.

2 Theory and method

2.1 YPD approximation

Young's modulus and Poisson's ratio are the central parameters in predicting the brittleness of the subsurface layers. Young's modulus can characterize the rigidity or brittleness of rocks, and Poisson's ratio can be regarded as a kind of fluid factor which can be used for pore fluid identification. The reflection coefficients approximate equation was derived in terms of Young's modulus, Poisson's ratio, and density (YPD approximation) with the hypothesis of planar incident wave by Zong et al. (2012c):

$$R(\theta) = \left(\frac{1}{4}\sec^2\theta - 2k\sin^2\theta\right)\frac{\Delta E}{E}$$
$$+ \left(\frac{1}{4}\sec^2\theta\frac{(2k-3)(2k-1)^2}{k(4k-3)} + 2k\sin^2\theta\frac{1-2k}{3-4k}\right)$$
$$\times \frac{\Delta\sigma}{\sigma} + \left(\frac{1}{2} - \frac{1}{4}\sec^2\theta\right)\frac{\Delta\rho}{\rho}, \tag{1}$$

where θ is the incident angle; $R(\theta)$ is the reflection coefficients; k stands for the square of the average S-to-P velocity ratio; and $\Delta E/E$, $\Delta\sigma/\sigma$, and $\Delta\rho/\rho$ represent the reflection coefficients of Young's modulus, Poisson's ratio, and density, respectively.

In order to perform the inversion for Young's modulus, Poisson's ratio and density, we should firstly build the forward model. Combining the convolution model and YPD approximation, we can get the pre-stack data in angle domain shown as Eq. (2).

$$\begin{bmatrix}\mathbf{d}(\theta_1)\\\mathbf{d}(\theta_2)\\\vdots\\\mathbf{d}(\theta_n)\end{bmatrix} = \begin{bmatrix}\mathbf{W}(\theta_1)\mathbf{C}_E(\theta_1) & \mathbf{W}(\theta_1)\mathbf{C}_\sigma(\theta_1) & \mathbf{W}(\theta_1)\mathbf{C}_\rho(\theta_1)\\\mathbf{W}(\theta_2)\mathbf{C}_E(\theta_2) & \mathbf{W}(\theta_2)\mathbf{C}_\sigma(\theta_2) & \mathbf{W}(\theta_2)\mathbf{C}_\rho(\theta_2)\\\vdots & \vdots & \vdots\\\mathbf{W}(\theta_n)\mathbf{C}_E(\theta_n) & \mathbf{W}(\theta_n)\mathbf{C}_\sigma(\theta_n) & \mathbf{W}(\theta_n)\mathbf{C}_\rho(\theta_n)\end{bmatrix}$$
$$\times \begin{bmatrix}\mathbf{r}_E\\\mathbf{r}_\sigma\\\mathbf{r}_\rho\end{bmatrix} \tag{2}$$

where $\mathbf{W}(\theta_i)$ represents the wavelet matrix and $\mathbf{C}_E(\theta_i)$, $\mathbf{C}_\sigma(\theta_i)$, and $\mathbf{C}_\rho(\theta_i)$ represent the weighting coefficients of Young's modulus reflectivity vector $\mathbf{r}_E$, Poisson's ratio reflectivity vector $\mathbf{r}_\sigma$, and density reflectivity vector $\mathbf{r}_\rho$, respectively. The product of the wavelet matrix and weighting coefficients makes up the kernel matrix $\mathbf{G}$. Setting the reflectivities as model vector $\mathbf{r}$, the forward model equation can be written in a linear equation as $\mathbf{d} = \mathbf{Gr}$.

2.2 Model parameters decorrelation

Decorrelation of model space parameters can enhance the stability of the three-parameter AVA inversion (Downton 2005; Zong et al. 2012b). We took the singular value decomposition (SVD) for covariance matrix $\mathbf{C}_r$ of model elastic parameters:

$$\mathbf{C}_r = \begin{bmatrix}\sigma_E^2 & \sigma_{E\sigma} & \sigma_{E\rho}\\\sigma_{\sigma E} & \sigma_\sigma^2 & \sigma_{\sigma\rho}\\\sigma_{\rho E} & \sigma_{\rho\sigma} & \sigma_\rho^2\end{bmatrix} = \mathbf{v}\mathbf{S}\mathbf{v}^T \tag{3}$$

where σ_E^2 is the variance of Young's modulus; $\sigma_{E\sigma}$ is the covariance of Young's modulus and Poisson's ratio, and so on; $\mathbf{v}$ is a matrix made up of three eigenvectors; and $\mathbf{S}$ is the diagonal matrix made up of the positive decreasing singular values.

For N samples, the inverse of the decorrelation matrix $\mathbf{V}$ can be expressed as the Kronecker product of $\mathbf{v}^{-1}$ and an N-order identity matrix. Therefore, the inverse of decorrelation matrix $\mathbf{V}$ is expressed as Eq. (4):

$$\mathbf{V}^{-1} = kron(\mathbf{v}^{-1}, \mathbf{I}) \tag{4}$$

In this case, the kernel matrix $\mathbf{G}$ becomes $\tilde{\mathbf{G}} = \mathbf{GV}^{-1}$, and the model vector becomes $\tilde{\mathbf{r}} = \mathbf{Vr}$. Therefore, the forward modeling equation should be written as

$$\mathbf{d} = \mathbf{Gr} = (\tilde{\mathbf{G}}\mathbf{V})(\mathbf{V}^{-1}\tilde{\mathbf{r}}) = \tilde{\mathbf{G}}\tilde{\mathbf{r}}. \tag{5}$$

2.3 Dipole decomposition and forward model

Then we used the dipole reflection coefficient decomposition method to update the forward model of pre-stack BPI inversion for Young's modulus, Poisson's ratio, and density. The reflection coefficient decomposition method is shown in Fig. 1. In this case, the vector of reflection coefficients containing Young's modulus, Poisson's ratio, and density can be written as Eq. (6).

$$\tilde{\mathbf{r}} = \begin{bmatrix}\tilde{\mathbf{r}}_E\\\tilde{\mathbf{r}}_\sigma\\\tilde{\mathbf{r}}_\rho\end{bmatrix} = \begin{bmatrix}\mathbf{r}_e & \mathbf{r}_o & 0 & 0 & 0 & 0\\0 & 0 & \mathbf{r}_e & \mathbf{r}_o & 0 & 0\\0 & 0 & 0 & 0 & \mathbf{r}_e & \mathbf{r}_o\end{bmatrix}\begin{bmatrix}\mathbf{a}_E\\\mathbf{b}_E\\\mathbf{a}_\sigma\\\mathbf{b}_\sigma\\\mathbf{a}_\rho\\\mathbf{b}_\rho\end{bmatrix}$$
$$= \begin{bmatrix}\mathbf{D} & & \\ & \mathbf{D} & \\ & & \mathbf{D}\end{bmatrix}\begin{bmatrix}\mathbf{m}_E\\\mathbf{m}_\sigma\\\mathbf{m}_\rho\end{bmatrix} = \tilde{\mathbf{D}}\mathbf{m} \tag{6}$$

where $\mathbf{D}$ stands for the dipole reflectivity decomposition operator. $\mathbf{m}_E$, $\mathbf{m}_\sigma$ and $\mathbf{m}_\rho$ are the sparse coefficients of Young's modulus, Poisson's ratio, and density, respectively, under the reflectivity decomposition, $\mathbf{m}_E = [\mathbf{a}_E^T\ \mathbf{b}_E^T]^T$, $\mathbf{m}_\sigma = [\mathbf{a}_\sigma^T\ \mathbf{b}_\sigma^T]^T$, and $\mathbf{m}_\rho = [\mathbf{a}_\rho^T\ \mathbf{b}_\rho^T]^T$. $\tilde{\mathbf{D}}$ is a large scale matrix consisting of three reflectivity decomposition operators. $\mathbf{m}$ is the sparse coefficients vector consisting of $\mathbf{m}_E$, $\mathbf{m}_\sigma$, and $\mathbf{m}_\rho$.

We can obtain the forward model shown as Eq. (7) by substituting Eq. (6) into Eq. (5):

$$\mathbf{d} = \tilde{\mathbf{G}}\tilde{\mathbf{D}}\mathbf{m}. \tag{7}$$

2.4 Bayesian inference and model constrained BPI

In this study, we constructed the objective function of the AVA inversion under the Bayesian framework (Ulrych et al. 2001). The Bayesian theorem is given by

$$p(\mathbf{m}|\mathbf{d}) = \frac{p(\mathbf{d}|\mathbf{m})p(\mathbf{m})}{p(\mathbf{d})}, \tag{8}$$

where $p(\mathbf{m}|\mathbf{d})$ is the posterior probability distribution function (PDF), $p(\mathbf{d}|\mathbf{m})$ is the likelihood function that is the PDF of noise, $p(\mathbf{m})$ is the prior information of the parameter $\mathbf{m}$,

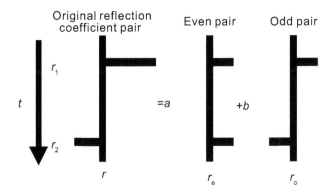

Fig. 1 The reflectivity decomposition (Zhang et al. 2013)

and $p(\mathbf{d})$ is the normalization factor which can be ignored as it is a constant value. Therefore, the Bayesian theorem is often expressed as Eq. (9) without the scaling factor:

$$p(\mathbf{m}|\mathbf{d}) \propto p(\mathbf{d}|\mathbf{m})p(\mathbf{m}). \qquad (9)$$

We suppose that the noise obeys a Gaussian distribution; hence, the likelihood function should be

$$p(\mathbf{d}|\mathbf{m}) \propto \exp\left((\tilde{\mathbf{G}}\tilde{\mathbf{D}}\mathbf{m} - \mathbf{d})^{\mathrm{T}} \mathbf{X}_{\mathrm{d}}^{-1} (\tilde{\mathbf{G}}\tilde{\mathbf{D}}\mathbf{m} - \mathbf{d})\right), \qquad (10)$$

where $\mathbf{X}_{\mathrm{d}}$ is the noise covariance matrix. For simplicity, we suppose that the noise is uncorrelated, so the covariance matrix should be $\mathbf{X}_{\mathrm{d}} = \sigma_{\mathrm{d}}^2 \mathbf{I}_{\mathrm{d}}$, where $\mathbf{I}_{\mathrm{d}}$ is the identity matrix and σ_{d}^2 is the variance of the Gaussian distributed noise.

In this paper, we assume that the prior distribution is constructed by two terms:

$$p(\mathbf{m}) = p_{\mathrm{t}}(\mathbf{m})p_{\mathrm{l}}(\mathbf{m}), \qquad (11)$$

where the first term, $p_{\mathrm{t}}(\mathbf{m})$, is the probability distribution of $\mathbf{m}$ which represents the sparsity of the coefficients and the second term, $p_{\mathrm{l}}(\mathbf{m})$, is the low-frequency model information that can enhance the lateral continuity. The prior information stabilized the inversion process and provided a principle to choose the "best" solution that can adequately fit the observed data.

In order to recover the discontinuity of the layer properties, the minimum l_1 norm that works well in selecting the sparse solution practically should be taken into consideration to constrain the inversion. This l_1 norm regularization can be incorporated into the Bayesian approach as the Laplacian distribution with a mean of zero:

$$p_{\mathrm{t}}(\mathbf{m}) \propto \exp\left(-\frac{1}{\sigma_{\mathrm{m}}}\sum_{i=1}^{M}|m_i|\right) \qquad (12)$$

In the lateral term, we suppose that the error between the inversion and low-frequency model obeys a normal distribution.

$$p_{\mathrm{l}}(\mathbf{m}) \propto \exp\left(\frac{1}{2}(\tilde{\mathbf{C}}\tilde{\mathbf{D}}\mathbf{m} - \boldsymbol{\xi})\mathbf{X}_{(\mathrm{m},\xi)}^{-1}(\tilde{\mathbf{C}}\tilde{\mathbf{D}}\mathbf{m} - \boldsymbol{\xi})\right)^{\mathrm{T}} \qquad (13)$$

where $\mathbf{X}_{(\mathrm{m},\xi)}$ is the covariance matrix associated with the three elastic properties: Young's modulus, Poisson's ratio, and density. Here we assume that the parameters at each sample are independent as we took decorrelation of the model parameters, and then $\mathbf{X}_{(\mathrm{m},\xi)} = \sigma_{(\mathrm{m},\xi)}^2 \mathbf{I}_{(\mathrm{m},\xi)}$, where $\sigma_{(\mathrm{m},\xi)}^2$ is the variance of the coefficients for estimation, and $\mathbf{I}_{(\mathrm{m},\xi)}$ is the identity matrix. $\boldsymbol{\xi}$ is the vector made up of the relevant Young's modulus, Poisson's ratio, and density; $\tilde{\mathbf{C}}$ is made up of diagonal integrated matrix $\mathbf{C} = \int_{t_0}^{t} \mathrm{d}\tau$. The expression of $\boldsymbol{\xi}$ and $\tilde{\mathbf{C}}$ can be written as Eq. (14):

$$\boldsymbol{\xi} = \begin{bmatrix} \ln(\mathbf{E}/E_0) \\ \ln(\boldsymbol{\sigma}/\sigma_0) \\ \ln(\boldsymbol{\rho}/\rho_0) \end{bmatrix} \quad \tilde{\mathbf{C}} = \begin{bmatrix} \mathbf{C} & & \\ & \mathbf{C} & \\ & & \mathbf{C} \end{bmatrix}. \qquad (14)$$

Under Bayes' framework, we can estimate the solution as the maximum a posteriori (MAP) solution. Substituting Eq. (12) and Eq. (13) into Eq. (11), the prior information can be written as

$$p(\mathbf{m}) \propto \exp\left[-\left(\frac{1}{\sigma_{\mathrm{m}}}\|\mathbf{m}\|_1 + \frac{1}{\sigma_{(\mathrm{m},\xi)}^2}\|\tilde{\mathbf{C}}\tilde{\mathbf{D}}\mathbf{m} - \boldsymbol{\xi}\|_2^2\right)\right] \qquad (15)$$

Substituting the likelihood function Eq. (10) and prior distribution Eq. (15) into the Bayesian theorem Eq. (9), we get the objective function shown as Eq. (16) under the Bayesian inference:

$$J(\mathbf{m}) = \|\tilde{\mathbf{G}}\tilde{\mathbf{D}}\mathbf{m} - \mathbf{d}\|_2^2 + \lambda\|\mathbf{m}\|_1 + \mu\|\tilde{\mathbf{C}}\tilde{\mathbf{D}}\mathbf{m} - \boldsymbol{\xi}\|_2^2, \qquad (16)$$

θwhere, λ and μ are the trade-off factors which balance the overall impact of the regularization, and $\lambda = \sigma_{\mathrm{d}}^2/\sigma_{\mathrm{m}}$, and $\mu = \sigma_{\mathrm{d}}^2/\sigma_{(\mathrm{m},\xi)}^2$.

The objective function Eq. (16) can be viewed as the normal expression for a basis pursuit problem via an augmented matrix:

$$J(\mathbf{m}) = \left\|\begin{pmatrix} \tilde{\mathbf{G}} \\ \sqrt{\mu}\tilde{\mathbf{C}} \end{pmatrix}\tilde{\mathbf{D}}\mathbf{m} - \begin{pmatrix} \mathbf{d} \\ \sqrt{\mu}\boldsymbol{\xi} \end{pmatrix}\right\|_2^2 + \lambda\|\mathbf{m}\|_1. \qquad (17)$$

Accordingly, utilizing the Gradient Projection for Sparse Reconstruction (GPSR) method (Figueiredo et al. 2007), we minimized the objective function $J(\mathbf{m})$ to obtain the sparse estimates. After that, the reflection coefficients with isolated spikes can be obtained by Eq. (6). The inverted results contain low frequencies as the low-frequency trend model data were added into the objective function as a penalty function. Therefore, the output Young's modulus, Poisson's ratio, and density with blocky boundaries can be obtained by Eq. (18):

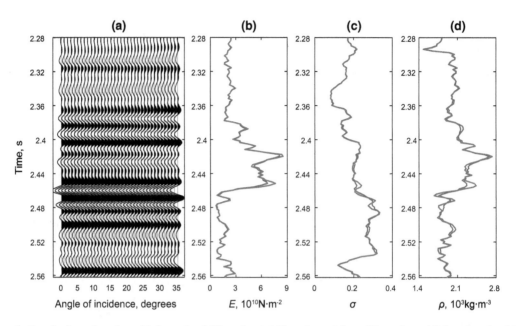

Fig. 2 a The synthetic seismic angle gather with free noise; **b** The estimated Young's modulus; **c** The estimated Poisson's ratio; **d** The estimated density. **b–d** the *red lines* mean the inverted results and the *blue lines* represent the real model curves

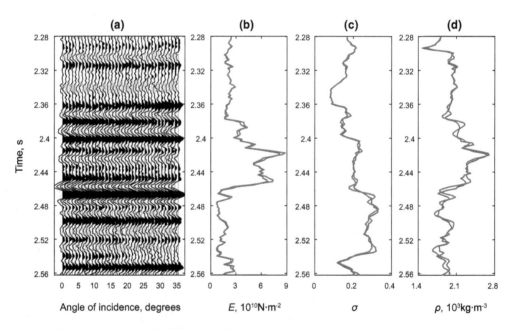

Fig. 3 a The synthetic seismic angle gather with SNR = 4:1; **b** The estimated Young's modulus; **c** The estimated Poisson's ratio; **d** The estimated density. **b–d** the *red lines* mean the inverted results, and the *blue lines* represent the real model curves

$$E(t) = E(t_0) \exp \int_{t_0}^{t} r_E(\tau) \, d\tau$$

$$\sigma(t) = \sigma(t_0) \exp \int_{t_0}^{t} r_\sigma(\tau) \, d\tau \,. \tag{18}$$

$$\rho(t) = \rho(t_0) \exp \int_{t_0}^{t} r_\rho(\tau) \, d\tau$$

3 Model test

We tested the validity of our proposed inversion method with well log data. The angle gather data with free noise (Fig. 2a) were synthesized with Zoeppritz equations in the time domain by utilizing the real well logs of P-wave

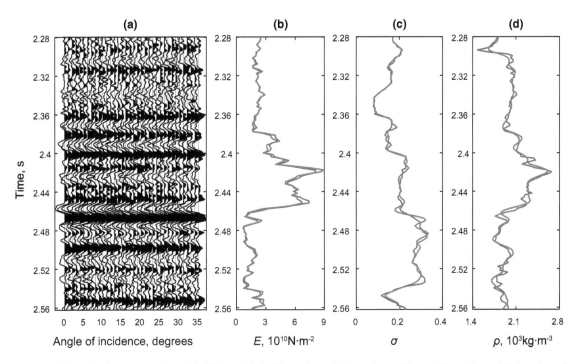

Fig. 4 a The synthetic seismic angle gather with SNR = 2:1; **b** The estimated Young's modulus; **c** The estimated Poisson's ratio; **d** The estimated density. **b–d** the *red lines* mean the inverted results, and the *blue lines* represent the real model curves

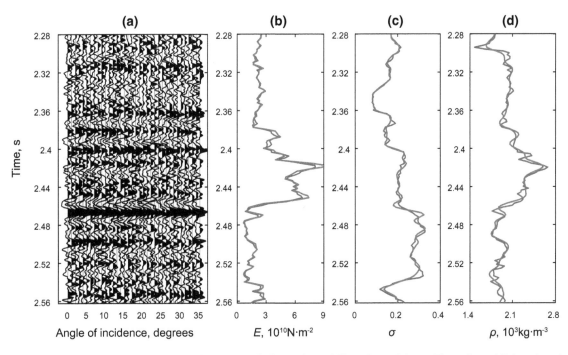

Fig. 5 a The synthetic seismic angle gather with SNR = 1:1; **b** The estimated Young's modulus; **c** The estimated Poisson's ratio; **d** The estimated density. **b–d** the *red lines* mean the inverted results, and the *blue lines* represent the real model curves

velocity, S-wave velocity, and density and a 35 Hz Ricker wavelet for incident angles ranging from 0° to 35°. Figure 2b–d displays the log curves of Young's modulus, Poisson's ratio, and density. The blue curves shown in

Fig. 2b–d are the real models and the red curves are the inversion results. From Fig. 2b–d, we can clearly see that the Young's modulus, Poisson's ratio, and density can be inverted reasonably with free noise. The error of the

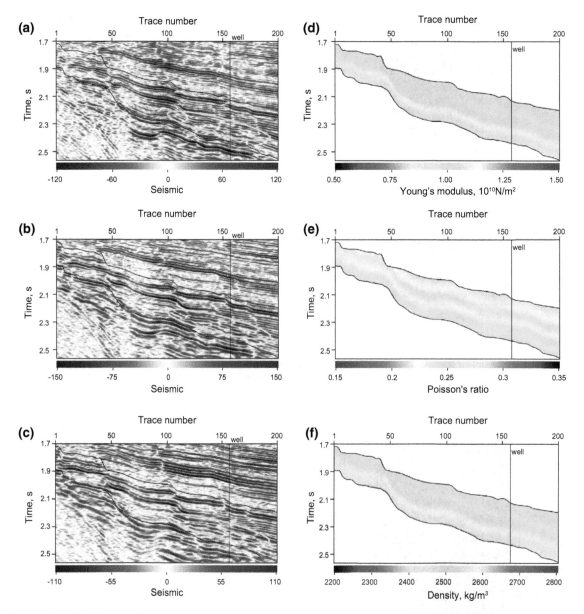

Fig. 6 The partial stack seismic profile and initial model of Young's modulus, Poisson's ratio, and density. **a** Partial stack seismic data with a small incident angle range, the mean angle is 8°; **b** Partial stack seismic data with a medium incident angle range, the mean angle is 16°; **c** Partial stack seismic data with a large incident angle range, the mean angle is 24°; **d–f** The profiles of initial model of Young's modulus, Poisson's ratio, and density, respectively

inverted density is a little bigger than that of the other two parameters. In order to verify the stability of the inversion method, we added random Gaussian noise to the synthetic gather data with different signal-to-noise ratios (SNRs). The gather traces are displayed in Figs. 3a, 4a and 5a, and the SNRs are 4:1, 2:1, and 1:1, respectively. The inversion results in different gather traces are shown in Figs. 3b–d, 4b–d and 5b–d, respectively. It is very clear that the inversion results estimated by the proposed inversion method match well with the real models as the low-frequency model enables the inversion results to approximate the real models. Although the inversion results are

influenced by the noise, especially the density, the Young's modulus and Poisson's ratio can match well with the real models to some degree so that the inversion results are helpful for us to evaluate the brittleness of the layer.

4 Real data example

The inversion method was applied to real partial angle stack seismic data, and the sampling interval of the seismic data is 2 ms. Figure 6 displays the used three partial angle stack seismic sections, and the mean angles of the seismic

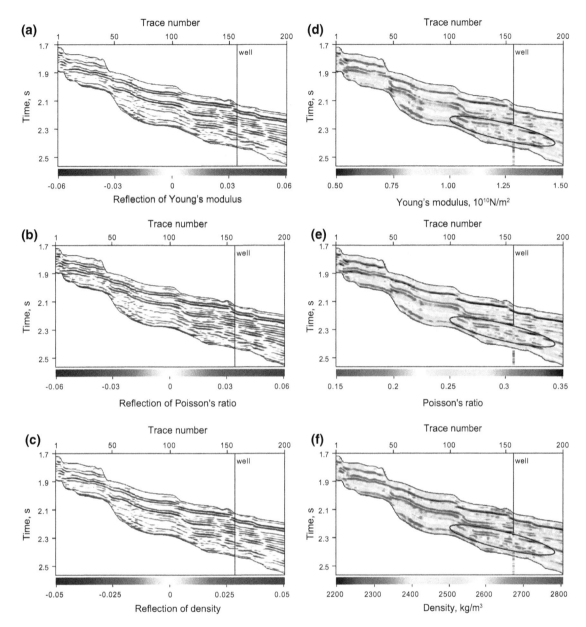

Fig. 7 The inverted reflection properties of Young's modulus, Poisson's ratio, and density and the properties generated by the reflections using Eq. (18). **a** Reflection of Young's modulus, **b** Reflection of Poisson's ratio, **c** Reflection of density, **d** Young's modulus, **e** Poisson's ratio, **f** Density

data in Fig. 6a–c are 8°, 16°, and 24°, respectively. In Fig. 6a–c, the green ellipse circles the target reservoir, and a well is drilled through the target at CDP 156. Figure 6d–f displays the initial models of Young's modulus, Poisson's ratio, and density, respectively. The initial models are established by spatial interpolation and extrapolation and low-pass filtering. The inverted isolated reflectivity spikes of r_E, r_σ, and r_ρ are shown in Fig. 7a–c, respectively. The structure of the reflectivity estimates is similar to the partial angle stack seismic profile and appears to have a better resolving power for the layer boundaries. The blocky Young's modulus, Poisson's ratio, and density displayed in

Fig. 7d–f are obtained from the estimated sparse reflectivity by using Eq. (18). The blocky results from model constrained BPI focus on the layer boundaries well, which is useful for us to interpret the inversion results. To some extent, the lateral continuity is improved because the low-frequency trend model is continuous laterally. The calculated logs of Young's modulus, Poisson's ratio and density are inserted into the sections. Figure 8a–c plot the inverted Young's modulus, Poisson's ratio, and density (red lines) at the well location, aligning with the well logs (dark lines). From Fig. 8, we can draw a conclusion that the inverted results of Young's modulus and Poisson's ratio have a

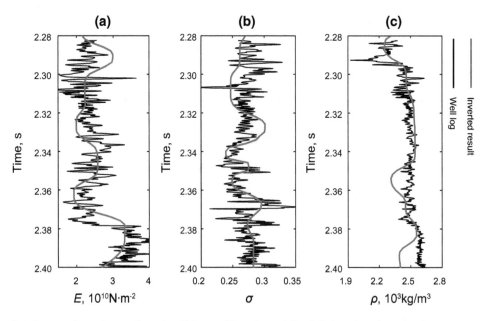

Fig. 8 The comparison between inversion results and well logs. **a** Young's modulus, **b** Poisson's ratio, **c** density

good fit with the logs, while the density inversion does not match as well as the other two elastic parameters as the maximum angle is not large enough for us to invert the density information. From the inverted results shown in Fig. 7d–f, we can clearly see that the Young's modulus exhibits high anomalous values and Poisson's ratio shows low anomalous values in the target circled with dark ellipses, which means that the brittleness of the circled target is high. The drilling shows the high brittleness at 2.3 s. Therefore, the brittleness evaluated by the inverted Young's modulus and Poisson's ratio using the proposed inversion method is consistent with the drilling.

5 Conclusions

In this paper, we presented a novel stable inversion method to estimate the Young's modulus and Poisson's ratio for brittleness evaluation from pre-stack seismic data with the YPD approximation. We introduced the low-frequency trend model into the basis pursuit inversion implementation. Therefore, we can call this method model constrained BPI. We derived the objective function of model constrained BPI by the Bayesian theorem. In the improved method, the low-frequency trend model as prior knowledge stabilized the inversion and improved the lateral continuity because the low-frequency trend model is continuous in the space axis. The l_1 norm of the model parameters and the GPSR algorithm kept the sparsity of inversion results so that we can obtain the isolated reflectivities and the elastic parameters with discontinuous jumps in the time axis. The model test results showed that we can obtain the high-precision

Young's modulus, Poisson's ratio, and density. The real data application is performed to confirm the validity of the proposed method, and it showed that the high anomalous value of Young's modulus and low anomalous value of Poisson's ratio which mean high brittleness matched well with the brittleness interpretation of drilling.

Acknowledgments We would like to acknowledge the sponsorship of the National "973 Program" of China (2013CB228604) and the National Grand Project for Science and Technology (2011ZX05030-004-002), China Postdoctoral Science Foundation (2014M550379), Natural Science Foundation of Shandong (2014BSE28009), Science Foundation for Post-doctoral Scientists of Shandong (201401018), and Science Foundation for Post-doctoral Scientists of Qingdao and Science Foundation from SINOPEC Key Laboratory of Geophysics (33550006-14-FW2099-0038). We also acknowledge the support of the Australian and Western Australian governments and the North West Shelf Joint Venture partners, as well as the Western Australian Energy Research Alliance (WA:ERA).

References

Aki K, Richards PG. Quantitative seismology. San Francisco: W. H. Freeman; 1980.

Alemie W, Sacchi MD. High-resolution three-term AVO inversion by means of a trivariate cauchy probability distribution. Geophysics. 2011;76(3):R43–55.

Beck A, Teboulle M. A fast iterative shrinkage-thresholding algorithm for linear inverse problems. SIAM J Imaging Sci. 2009;2(1):183–202.

Buland A, Omre H. Bayesian linearized AVO inversion. Geophysics. 2003;68(1):185–98.

Chopra S, Castagna J, Portniaguine O. Seismic resolution and thin-bed reflectivity inversion. CSEG Rec. 2006;31(1):19–25.

Downton JE. Seismic parameter estimation from AVO inversion. Ph.D. Thesis, University of Calgary. 2005.

Duijndam AJW. Detailed Bayesian inversion of seismic data. Ph.D. Thesis, Technische Universiteit Delft. 1987.

Figueiredo MAT, Nowak RD, Wright SJ. Gradient projection for sparse reconstruction: application to compressed sensing and other inverse problems. IEEE J Sel Top Signal Process. 2007;1(4):586–97.

Gray D, Goodway B, Chen T. Bridging the gap: using AVO to detect changes in fundamental elastic constants. SEG Technical Program Expanded Abstracts. 1999.

Grieser WV, Bray JM. Identification of production potential in unconventional reservoirs. Production and Operations Symposium. 2007.

Hansen PC. Analysis of discrete ill-posed problems by means of the L-curve. Soc Ind Appl Math. 1992;34(4):561–80.

Harris NB, Miskimins JL, Mnich CA. Mechanical anisotropy in the Woodford Shale, Permian Basin: origin, magnitude, and scale. Lead Edge. 2011;30(3):284–91.

Levy S, Fullagar PK. Reconstruction of a sparse spike train from a portion of its spectrum and application to high-resolution deconvolution. Geophysics. 1981;46(9):1235–43.

Pérez DO, Velis DR, Sacchi MD. High-resolution prestack seismic inversion using a hybrid FISTA least-squares strategy. Geophysics. 2013;78(5):R185–95.

Rickman R, Mullen MJ, Petre J E, et al. A practical use of shale petrophysics for stimulation design optimization: All shale plays are not clones of the Barnett shale. SPE Annual Technical Conference and Exhibition. 2008.

Russell BH, Gray D, Hampson DP. Linearized AVO and poroelasticity. Geophysics. 2011;76(3):C19–29.

Sacchi MD. Reweighting strategies in seismic deconvolution. Geophys J Int. 1997;129(3):651–6.

Sena A, Castillo G, Chesser K, et al. Seismic reservoir characterization in resource shale plays: "Sweet spot" discrimination and optimization of horizontal well placement. SEG Technical Program Expanded Abstracts. 2011.

Tarantola A. Inverse problem theory and methods for model parameter estimation. Soc Ind Appl Math. 2005;9(5):1597–1620. doi:10.1137/1.9780898717921.

Theune U, Jensås IØ, Eidsvik J. Analysis of prior models for a blocky inversion of seismic AVA data. Geophysics. 2010;75(3):C25–35.

Tikhonov AN. Solution of incorrectly formulated problems and the regularization method. Sov Math Dokl. 1963;4:1035–8.

Ulrych TJ, Sacchi MD, Woodbury A. A Bayes tour of inversion: a tutorial. Geophysics. 2001;66(1):55–69.

Yin XY, Zhang SX. Bayesian inversion for effective pore-fluid bulk modulus based on fluid-matrix decoupled amplitude variation with offset approximation. Geophysics. 2014;79(5):R221–32.

Yin XY, Yang PJ, Zhang GZ. A novel prestack AVO inversion and its application. SEG Technical Program Expanded Abstracts. 2008.

Yin XY, Zong ZY, Wu GC. Seismic wave scattering inversion for fluid factor of heterogeneous media. Sci China Earth Sci. 2014;57(3):542–9.

Yin XY, Zong ZY, Wu GC. Research on seismic fluid identification driven by rock physics. Sci China Earth Sci. 2015;58(2):159–71.

Yuan SY, Wang SX. Spectral sparse Bayesian learning reflectivity inversion. Geophys Prospect. 2013;61(4):735–46.

Yuan SY, Wang SX, Luo CM, et al. Simultaneous multitrace impedance inversion with transform-domain sparsity promotion. Geophysics. 2015;80(2):R71–80.

Zhang R, Castagna J. Seismic sparse-layer reflectivity inversion using basis pursuit decomposition. Geophysics. 2011;76(6):R147–58.

Zhang SX, Yin XY, Zhang FC. Fluid discrimination study from fluid elastic impedance (FEI). SEG Technical Program Expanded Abstracts. 2009.

Zhang R, Sen MK, Srinivasan S. A prestack basis pursuit seismic inversion. Geophysics. 2013;78(1):R1–11.

Zoeppritz K. VII b. Über Reflexion und Durchgang seismischer Wellen durch Unstetigkeitsflächen. Nachrichten von der Königlichen Gesellschaft der Wissenschaften zu Göttingen, Mathematisch-physikalische Klasse. 1919; 66–84.

Zong ZY, Yin XY, Wu GC. AVO inversion and poroelasticity with P- and S-wave moduli. Geophysics. 2012a;77(6):N17–24.

Zong ZY, Yin XY, Wu GC. Elastic impedance variation with angle inversion for elastic parameters. J Geophys Eng. 2012b;9(3):247–60.

Zong ZY, Yin XY, Zhang F, et al. Reflection coefficient equation and pre-stack seismic inversion with Young's modulus and Poisson ratio. Chin J Geophys. 2012c;55(11):3786–94 (in Chinese).

Zong ZY, Yin XY, Wu GC. Elastic impedance parameterization and inversion with Young's modulus and Poisson's ratio. Geophysics. 2013;78(6):N35–42.

8

Layer regrouping for water-flooded commingled reservoirs at a high water-cut stage

Chuan-Zhi Cui[1] · Jian-Peng Xu[1] · Duan-Ping Wang[2] · Zhi-Hong Liu[2] ·
Ying-song Huang[2] · Zheng-Ling Geng[3]

Abstract Layer regrouping is to divide all the layers into several sets of production series according to the physical properties and recovery percent of layers at high water-cut stage, which is an important technique to improve oil recovery for high water-cut multilayered reservoirs. Different regroup scenarios may lead to different production performances. Based on unstable oil–water flow theory, a multilayer commingled reservoir simulator is established by modifying the production split method. Taking into account the differences of layer properties, including permeability, oil viscosity, and remaining oil saturation, the pseudo flow resistance contrast is proposed to serve as a characteristic index of layer regrouping for high water-cut multilayered reservoirs. The production indices of multilayered reservoirs with different pseudo flow resistances are predicted with the established model in which the data are taken from the Shengtuo Oilfield. Simulation results show that the pseudo flow resistance contrast should be less than 4 when the layer regrouping is implemented. The K-means clustering method, which is based on the objective function, is used to automatically carry out the layer regrouping process according to pseudo flow resistances. The research result is applied to the IV–VI sand groups of the second member of the Shahejie Formation in the Shengtuo Oilfield, a favorable development performance is obtained, and the oil recovery is enhanced by 6.08 %.

Keywords Water-flooded reservoirs · Layer regrouping · Flow resistance · High water cut · Reservoir simulation

1 Introduction

For multilayer commingled reservoirs, the difference in oil recovery among different layers will become increasingly larger along the development process due to the interlayer heterogeneity (Ehlig-Economides and Joseph 1987; Jackson and Banerjee 2000). Layer regrouping is to divide all the layers into several sets of production series according to the physical properties and recovery percent of layers, which is an important technique to eliminate the differences in oil recovery of multilayered reservoirs at a high water-cut stage (Shi et al. 2006; Cui and Zhao 2010; Hu et al. 2010). The current research about layer regrouping is mostly focused on technical limits using the ordinary static parameters (Fu et al. 2002; Zhang et al. 2005; Chen et al. 2007; Liu et al. 2007). Chen et al. (2007) proposed that the principles of layer regrouping are that the permeability contrast (defined as a ratio of maximum to minimum permeability in the same development unit) should be less than 10, the layer number should be no more than 10 in the same development unit, and the thickness of the commingled production layers should be less than 20 m. In the Lasaxing oilfield, the layer permeability contrast (max/min permeability ratio) is suggested to be around 2.5 (Fu et al. 2002). However, the reservoir permeability, oil viscosity, and oil saturation change along with the reservoir development (Sun et al. 1996; Zhang et al. 1997; Li et al. 2009), and there are many factors influencing the production performance of layer regrouping.

✉ Chuan-Zhi Cui
   cuichuanzhi@126.com

[1] College of Petroleum Engineering, China University of Petroleum, Qingdao 266580, Shandong, China

[2] Exploration and Development Research Institution, Shengli Oilfield of SINOPEC, Dongying 257015, Shandong, China

[3] Center for Educational Development, China University of Petroleum, Qingdao 266580, Shandong, China

Edited by Yan-Hua Sun

Conventional indexes and technical limits of layer regrouping cannot meet the current demand at the high water-cut stage, and therefore, it is necessary to propose a new comprehensive index and limit for layer regrouping.

The existing methods of layer regrouping are usually based on the weighting of parameters to achieve a comprehensive index by a fuzzy evaluation, and then layer regrouping is conducted via clustering (Wang and Zhang 2001; Geng et al. 2006; Bao et al. 2010). However, the weightings of parameters are usually assigned on the basis of development experience and discussion among experts, so they are not objective enough.

Numerical reservoir simulation is usually used to study the technical limits of layer regrouping and to predict the development indexes of multilayered reservoirs (Lang 1991; Cheng et al. 2004; Mallison et al. 2004; Bokhari and Islam 2005; Jiang et al. 2006; Mustafiz and Islam 2008; Kasiri and Bashiri 2010). However, this method showed a significant deficiency in simulating multilayer commingled reservoirs at the high water-cut stage. There are marked differences between simulation results and actual measured data of the absorption rate in each layer (Ji et al. 2009). In the development of water-flooded reservoirs, interference exists objectively among different oil layers. The interlayer interference increases the water absorption capacity of high permeable layers and decreases that of low permeable ones. The water absorption capacity of the high permeable layers becomes stronger and stronger, and that of the low permeable layer becomes weaker and weaker. Simulation results obtained from the conventional numerical simulators cannot actually reflect the situation mentioned above at the high water-cut stage. The time-varying characteristics of permeability and oil viscosity are not considered in the current reservoir simulators (Wolcott et al. 1996; Choi et al. 1997; Vaziri et al. 2002; Maschio and Jose Schiozer 2003; Bhambri and Mohanty 2008; Lolon et al. 2008). Therefore, it is necessary to establish a numerical simulator for a multilayer commingled reservoir which can reflect the actual interlayer differences at the high water-cut stage.

In this paper, a new comprehensive characteristic index of layer regrouping and a method for automatic layer regrouping is presented. A new numerical simulator was also established. The data from the second member of the Shahejie Formation in the second district of the Shengtuo Oilfield were used in the calculation of the model and in the layer regrouping to validate the effectiveness of this technique.

2 Simulation of multilayer commingled reservoirs at the high water-cut stage

We assume that one production unit consists of n single layers varying in permeability, thickness, porosity, crude oil viscosity, etc. Two-phase flow of oil and water exists in each layer, and the fluid flow follows Darcy's law. Both rock and fluids are slightly compressible, and we assume that no vertical flow occurs between layers. The influence of capillary force and gravity is ignored. Thus, considering the time-varying characteristics of the reservoir permeability and crude oil viscosity, a mathematic model for injection–production allocation in a multilayer commingled reservoir is established.

For convenience, the fluid flow in the multilayer model is regarded as a combination of one-dimensional one-way flow in n layers. All layers are linked through an oil well and a water well. The method for allocating the production rate for each layer is improved, and the water adsorption rate or liquid production rate of a single layer is calculated. The pressure, oil saturation distribution, and production indices of each layer are calculated. The production indices of each layer in the same production unit at the same moment are added up. Thus, the production indices of one production unit can be obtained.

2.1 Mathematical model

(1) *Differential equations*

The differential equations describing one-dimensional flow of the oil and water phases are as follows:

Oil phase:

$$-\frac{\partial}{\partial x}(\rho_o v_{ox}) + q_o = \frac{\partial}{\partial t}(\rho_o \phi S_o). \tag{1}$$

Water phase:

$$-\frac{\partial}{\partial x}(\rho_w v_{wx}) + q_w = \frac{\partial}{\partial t}(\rho_w \phi S_w). \tag{2}$$

(2) *Motion equation*

The motion equations of oil and water phases are as follows:

Oil phase:

$$v_{ox} = -\frac{kk_{ro}\partial p}{\mu_o \partial x}. \tag{3}$$

Water phase:

$$v_{wx} = -\frac{kk_{rw}\partial p}{\mu_w \partial x}. \tag{4}$$

(3) *Auxiliary equation*

The auxiliary equation is the saturation equation:

$$S_o + S_w = 1. \tag{5}$$

In the above formulas, ρ, v, S, q, μ, and k_r represent density, flow velocity, saturation, production rate, viscosity, and relative permeability, respectively; k is the absolute permeability; p is the formation pressure; ϕ is porosity; and x is the flow direction. Subscripts o and w represent the oil and water phases, respectively.

2.2 The solution to the model

2.2.1 Solution method for the mathematical model

The finite difference method, IMPES, is used to solve the equations. In the computation of each time step, what are needed are the following: the water saturation of each grid at the last time step is used to achieve the water cut by using the fractional flow equation; the changed permeability and crude oil viscosity of each grid are obtained according to the rules that the permeability and viscosity change along with the water cut; and the water adsorption rate and the liquid production rate at each layer are calculated. A program is made to realize the solution process.

2.2.2 Introduction of time-varying parameters

The changes of permeability and crude oil viscosity along with the water cut are introduced into the model. Cui and Zhao (Cui and Zhao 2004) reported the changes of permeability and crude oil viscosity along with the water cut in the second member of the Shahejie Formation in the second district of the Shengtuo Oilfield. The expression describing the permeability multiplier changing with the water cut is given by

$$k_{\mathrm{c}} = 1.0733 + 0.0034 f_{\mathrm{w}}, \tag{6}$$

and the crude oil viscosity changing with water cut may be described by

$$\mu_{\mathrm{o}} = \mu_{\mathrm{oi}} e^{0.0122 f_{\mathrm{w}}}, \tag{7}$$

where f_{w} is the water cut, %; k_{c} is the permeability multiplier; μ_{oi} is the initial viscosity of the formation crude oil, and μ_{oi} is equal to 18 mPa s in the second district of the Shengtuo Oilfield.

The relationships of the changes in permeability and oil viscosity with the water cut may be different in different reservoirs. The different relationships can lead to different calculation results. In this paper, Eqs. (6) and (7) are used in the calculation.

2.2.3 Allocation method for the water injection rate at each layer

In conventional numerical simulation, when the liquid production rate or the water injection rate is fixed, injection–production allocation at each layer is based on the parameters of the grids which the oil and water wells are located in, without considering the influence of the flow resistance

from the water well to the oil well. In this paper, the flow resistance in each layer between the water and oil wells is taken into consideration in the rate allocation at each layer.

If there are n layers in one set of production series, and it is one-dimensional flow from the injector to the producer, each layer is discretized into m grids. The flow resistance of the kth layer is expressed as

$$R_k = \sum_{i=1}^{m} \left[\frac{dx}{A} \times \frac{1}{k(k_{\mathrm{ro}}/\mu_{\mathrm{o}} + k_{\mathrm{rw}}/\mu_{\mathrm{w}})} \right]_i \tag{8}$$

If Q_{N} is the total water injection rate of the water well in one production unit, the water injection rate of the kth layer is

$$Q_{\mathrm{N}k} = \left(\frac{1}{R_k} / \sum_{j=1}^{n} \frac{1}{R_j} \right) Q_{\mathrm{N}} \tag{9}$$

Similarly, the liquid production rate of each layer can be obtained. The oil production rate and the water production rate of one well at the kth layer are allocated according to the water and oil mobility of the grid where the well is located, and the expressions are

$$\begin{cases} Q_{\mathrm{o}k} = \left(\dfrac{\lambda_{\mathrm{mo}}}{\lambda_{\mathrm{mo}} + \lambda_{\mathrm{mw}}} \right)_k Q_{\mathrm{l}k} \\ Q_{\mathrm{w}k} = Q_{\mathrm{l}k} - Q_{\mathrm{o}k} \end{cases} \tag{10}$$

where λ_{mo} and λ_{mw} represent the oil mobility and water mobility, respectively; $Q_{\mathrm{o}k}$, $Q_{\mathrm{w}k}$, and $Q_{\mathrm{l}k}$ represent the oil production rate, water production rate, and the liquid production rate at the kth layer, respectively.

2.2.4 Analysis of calculation results

In a case study, the oil-producing zone is composed of two layers with different permeability. When the permeability contrast is 3, 5, and 12, respectively, the water injection rate for each layer, obtained from the conventional reservoir simulation, is shown in Fig. 1. According to Fig. 1, the difference in the allocation proportion of the water injection rate between two layers increases at the early stage, but at the later stage, the difference tends to decrease. This is not in accordance with the field situation. The reason is that, in the conventional numerical simulation, the allocation results are calculated using the parameters of the grids where the well is located without consideration of the oil saturation distribution from the injector to the producer.

Figure 2 shows allocation results calculated from the reservoir simulation model built in this paper. In Fig. 2, from the early to mid-stage, the difference in the allocation proportion between two layers keeps increasing. In the later stage, the difference in the allocation proportion tends to be stable. This is in accordance with the actual case and proves the validity of the model.

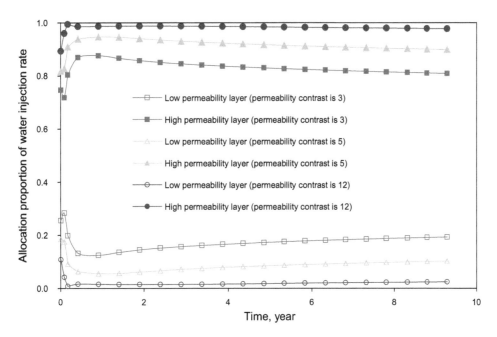

Fig. 1 Calculation results from the conventional reservoir simulator

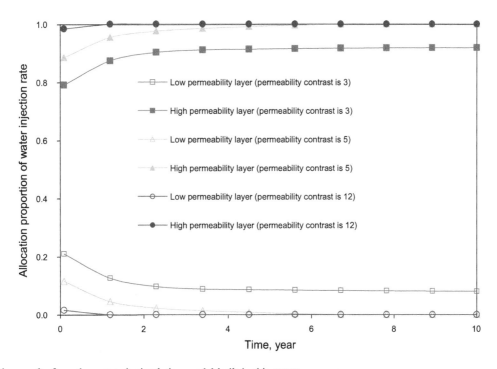

Fig. 2 Calculation results from the reservoir simulation model built in this paper

3 The comprehensive index and technical limit of layer regrouping

3.1 The comprehensive index of layer regrouping

For high water-cut reservoirs, representation indexes of layer regrouping, such as permeability contrast and crude oil viscosity contrast, cannot meet the needs of layer regrouping. In this paper, the pseudo flow resistance contrast is taken as a comprehensive index of layer regrouping at the high water-cut stage.

When oil and water phases flow simultaneously, the liquid production rate is derived from Darcy's law:

$$Q_t = \left(\frac{k_o}{\mu_o(f_w)} + \frac{k_w}{\mu_w} \right) A \frac{\Delta p}{L} = \frac{\Delta p}{\frac{\mu_o(f_w)\mu_w}{k_o\mu_w + k_w\mu_o(f_w)} \frac{L}{A}} = \frac{\Delta p}{R}. \quad (11)$$

The pseudo flow resistance is defined as

$$R' = \frac{\mu_o(f_w)\mu_w}{k_o\mu_w + k_w\mu_o(f_w)}. \quad (12)$$

Equation (12) shows that the pseudo flow resistance is related to the crude oil viscosity and effective permeability of oil and water phases. The effective permeability reflects the effect of absolute permeability and remaining oil saturation, and the expressions are given by

$$k_o = k(f_w)k_{ro}(S_w), \ k_w = k(f_w)k_{rw}(S_w). \quad (13)$$

The pseudo flow resistance contrast is a ratio of maximum to minimum pseudo flow resistance among layers in the same one set of production series.

3.2 The technical limit of layer regrouping

An 18-layer commingled reservoir model is established to study the technical limit of layer regrouping at the high water-cut stage. The initial permeability values of 18 layers are listed in Table 1. The producer–injector spacing is 300 m, the crude oil viscosity is 18 mPa s, and the porosity is 0.28. The reservoir simulation model built above is used for calculations. When the water cut is 95 %, the conditions of each layer are shown in Table 1.

This research on layer regrouping is conducted when the comprehensive water cut is 95 %. Two layers with different pseudo flow resistance are recombined separately, and the production continues until the water cut is 98 %. The curve between the oil recovery factor (the proportion of the oil in a reservoir which is recovered) and the pseudo flow resistance contrast is shown in Fig. 3. Figure 4 shows the curves between cumulative water–oil ratio and pseudo flow resistance contrast when the commingled production time is 5, 10, 15, 20, and 25 years respectively.

Figures 3 and 4 indicate that the production performance becomes worse when the pseudo flow resistance contrast increases. When the pseudo flow resistance contrast is greater than 4, the oil recovery dramatically decreases and the cumulative water–oil ratio increases rapidly. Therefore, it is appropriate to limit pseudo flow resistance contrast within 4 in layer regrouping at the high water-cut stage.

4 Optimizing method of layer regrouping

If there are many layers in one reservoir, there may be many different recombination scenarios of layers even we set the pseudo flow resistance contrast less than 4. So it is

Table 1 Parameters of each layer at water cut 95 %

Layers number	Initial permeability, 10^{-3} μm^2	Water cut, %	Oil recovery, %	Pseudo flow resistance, mPa s/μm^2
1	200	0	0.03	46.59
2	400	7.34	2.33	22.57
3	500	23.57	7.89	16.33
4	600	46.55	16.39	11.00
5	700	75.71	20.15	7.82
6	800	84.17	22.27	6.11
7	900	87.86	23.84	5.02
8	1000	89.22	25.2	4.22
9	1100	90.83	26.45	3.61
10	1200	92.28	27.55	3.14
11	1400	94.39	29.35	2.48
12	1600	95.36	30.81	2.04
13	1800	95.56	32.2	1.71
14	2000	95.97	33.58	1.45
15	2200	96.46	34.87	1.24
16	2400	96.91	36.06	1.08
17	2600	97.29	37.15	0.96
18	2800	97.58	38.14	0.85

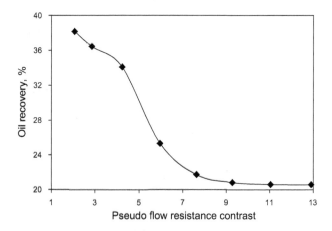

Fig. 3 Curve between oil recovery and pseudo flow resistance contrast

required to find an optimal one to obtain the highest oil recovery.

Based on the pseudo flow resistance of one single layer, using the K-means clustering method (Wang and Niu 2004; Kong et al. 2004), layer regrouping is carried out to obtain the optimal production performance. This is the basic principle of the K-means clustering method. Assume that there are n samples $(x_1, x_2, \ldots, x_n)$, and they are classified into p types:$(C_1, C_2, \ldots, C_p)$. Assume that the number of the ith type is N_i, and the mean of the types is $(m_1, m_2, \ldots, m_p)$, and then $m_i = \frac{1}{N_i} \sum_{i=1}^{N_i} x_i (i=1, 2,$

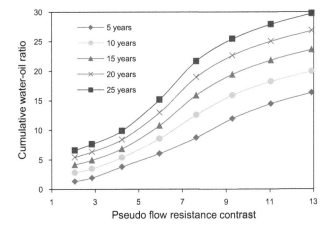

Fig. 4 Curves between cumulative water–oil ratio and pseudo flow resistance contrast

..., p). K-means clustering is on the basis of the least square method, and its object function is

$$\min J = \sum_{i=1}^{p} \sum_{j=1}^{N_i} ||x_j - m_i||^2 \qquad (14)$$

In the layer regrouping, supposing that the commingled reservoir of n layers comes to high water-cut stage, p sets of production series are divided according to the pseudo flow resistance of each layer. p single layers are randomly selected to be the initial center of the p sets of production series, and the rest of layers are assigned into the production series where the nearest production series center

locates. This is the initial division of the p sets of production series. Calculation of new centers is done to the newly allocated production series, and then the allocation of other layers is continued. After several cycles of iteration, the centers of the p sets of production series do not change any more, which means that all single layers have been allocated to their own production series. Correct clustering leads to a convergence function; otherwise, the iteration continues. A program is made to realize the optimal process.

5 Case analysis

The IV–VII sand groups of the second member of the Shahejie Formation in the second district of the Shengtuo Oilfield are located at the southwest of the structural high position of the Shengtuo Oilfield and are delta plain sedimentary subfaces. The IV–VI sand groups consist of 18 oil-bearing layers. Among them, there are 5 layers in the 4th sand group, 6 layers in the 5th sand group, and 7 layers in the 6th sand group. These sand groups have been put into production as one production series since 1975, and a high water cut was observed, and the oil recovery percent (a ratio of the produced oil to OIPP) was 35.5 %. Table 2 shows the parameters and the calculation results of pseudo flow resistance of each layer.

Table 3 shows the recombination results from the K-means clustering method. When these layers are

Table 2 Parameters of the IV–VI sand groups in the Shengtuo Oilfield

Sublayer name	Permeability, 10^{-3} μm^2	Net thickness, m	Porosity	Remaining oil saturation	Remaining reserves, 10^4 t	Pseudo flow resistance, mPa s/μm^2
$S_2 4^1$	3041.5	3.44	0.274	0.445	17.88	1.252
$S_2 4^2$	3796.0	3.09	0.286	0.421	15.69	0.905
$S_2 4^3$	1084.0	1.37	0.278	0.574	5.08	6.525
$S_2 4^4$	925.0	1.97	0.271	0.587	7.28	8.538
$S_2 4^5$	1427.9	2.64	0.261	0.534	7.12	4.093
$S_2 5^1$	1100.9	0.86	0.276	0.571	0.93	6.278
$S_2 5^2$	2790.4	1.08	0.267	0.453	3.68	1.415
$S_2 5^3$	1329.7	2.06	0.277	0.550	9.97	4.681
$S_2 5^4$	1445.6	2.38	0.270	0.537	6.11	4.093
$S_2 5^5$	3512.0	3.36	0.288	0.432	17.26	1.025
$S_2 5^6$	1735.1	1.10	0.270	0.512	1.68	3.022
$S_2 6^1$	1581.6	1.44	0.288	0.534	5.52	3.695
$S_2 6^2$	615.8	1.73	0.230	0.609	6.55	14.769
$S_2 6^3$	903.7	0.89	0.255	0.583	1.76	8.432
$S_2 6^4$	1276.8	1.34	0.279	0.556	4.71	4.994
$S_2 6^5$	2767.1	1.61	0.275	0.458	5.08	1.458
$S_2 6^6$	1903.7	1.49	0.284	0.506	5.49	2.681
$S_2 6^7$	1443.0	2.54	0.268	0.536	13.71	4.084

Table 3 Layer regrouping programs of the IV–VI sand groups in the Shengtuo Oilfield

Layer regrouping program	Layer names	Pseudo flow resistance contrast	Remaining reserves, 10^4 t
One set	All layers	16.327	135.51
Two sets	S_24^3, S_24^4, S_25^1, S_26^2, S_26^3	2.352	21.60
	S_24^1, S_24^2, S_24^5, S_25^2, S_25^3, S_25^4, S_25^5, S_25^6, S_26^1, S_26^4, S_26^5, S_26^6, S_26^7	5.521	113.90
Three sets	S_24^4, S_26^2, S_26^3	1.752	15.59
	S_24^1, S_24^2, S_25^2, S_25^5, S_25^6, S_26^5, S_26^6	3.341	66.76
	S_24^3, S_24^5, S_25^1, S_25^3, S_25^4, S_26^1, S_26^4, S_26^7	1.766	53.16
Four sets	S_26^2	1.000	6.55
	S_24^3, S_24^4, S_25^1, S_26^3	1.360	15.05
	S_24^1, S_24^2, S_25^2, S_25^5, S_26^5	1.612	59.59
	S_24^5, S_25^3, S_25^4, S_26^5, S_26^1, S_26^4, S_26^6, S_26^7	1.862	54.32

recombined to two sets of production series, the pseudo flow resistance contrast of one set is greater than 4. When these layers are recombined to three or four production series, the pseudo flow resistance contrasts are smaller than 4, which meet the technical limit of layer regrouping. According to the requirement of individual-well control reserves at the high water-cut stage (Wang and Niu 2004), the remaining reserves in S_26^2 layer in the four sets of production series are only 6.552×10^4 tons, and it is not economical to set it as one set of production series. Therefore, the 4th–6th sand groups are recombined to three sets of production series.

After layer regrouping, the development indexes were predicted with the reservoir simulation model built above (Fig. 5). Compared with the scenario without layer regrouping (one set of production series), the oil recovery of two sets of production series increases by 2.14 %, and the oil recovery of three sets of production series increases by 6.08 %. Hence, recombination of three sets of production series can achieve better performance and are recommended.

6 Conclusions

(1) A numerical simulator was established for a multi-layer commingled reservoir, which considers the changes of permeability and oil viscosity during oil production. A method for allocating the water injection rate and the liquid production rate of wells at each layer was modified. The results of the simulator can actually reflect the characteristics of fluid flow in different producing layers in the multilayer commingled reservoir at the high water-cut stage.

(2) The pseudo flow resistance contrast was proposed to be a characteristic index of layer regrouping at the high water-cut stage, which considers each single

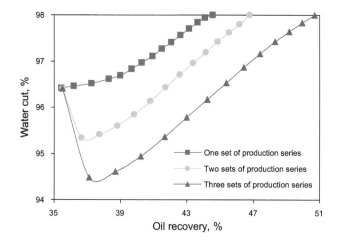

Fig. 5 Curves between water cut and oil recovery

layer's permeability, crude oil viscosity, and the remaining oil saturation. By analyzing the simulation results from the numerical simulator built in this paper, the pseudo flow resistance contrast in one set of production series should be controlled within 4 in layer regrouping at the high water-cut stage. A K-means clustering method was used to implement the automatic optimization of layer regrouping.

(3) The limit of the pseudo flow resistance contrast in this paper was obtained through the data from the second district of the Shengtuo Oilfield. The limit may be different in different reservoirs. In the layer regrouping at the high water-cut stage, in addition to the pseudo flow resistance contrast, the remaining reserves, well pattern, well spacing, etc., are needed to be considered.

Acknowledgments This work was supported by the Program for Changjiang Scholars and Innovative Research Team in University

(IRT1294) and the China National Science and Technology Major Projects (Grant No: 2016ZX05011).

References

Bao JW, Song XM, Ye JG, et al. Recombination of layer series of development in a high water-cut oil field. Xinjiang Pet Geol. 2010;31(3):291–4 (in Chinese).

Bhambri P, Mohanty KK. Two and three-hydrocarbon phase streamline-based compositional simulation of gas injections. J Pet Sci Eng. 2008;62(1):16–27. doi:10.1016/j.petrol.2008.06.003.

Bokhari K, Islam MR. Improvement in the time accuracy of numerical methods in petroleum engineering problems: a new combination. Energy Sources. 2005;27(1–2):45–60. doi:10.1080/00908310490448109.

Chen MF, Jiang HQ, Zeng YX. A study of maximum/minimum permeability ratio boundaries for reasonable developed-layer recombination in extremely heterogeneous reservoirs. China Offshore Oil Gas. 2007;19(5):319–22 (in Chinese).

Cheng H, Kharghoria A, Zhong H, et al. Fast history matching of finite difference models using streamline-derived sensitivities. In: SPE/DOE symposium on improved oil recovery, 17–21 April, Tulsa, Oklahoma; 2004. doi:10.2118/89447-PA.

Choi ES, Cheema T, Islam MR. A new dual-porosity/dual permeability model with non-Darcian flow through fractures. J Pet Sci Eng. 1997;17(3):331–44. doi:10.1016/S0920-4105(96)00050-2.

Cui CZ, Zhao XY. A numerical reservoir simulation study with the variety of reservoir parameters. J Hydrodyn. 2004;19(A):912–5 (in Chinese).

Cui CZ, Zhao XY. Method for calculating production indices of multilayer water drive reservoirs. J Pet Sci Eng. 2010;75(1):66–70. doi:10.1016/j.petrol.2010.10.003.

Ehlig-Economides CA, Joseph J. A new test for determination of individual layer properties in a multilayered reservoir. SPE Form Eval. 1987;2(3):261–83. doi:10.2118/14167-PA.

Fu BZ, Kong XT, Chen ZM, et al. Principles for combining series of polymer flooding in Lamadian-Saertu-Xingshugang oilfields. Pet Geol Oilfield Dev Daqing. 2002;21(6):51–4 (in Chinese).

Geng ZL, Jiang HQ, Sun MR, et al. Model of rhythmite reconstruction in the reservoirs with extra high water cut. J Oil Gas Technol. 2006;28(6):102–6 (in Chinese).

Hu DD, Tang W, Chang YW, et al. A study for redevelopment trends after polymer flooding and its field application in La-sa-xing oilfield. In: International oil and gas conference and exhibition in China, 8–10 June, Beijing, China; 2010. doi:10.2118/130903-MS.

Jackson RR, Banerjee R. Advances in multilayer reservoir testing and analysis using numerical well testing and reservoir simulation. In: SPE annual technical conference and exhibition, 1–4 October, Dallas, Texas; 2000. doi:10.2118/62917-MS.

Ji BY, Zhao GZ, Li H, et al. Multidisciplinary integrated reservoir research methods and application. Beijing: Petroleum Industry Press; 2009 (in Chinese).

Jiang HQ, Yao J, Jiang RZ. Principles and methods of reservoir engineering. Dongying: China University of Petroleum Press; 2006 (in Chinese).

Kasiri N, Bashiri A. Comparative study of different techniques for numerical reservoir simulation. Pet Sci Technol. 2010;28(5):494–503. doi:10.1080/10916460903515532.

Lang ZX. Reservoir engineering. Beijing: Petroleum Industry Press; 1991 (in Chinese).

Li H, Wang XW, Liu SL. Variation law of parameters of reservoir physical property in old oilfield. J Southwest Pet Univ (Sci Technol Ed). 2009;31(2):85–9 (in Chinese).

Liu YP, Chen YM, Yuan SB, et al. Well pattern reconstruction of a refined cyclothem in the 8th member of Es3 in fault block Tuo21. J Oil Gas Technol. 2007;29(5):116–20 (in Chinese).

Lolon E, Archer RA, Ilk D, et al. New semianalytical solutions for multilayer reservoirs. In: CIPC/SPE gas technology symposium 2008 joint conference, 16–19 June, Calgary, Alberta; 2008. doi:10.2118/114946-MS.

Kong R, Zhang GX, Shi ZS. Kernel-based K-means clustering. Comput Eng. 2004;30(11):12–3 (in Chinese).

Mallison BT, Gerritsen MG, Matringe SF. Improved mappings for streamline based simulation. In: SPE/DOE symposium on improved oil recovery, 17–21 April, Tulsa, Oklahoma; 2004. doi:10.2523/89352-MS.

Maschio C, Jose Schiozer D. A new upscaling technique based on Dykstra-Parsons coefficient: evaluation with streamline reservoir simulation. J Pet Sci Eng. 2003;40(1):27–36. doi:10.1016/S0920-4105(03)00060-3.

Mustafiz S, Islam MR. State-of-the-art petroleum reservoir simulation. Pet Sci Technol. 2008;26(10–11):1303–29. doi:10.1080/10916460701834036.

Shi CF, Du QL, Zhu LH, et al. Research on remaining oil distribution and further development methods for different kinds of oil layers in Daqing oilfield at high water-cut stage. In: SPE Asia Pacific oil & gas conference and exhibition, 11–13 September, Adelaide; 2006. doi:10.2118/101034-MS.

Sun SX, Han JW, Guo YR, et al. Laboratory experiment on physical properties of flooding sandstone in Shengtuo oilfield. J Univ Pet. 1996;20(Suppl):33–5 (in Chinese).

Vaziri HH, Xiao Y, Islam R, et al. Numerical modeling of seepage-induced sand production in oil and gas reservoirs. J Pet Sci Eng. 2002;36(1):71–86. doi:10.1016/S0920-4105(02)00264-4.

Wang J, Zhang JS. Comparing several methods of assuring weight vector in synthetic evaluation. J Hebei Univ Technol. 2001;30(2):52–7 (in Chinese).

Wang SB, Niu SW. Layer subdivision in the late high water cut stage in the complex fault block reservoirs, Dongxin oilfield. Pet Explor Dev. 2004;31(3):116–8 (in Chinese).

Wolcott DS, Kazemi H, Dean RH. A practical method for minimizing the grid orientation effect in reservoir simulation. In: SPE annual technical conference and exhibition, 6–9 Oct, Denver, Colorado; 1996. doi:10.2118/36723-MS.

Zhang HX, Liu QN, Li FQ, et al. Variations of petrophysical parameters after sandstone reservoirs watered out in Daqing oilfield. SPE Adv Technol Ser. 1997;5(1):128–39. doi:10.2118/30844-PA.

Zhang SM, Liu ZH, Wan HY, et al. Technical limits for pattern rearrangement in the late period of high water-cut in an uncompartmentalized oilfield. Spec Oil Gas Reserv. 2005;12(2):57–62 (in Chinese).

9

Stress redistribution in multi-stage hydraulic fracturing of horizontal wells in shales

Yi-Jin Zeng[1] · **Xu Zhang**[1] · **Bao-Ping Zhang**[1]

Abstract Multi-stage hydraulic fracturing of horizontal wells is the main stimulation method in recovering gas from tight shale gas reservoirs, and stage spacing determination is one of the key issues in fracturing design. The initiation and propagation of hydraulic fractures will cause stress redistribution and may activate natural fractures in the reservoir. Due to the limitation of the analytical method in calculation of induced stresses, we propose a numerical method, which incorporates the interaction of hydraulic fractures and the wellbore, and analyzes the stress distribution in the reservoir under different stage spacing. Simulation results indicate the following: (1) The induced stress was overestimated from the analytical method because it did not take into account the interaction between hydraulic fractures and the horizontal wellbore. (2) The hydraulic fracture had a considerable effect on the redistribution of stresses in the direction of the horizontal wellbore in the reservoir. The stress in the direction perpendicular to the horizontal wellbore after hydraulic fracturing had a minor change compared with the original in situ stress. (3) Stress interferences among fractures were greatly connected with the stage spacing and the distance from the wellbore. When the fracture length was 200 m, and the stage spacing was 50 m, the stress redistribution due to stage fracturing may divert the original stress pattern, which might activate natural fractures so as to generate a complex fracture network.

Keywords Shale gas · Horizontal well · Multi-stage fracturing · Complex fracture · Stage spacing · Induced stress

1 Introduction

Multi-stage fracturing in horizontal wells is an important and effective completion method for compact, low permeability reservoirs (Cai et al. 2009; Yao et al. 2013; Koløy et al. 2014; Li et al. 2014; Sorek et al. 2014; Zhang and Li 2014). The essence of hydraulic fracturing is to inject high-pressure fluids into a reservoir to create induced fractures around the wellbore. As a reliable and economic formation stimulation technology, it has been successfully used in shale gas reservoirs. Due to low matrix porosity and low permeability of shale gas reservoirs, to make long well lengths for effective drainage areas in traditional designs, it is not practical for tight shales (Zhou et al. 2010; Zeng et al. 2010; Wu et al. 2011, 2012; Wang et al. 2014a). In order to create complex fractures in shale reservoirs through multi-stage hydraulic fracturing in horizontal wells, analysis of stage spacing and stress interference among fractures needs to be done.

Stress distribution around a horizontal wellbore is very complex, affected by filtration of fracturing fluids, pore pressure, etc. (Fischer et al. 1994; Economides 2006; Civan 2010; Fang and Khaksar 2011; Zhang et al. 2012; Ziarani and Aguilera 2012; Hou et al. 2013; Pan et al. 2014; Chen et al. 2015). Meanwhile, previous stages of hydraulic fracturing would affect later stages of fracturing (Xu 2009; Wei et al. 2011; Chuprakov et al. 2011; Wang et al. 2014b).

Some researchers investigated optimization of hydraulic fracturing based on a one-factor analysis of fracture parameters and evaluated effects of fracture parameters on

✉ Yi-Jin Zeng
zengyj.sripe@sinopec.com

[1] Sinopec Research Institute of Petroleum Engineering, Beijing 100101, China

Edited by Yan-Hua Sun

productivity of horizontal wells (Zhu et al. 2013). A new grid refinement was used to optimize fracturing parameters. Effects of the horizontal-section length, number of fractures, fracture length, and conductivity on the productivity of horizontal wells were also studied (Chen et al. 2013). An optimization method for perforation spacing was established, it included a mathematical model of the induced stress field on the basis of a homogeneous and isotropic 2D plane strain model and the shift between the maximum and minimum horizontal principal stress (Yin et al. 2012). Qu et al. put forward a design method for optimizing horizontal well fracturing parameters (Qu et al. 2012). Shang et al. established a wellbore stress distribution model and a fracture pressure calculation model (Shang et al. 2009).

In the above-mentioned research, an analytical method based on classic fracture mechanics was used to analyze characteristics of the fracture-induced stress field in hydraulic fracturing, but this model did not consider interaction between fractures and the horizontal wellbore, so did not explicitly give characteristics of different stages of spacing and stress interference among fractures.

On the basis of analyzing limitations of the classical analytical method, a comparative analysis of the numerical method is carried out. With the numerical model, we analyze in situ stress change and interference among fractures.

2 Fracture geometry and net pressure for multi-stage fractured horizontal wells

2.1 Fracture geometry

During hydraulic fracturing, a fracturing fluid is injected continuously into the formation, and fractures will propagate dynamically. However, it is difficult to predetermine the fracture geometry using the pressure distribution function. While using the criterion of $K_I = K_{Ic}$ at each moment, the hydraulic mechanical coupling problem can be solved.

From the theory of fracture mechanics and the Castigliano theorem, the width of a fracture under plain strain conditions can be calculated from the following equation:

$$w(y) = \frac{4}{\pi E} \int_y^a \left[\int_{-\xi}^{\xi} \Delta p(y) \sqrt{\frac{\xi + y}{\xi - y}} dy \right] \frac{1}{\sqrt{\xi^2 - y^2}} d\xi, \quad (1)$$

where $\Delta p(y) = p(y) - \sigma_h$ is the net pressure in the fracture; $p(y)$ is the fluid pressure; σ_h is the minimum horizontal in situ stress perpendicular to the fracture plane; E is the elastic modulus of the rock; a is the fracture half length; ξ is the temporary fracture half length during integration, as shown in Fig. 1.

When the net pressure is distributed smoothly and continuously along the fracture, it can be defined by a

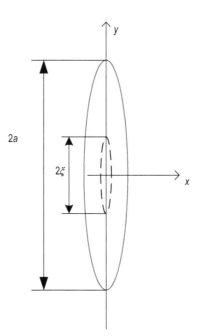

Fig. 1 Geometry of a hydraulic fracture

continuous function $p(y) = p_w f(y)$. The stress intensity factor K_I at the fracture tip can be expressed as

$$K_I = \frac{1}{\sqrt{\pi a}} \int_{-a}^{a} [p(y) - \sigma_h] \sqrt{\frac{a + y}{a - y}} dy$$
$$= \frac{1}{\sqrt{\pi a}} \int_{-a}^{a} [p_w f(y) - \sigma_h] \sqrt{\frac{a + y}{a - y}} dy, \quad (2)$$

where p_w is the wellbore pressure; $f(x)$ is the eigen function of the pressure distribution.

According to the mean value theorem of integral, Eq. (2) becomes

$$K_I = A p_w \sqrt{a} - \sigma_h \sqrt{\pi a}, \quad (3)$$

where $A = \frac{2}{\pi} f(b) \sqrt{\frac{a+b}{a-b}}$ $(-a < b < a)$ is a constant, and $A = \frac{\sigma_h \sqrt{\pi a} + K_{Ic}}{p_w \sqrt{a}}$ considering the fracture propagation criterion $K_I = K_{Ic}$.

Substituting Eq. (3) into (1) and integrating it give

$$w(y) = \frac{4}{E} K_{Ic} \sqrt{\frac{a^2 - y^2}{\pi a}}, \quad (4)$$

where K_{Ic} is the fracture toughness of the target zone. From Eq. (4), it can be seen that the fracture cross section is a slim ellipse determined by the fracture half length a, E, K_{Ic}, and independent of the pressure distribution in it.

2.2 Net pressure

We assume that the fracturing fluid is incompressible, and the formation is impermeable, then the fracture volume is equal to the volume of the fluid injected into the formation.

$$Q = V(a) = \frac{2K_{Ic}ha\sqrt{\pi a}}{E}, \tag{5}$$

$$Q = qt, \tag{6}$$

where q is the injection rate; t is the injection time.

The fracture parameters can be obtained as

$$a = \left(\frac{qEt}{2\sqrt{\pi}hK_{Ic}}\right)^{2/3}, \tag{7}$$

$$w(0) = \frac{4K_{Ic}}{E\sqrt{\pi}}\left(\frac{qEt}{2\sqrt{\pi}hK_{Ic}}\right)^{1/3}. \tag{8}$$

Using fluid mechanics theory, the distribution of the net pressure can be approximated by a linear equation:

$$p(y) = p_w\left(1 - 0.42\frac{y}{a}\right). \tag{9}$$

Substituting Eq. (9) into (2) gives the wellbore pressure p_w:

$$p_w = \frac{0.77K_{Ic}}{\sqrt{a}} + 1.365\sigma_h. \tag{10}$$

The fluid pressure distribution $p(y)$ and the tip pressure p_a are

$$p(y) = \left(\frac{0.77K_{Ic}}{\sqrt{a}} + 1.365\sigma_h\right)\left(1 - 0.42\frac{y}{a}\right), \tag{11}$$

$$p_a = \frac{0.45K_{Ic}}{\sqrt{a}} + 0.79\sigma_h = 0.58p_w. \tag{12}$$

Equations (11) and (12) indicate that the tip pressure p_a is not 0, and the wellbore pressure p_w should be always greater than the minimum horizontal stress σ_h to maintain the fracture propagation considering the pressure drop.

3 In situ stress model for multi-stage fractured horizontal wells

Due to shale formations of low permeability, multi-stage hydraulic fracturing of horizontal wells is used to interact with natural fractures and weak bedding planes, to create as many crossing fractures as possible to maximize the stimulated reservoir volume. During stage fracturing, the main hydraulic fractures are created one after another. When a main fracture is formed, there will exist an induced stress field around the fracture, which influences the in situ stress field around it. The superposition of the induced stress field and the in situ stress field will affect the initiation and propagation of subsequent fractures. At present, classical fracture mechanics theory is used to calculate the stress field after fracturing.

3.1 Limitation of the classical analytical method

According to the theory of fracture mechanics, a two-dimensional model was established, based on the assumption of homogeneity, isotropy, and plane strain conditions, to calculate the induced stress field, as shown in Fig. 2.

We assume that the fracture is vertical, its longitudinal section is elliptical, and the height is H. z-axis is along the fracture height direction. x-axis is along the horizontal wellbore, and y-axis is along the direction of the maximum horizontal stress. It is assumed that the tensile stress is positive and the compressive stress is negative. The induced stresses at an arbitrary point (x, y, z) are as follows (Yin et al. 2012):

$$\begin{cases} \sigma_x = -p\frac{r}{c}\left(\frac{c^2}{r_1r_2}\right)^{\frac{3}{2}}\sin\beta\sin\left[\frac{3}{2}(\beta_1 + \beta_2)\right] \\ \qquad + p\left[\frac{r}{(r_1r_2)^{\frac{1}{2}}}\cos\left(\beta - \frac{1}{2}\beta_1 - \frac{1}{2}\beta_2\right) - 1\right] \\ \sigma_z = p\frac{r}{c}\left(\frac{c^2}{r_1r_2}\right)^{\frac{3}{2}}\sin\beta\sin\left[\frac{3}{2}(\beta_1 + \beta_2)\right] \\ \qquad + p\left[\frac{r}{(r_1r_2)^{\frac{1}{2}}}\cos\left(\beta - \frac{1}{2}\beta_1 - \frac{1}{2}\beta_2\right) - 1\right] \\ \sigma_y = \nu(\sigma_x + \sigma_z) \\ \tau_{xz} = -p\frac{r}{c}\left(\frac{c^2}{r_1r_2}\right)^{\frac{3}{2}}\sin\beta\sin\left[\frac{3}{2}(\beta_1 + \beta_2)\right] \end{cases} \tag{13}$$

with

$$c = H/2,$$

where σ_x, σ_y, and σ_z are the three normal stress components induced by a fracture, MPa; τ_{xz} is the shear component, MPa; p is the fluid pressure, MPa; ν is the Poisson ratio.

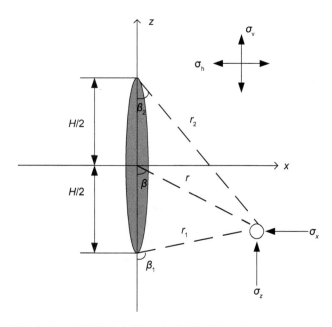

Fig. 2 Stress field induced by a hydraulic fracture

The relationships between these parameters are as follows:

$$\begin{cases} r = \sqrt{x^2 + z^2} \\ r_1 = \sqrt{x^2 + (c - z)^2} \\ r_2 = \sqrt{x^2 + (c + z)^2} \\ \beta = \arctan\left(-\dfrac{x}{c}\right) \\ \beta_1 = \arctan\left(\dfrac{x}{c - z}\right) \\ \beta_2 = \arctan\left(-\dfrac{x}{c + z}\right) \end{cases} \tag{14}$$

In situ stresses are composed of σ_x, σ_y, and σ_z. The stress field around a later fracture may be a summation of induced stresses of the former fractures and the in situ stresses. According to the principle of superposition, the stress field around the nth fracture was studied by Zhang and Chen (2010a, b, c):

$$\begin{cases} \sigma'_{H(n)} = \sigma_H + v\left(\displaystyle\sum_{i=1}^{n-1} \sigma_{x(in)} + \sum_{i=1}^{n-1} \sigma_{z(in)}\right) \\ \sigma'_{h(n)} = \sigma_h + \displaystyle\sum_{i=1}^{n-1} \sigma_{x(in)}, \\ \sigma'_{v(n)} = \sigma_v + \displaystyle\sum_{i=1}^{n-1} \sigma_{z(in)} \end{cases} \tag{15}$$

where $\sigma'_{H(n)}$, $\sigma'_{h(n)}$, and $\sigma'_{v(n)}$ are the three principal stresses around the nth fracture, MPa; $\sigma_{x(in)}, \sigma_{y(in)}$, and $\sigma_{z(in)}$ are the induced stress components around the nth fracture caused by the ith fracture, MPa.

When a new fracture is formed, it will produce an induced stress field which can be calculated by Eq. (13). Superimposing the newly induced stress field on the old one gives the final stress field (Eq. 15). This method just considers the effect of hydraulic fractures on the stress field but ignores the influence of the horizontal well. So it is limited and cannot provide the exact value of the stress field.

Because the stress field around the horizontal wellbore changes dramatically, the effect of the horizontal wellbore on the stress field distribution should be considered. Numerical calculation can achieve the above purpose.

3.2 Numerical method for in situ stress field in horizontal wells

The initial stress field around the wellbore will be disturbed and redistributed in the shale reservoir during drilling and completion operations. When the first hydraulic fracture is formed, the in situ stress field around the hydraulic fracture and the wellbore will be redistributed. So interaction

between fractures and the horizontal wellbore should be considered when calculating the variation of stress in horizontal well fracturing.

Considering interaction between fractures and the horizontal wellbore, analysis of the in situ stress field can be simplified as a plane strain problem. In this analysis, the stress induced by the change of reservoir temperature is ignored, and the fluid flowing in fractures is incompressible.

Figure 3 shows a model of stress fields considering interaction between fractures and the horizontal wellbore.

In order to reduce calculation time, we considered symmetrical geometry, in which we would calculate the stress field in the finite element model.

4 Comparison of classical analytical and numerical methods

The stress difference around a fracture before and after fracturing can be expressed as

$$\begin{cases} \sigma_{ax} = \sigma'_h - \sigma_h \\ \sigma_{ay} = \sigma'_H - \sigma_H \\ \sigma_{az} = \mu(\sigma_{ax} + \sigma_{ay}) \end{cases} \tag{16}$$

where σ'_h and σ'_H are the minimum and maximum horizontal stresses around the fracture after fracturing, respectively.

The main parameters used in numerical simulation are shown Tables 1 and 2.

The induced stress in the first stage of fracturing is closely related to the net pressure and the fracture length. The induced stress is a function of the distance away from

Table 1 Mechanical parameters of the shale reservoir

E_1, GPa	E_2, GPa	v_1	v_2	G_{12}, GPa
24.910	14.093	0.324	0.367	7.814

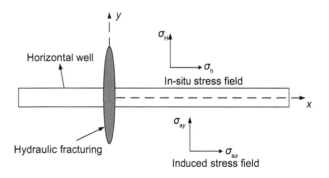

Fig. 3 Stress fields considering interaction between fractures and the horizontal wellbore

Table 2 Input parameters

Well depth, m	Wellbore length, m	Minimum horizontal stress, MPa	Maximum horizontal stress, MPa	Reservoir pressure, MPa	Net pressure, MPa	The model length, m	The model width, m	Fracture length, m	Facture width, mm
2410	1000	50	65	33	10	2000	2000	200	2

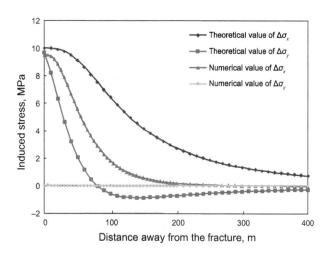

Fig. 4 Relationship between the induced stress field and the distance away from the fracture

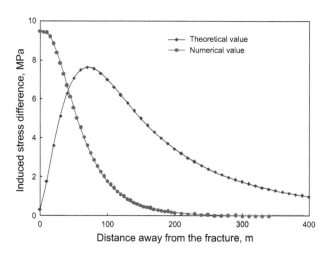

Fig. 5 Relationship between the induced stress difference and the distance away from the fracture

the first fracture. Figure 4 shows the induced stress from the classical analytical and numerical methods.

Ignoring the interaction between the fracture and the wellbore (in the analytical model), the stress induced by the first fracture is relatively high in the wellbore direction. The theoretical method overestimates the induced stress in the horizontal wellbore direction. The induced stress in the y-axis direction decreases rapidly. However, the induced stress around the wellbore is affected by the wellbore. Generally, the induced stress in the minimum horizontal stress direction is greater than that in the maximum horizontal stress direction, so the difference between the maximum and minimum stresses will reduce. The fracture may turn when the stress difference reaches a certain value. This would provide favorable conditions for creating complex fractures and communicating with natural fractures in the propagation of hydraulically induced fractures.

Figure 5 shows that the stress difference calculated from the classical analytical method is relatively low in the region close to the fracture, when not considering the interaction between fractures and the wellbore. On the other hand, the stress difference calculated from the analytical method is relatively high in the region far away from the fracture.

5 Distribution of the induced stress with different spacing among fracturing stages

As shown in Figs. 4 and 5, the induced stress is related to the distance away from the fracture. With an increase in the fracture spacing in two stages of fracturing, the stress difference decreases gradually.

Therefore, in order to create a fracture network to improve well productivity in shale reservoirs, we should try to reduce the fracture spacing between the two stages. When the fracture spacing is too small, not only would it waste fracturing materials and resources, but also it cannot achieve the expected economic benefits.

In this paper, the stress field is studied with the numerical method at different fracture spacing.

5.1 Fracture reorientation criterion

When the net pressure in the fracture is greater than the stress difference between the minimum and maximum horizontal stresses plus reservoir rock tensile strength, a branch of fractures may be initiated (Sam et al. 2011; Zhao et al. 2012; Jiang 2013).

$$\sigma_H + \sigma_y \leq \sigma_h + \sigma_x \tag{17}$$

Equation (7) can be changed to

$$\sigma_x - \sigma_y \geq \sigma_H - \sigma_h \qquad (18)$$

When Eq. (18) is satisfied, a branch fracture may deviate from the original extension path and extend along the horizontal wellbore as shown in Fig. 6.

Formation mechanisms of a complex fracture network are as follows:

(1) To form a plurality of main fractures in the formation, the stress induced by fracturing might change the initial stress pattern.

(2) Proper fracture spacing to create reasonably large induced stress so as to make the fracture reorient and to make connections with natural fractures.

5.2 Stress change under different fracture spacing

Figure 7 shows the change of stress with varied fracture spacing of 50, 100, and 150 m. This indicates that stress concentration around the fracture will be strengthen as fracture spacing reduces. When the fracture spacing is 50 m, the maximum stress along the wellbore direction is in the middle of the two fractures. When it increases to 100 m, the maximum stress will be in the region near the fractures, and the minimum is in the middle of fractures. The shape of distribution of stress along the wellbore direction is similar to a funnel. When it changes to 150 m, the maximum stress is in the region near the fractures. So, the stress along the wellbore direction increases remarkably, and the interference of two fractures is strong when the fracture was 50 m apart. When the fractures were 100 and 150 m apart, the stress will be reduced and the degree of interference is also weakened.

Figure 8 shows a change of stress in the direction perpendicular to the wellbore direction at different fracture spacing of 50, 100, to 150 m. This indicates that the existence of the wellbore plays a dominate role in the change of stress.

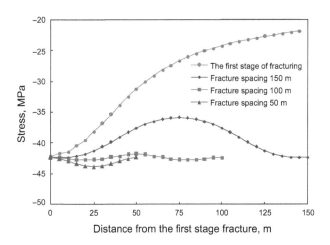

Fig. 7 Change of stress with fracture spacing in the wellbore direction

Figure 9 shows a change of difference between the maximum and minimum stresses in the wellbore direction at different fracture spacing. This indicates that the difference will be large when the fracture spacing is small. So it may change the horizontal stress pattern between horizontal stresses, which would be helpful in the formation of a complicated fracture network. However, the difference of horizontal stresses parallel and perpendicular to the wellbore would decrease as the fracture spacing increases.

Figure 10 shows a change of stress at the fracture tips parallel to the wellbore as the fracture spacing varies from 50, 100, to 150 m. This indicates that the smaller the fracture spacing, the higher the stress at the fracture tip and the stronger the interference between stresses. This would increase the likelihood of fracture reorientation and the formation of a complex fracture network. While with an increase in the fracture spacing, a high stress appears only near fractures and they have strong interference. In the

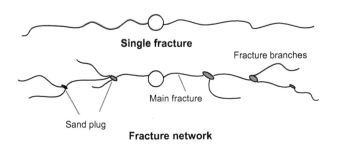

Fig. 6 Comparison of a single fracture and a complex fracture network

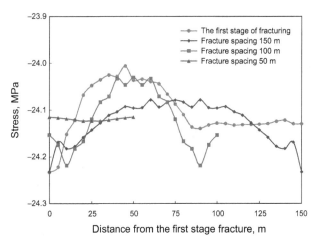

Fig. 8 Change of stress perpendicular to the wellbore direction at different fracture spacing

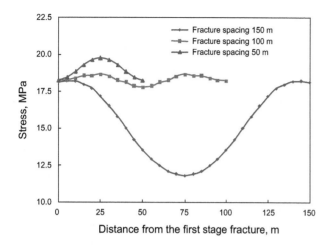

Fig. 9 Change of difference between the maximum and minimum horizontal stresses in the wellbore direction at different fracture spacing

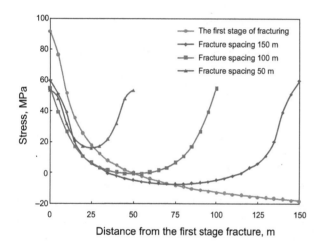

Fig. 10 Stress near fracture tips in the wellbore direction at different fracture spacing

central area between fractures, the stress interference is much less, so the creation of a complex fracture network is unlikely.

Figure 11 shows the change of stress at fracture tips in the direction perpendicular to the wellbore. This demonstrates that the influenced region is quite limited as the fracture spacing increases. When the fracture spacing extends the influenced region, the stress would be mainly influenced by the wellbore.

So the stress interference between fractures is closely related to the fracture spacing and the distance from the wellbore in stage fracturing of the horizontal well. The change of stress in the wellbore direction is mainly affected by the fracture spacing while the distance from the wellbore dominates the change of stress in the direction perpendicular to the wellbore. When the fracture spacing is 50 m, the horizontal stress pattern may be changed; the

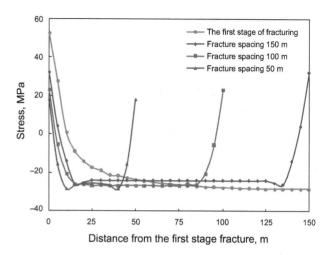

Fig. 11 Stress near fracture tips in the direction perpendicular to the wellbore at different fracture spacing

original maximum stress turns to the minimum and vice versa due to high induced stress, and it may connect natural fractures more effectively so as to create a complex fracture system more easily.

6 Conclusions

(1) A comparison was made between the traditional classic analytical and numerical methods in analysis of stress in multi-stage fracturing of horizontal wells. The analytical method did not take into account interaction among fractures and the horizontal wellbore when analyzing the induced stress, so a high stress was predicted in the minimum horizontal stress direction and the influence of the induced stress from fracturing was overestimated. In addition, the induced stress in the maximum horizontal stress direction was influenced by near wellbore effects. So the analytical method has its limitations, and the numerical method can provide more reasonable results.

(2) The induced stress along wellbore direction from fracturing may significantly influence the stress distribution around the horizontal wellbore. The influence will gradually reduce to its original stress in the far field. Stress redistribution in the direction perpendicular to the wellbore is mainly influenced by near wellbore effects.

(3) The stress interference is related to fractures and near wellbore effects in multi-stage fracturing of horizontal wells. The stress along the wellbore is mainly influenced by the fracture spacing. For a fracture with 200 m length, fracture reorientation may occur due to high induced stress so as to create complex fractures when the fracture spacing is around 50 m.

Acknowledgments This research was supported by the Natural Science Foundation of China (Grant No. 51490653, Basic Theoretical Research of Shale Oil and Gas Effective Development).

References

Cai WB, Li ZM, Zhang XL, et al. Horizontal well fracturing technology for reservoirs with low permeability. Pet Explor Dev. 2009;36(1):80–5.

Chen FJ, Tang Y, Liu SD, et al. Study of the optimization of staged fracturing of a horizontal well in a tight gas reservoir with low permeability. Spec Oil Gas Reserv. 2013;19(6):85–7 (in Chinese).

Chen JN, Li TT, Zhang Y. Application of the unstructured grids in the numerical simulation of fractured horizontal wells in ultra-low permeability gas reservoirs. J Nat Gas Sci Eng. 2015;22(1):580–90. doi:10.1016/j.jngse.2015.01.003.

Chuprakov DA, Akulich AV, Siebrits E, et al. Hydraulic-fracture propagation in a naturally fractured reservoir. SPE Prod Oper. 2011;26(1):88–97.

Civan F. Effective correlation of apparent gas permeability in tight porous media. Transp Porous Media. 2010;82(2):375–84.

Economides MJ. Reservoir stimulation. Beijing: Petroleum Industry Press; 2006. p. 90–125.

Fang Z, Khaksar A. Complexity of minifrac tests and implications for in situ horizontal stresses in coalbed methane reservoirs. In: International petroleum technology conference, 7–9 February, Bangkok, Thailand, 2011.

Fischer MP, Gross MR, Engelder T, et al. Finite-element analysis of the stress distribution around a pressurized crack in a layered elastic medium: implications for the spacing of fluid-driven joints in bedded sedimentary rock. Tectonophysics. 1994;247(1):49–64.

Hou B, Chen M, Wang Z, et al. Hydraulic fracture initiation theory for a horizontal well in a coal seam. Pet Sci. 2013;10(2):219–25. doi:10.1007/s12182-013-0270-9.

Jiang TX. The fracture complexity index of horizontal wells in shale oil and gas reservoirs. Pet Drill Tech. 2013;41(2):7–12 (in Chinese).

Koløy TR, Brække K, Sørheim T, et al. The evolution, optimization & experience of multistage frac completions in a North Sea environment. In: SPE annual technical conference and exhibition, 27–29 October, Amsterdam, The Netherlands, 2014. doi:10.2118/170641-MS.

Li LL, Yao J, Li Y, et al. Study of productivity calculation and distribution of staged multi-cluster fractured horizontal wells. Pet Explor Dev. 2014;41(4):504–8.

Pan BZ, Li D, Chen G, et al. Numerical simulation of wellbore and formation temperature fields in carbonate formations during drilling and shut-in in the presence of lost circulation. Pet Sci. 2014;11(2):293–9. doi:10.1007/s12182-014-0343-4.

Qu ZQ, Zhao YJ, Wen QZ, et al. Fracture parameter optimization in integral fracturing of horizontal wells. Pet Geol Recover Effic. 2012;19(4):106–11 (in Chinese).

Sam ZS, Zhou XY, Yang HJ, et al. Fractured reservoir modeling by discrete fracture network and seismic modeling in the Tarim Basin. China. Pet Sci. 2011;8(12):433–45.

Shang XT, He SL, Liu GF, et al. Breakdown pressure calculation of staged fracturing for horizontal wells. Drill Prod Technol. 2009;31(2):96–100 (in Chinese).

Sorek N, Moreno JA, Rice R, et al. Optimal hydraulic fracture angle in productivity maximized shale well design. In: Annual technical conference and exhibition, 27–29 October, Amsterdam, The Netherlands, 2014. doi:10.2118/170965-MS.

Wang T, Zhou WB, Chen JH, et al. Simulation of hydraulic fracturing using particle flow method and application in a coal mine. Int J Coal Geol. 2014a;121(5):1–13. doi:10.1016/j.coal.2013.10.012.

Wang H, Liao X, Zhao X, et al. Advances of technology study of stimulated reservoir volume in an unconventional reservoir. Spec Oil Gas Reserv. 2014b;21(2):8–15 (in Chinese).

Wei H, Li L, Wu X, et al. The analysis and theory research on the factor of multiple fractures during hydraulic fracturing of CBM wells. Proc Earth Planet Sci. 2011;22(3):231–7. doi:10.1016/j.proeps.2011.09.088.

Wu Q, Xu Y, Wang T, et al. The revolution of reservoir stimulation: an introduction to volume fracturing. Nat Gas Ind. 2011;31(4):7–11 (in Chinese).

Wu Q, Xu Y, Wang X, et al. Volume fracturing technology of unconventional reservoirs: connotation, design optimization and implementation. Pet Explor Dev. 2012;39(3):377–84.

Xu YB. Influential factors for fracture geometry as multi-fracture simultaneous extension. Pet Geol Oilfield Dev Daqing. 2009;28(3):89–92 (in Chinese).

Yao J, Sun H, Fan DY, et al. Numerical simulation of gas transport mechanisms in tight shale gas reservoirs. Pet Sci. 2013;10(5):528–37. doi:10.1007/s12182-013-0304-3.

Yin J, Guo JC, Zeng FH. Perforation spacing optimization for staged fracturing of horizontal well. Pet Drill Tech. 2012;40(5):67–71 (in Chinese).

Zeng BQ, Cheng LS, Li CL, et al. Development evaluation of fractured horizontal wells in ultra-low permeability reservoirs. Acta Pet Sin. 2010;31(5):791–6 (in Chinese).

Zhang GQ, Chen M. Dynamic fracture propagation in hydraulic re-fracturing. J Pet Sci Eng. 2010a;70(3–4):266–72.

Zhang GQ, Chen M. Study of damage mechanism by production testing after acid fracturing of carbonate reservoir. Pet Sci Technol. 2010b;28(2):125–34.

Zhang GQ, Chen M. The relationship between the production rate and initiation location of new fractures in a refractured well. Pet Sci Technol. 2010c;28(7):655–66.

Zhang WM, Meng G, Wei X. A review on slip models for gas microflows. Microfluid Nanofluid. 2012;13(6):845–82.

Zhang YJ, Li ZW. Electricity generation from enhanced geothermal systems by oilfield produced water circulating through a reservoir stimulated by staged fracturing technology for horizontal wells: a case study in the Xujiaweizi area in Daqing Oilfield, China. Energy. 2014;78(6):788–805. doi:10.1016/j.energy.2014.10.073.

Zhao HF, Chen M, Jin Y, et al. Rock fracture kinetics of the fracture mesh system in shale gas reservoirs. Pet Explor Dev. 2012;39(4):465–70.

Zhou CN, Dong DZ, Wang SJ, et al. Geological characteristics, formation mechanism and resource potential of shale gas in China. Pet Explor Dev. 2010;37(6):641–53.

Zhu SY, Li HT, Yang MJ, et al. Optimization of fracture laying patterns of horizontal wells in multi-stage fracturing of a low permeability reservoir. Fault-Block Oil Gas Field. 2013;20(3):373–6 (in Chinese).

Ziarani AS, Aguilera R. Knudsen's permeability correction for tight porous media. Transp Porous Media. 2012;91(1):239–60.

10

Inflationary effects of oil prices and domestic gasoline prices: Markov-switching-VAR analysis

Selin Ozdemir[1] · Işil Akgul[1]

Abstract The purpose of this study is to contribute to the literature by studying the effects of sudden changes both on crude oil import price and domestic gasoline price on inflation for Turkey, an emerging country. Since an inflation targeting regime is being carried out by the Central Bank of Turkey, determination of such effects is becoming more important. Therefore empirical evidence in this paper will serve as guidance for those countries, which have an inflation targeting regime. Analyses have been done in the period of October 2005–December 2012 by Markov-switching vector autoregressive (MS-VAR) models which are successful in capturing the nonlinear properties of variables. Using MS-VAR analysis, it is found that there are 2 regimes in the analysis period. Furthermore, regime changes can be dated and the turning points of economic cycles can be determined. In addition, it is found that the effect of the changes in crude oil and domestic gasoline prices on consumer prices and core inflation is not the same under different regimes. Moreover, the sudden increase in gasoline price is more important for consumer price inflation than crude oil price shocks. Another finding is the presence of a pass-through effect from oil price and gasoline price to core inflation.

Keywords Crude oil price · Domestic gasoline price · Consumer price index · Core inflation · MS-VAR model

✉ Selin Ozdemir
sozdemir@marmara.edu.tr

[1] Department of Econometrics, Faculty of Economics, Marmara University, Istanbul, Turkey

Edited by Xiu-Qin Zhu

1 Introduction

Oil price has acquired increasing attention from both academicians and politicians after the dramatic rise in oil price in 1973 (Robert and Tatom 1977; Santini 1986; Mork 1989; Dotsey and Reid 1992; Davis et al. 1996; Hamilton 1996; Cunado and Perez de Gracia 2003; Cologni and Manera 2005; Kilian and Vigfusson 2009; Du et al. 2011; Reboredo 2011; Kilian and Murphy 2013). The reason is that understanding the inflationary effects caused by an increase of oil price can assist politicians and central banks to implement policies to get inflation under control.

Oil is one of the most important sources of energy, and the impacts of oil price fluctuations are risky in terms of the economies of countries. Fluctuation in oil price affects both the oil-importing and the oil-exporting countries. Especially in the non-oil-producing countries, oil price fluctuation can have a great impact on economic variables such as consumer prices and core inflation. Moreover, it is generally accepted that instant and huge changes in oil price can cause a rise in consumer prices and core inflation, which will result in economic recession in oil-importing countries.

The impacts of oil price increases on high inflation are basically reflected in three ways: the first impact appears because oil constitutes a proportion of household consumption. This proportion comprises processed products such as gasoline used for transportation and fuel used for heating, which fall into the household consumption basket within the consumer price index. The second impact is reflected in the form of consumer prices through producer prices. Firms and factories pass on the increase in energy prices to the prices of final products. In turn, this creates an impact on the consumer price index, which is an indirect effect. The third impact is that there could be an expectation of higher inflation and higher wages. In order to compensate for the decrease in real income, a negotiation process is conducted for wages. Production costs increase because of the rising of oil prices, which is called as a second round impact.

Based on the above factors, the Central Bank of the Republic of Turkey confirmed that oil price uncertainty is a risk factor with regard to inflation for Turkey, which carries out an inflation targeting regime (CBRT Monetary Policy Report 2012). Furthermore, it has been indicated that the emergence of oil supply problems could lead to an increase in energy prices; consequently, it could exacerbate the expectation of inflation, and eventually necessary steps would need to be taken to prevent it. Therefore, this paper discusses a topic that is on the agenda, and aims to determine whether shocks in oil and gasoline prices constitute a risk to consumer price inflation and core inflation.

Various researchers have investigated the relationship between oil price and inflation in Turkey. But only a few studies have investigated whether the relationship is nonlinear. Oil price and inflation series may exhibit nonlinear behavior due to factors such as policy changes, energy crises, etc. Thus, if oil price and inflation data exhibit structural regime shifts, then a model assuming constant parameters, mean, and variance is likely to yield misleading results. Therefore, modeling the relationship between oil price and inflation within a nonlinear framework is more suitable.

This paper investigates the effect of a nonlinear relationship between crude oil import price and domestic gasoline price on consumer price inflation and core inflation, and differs from the existing literature using Markov-switching vector autoregressive (MS-VAR) models. The paper proceeds as follows: Section 2 presents a review of previous studies on the empirical evidence of oil price changes and their effects on inflation. Section 3 deals with data used in the analysis, methodological issues, and the empirical analysis. Section 4 concludes with a summary and policy implications.

2 Previous studies

Several researchers have investigated the relationship between oil prices and inflation using different econometric approaches, countries, and sample periods. The relevant literature includes the following studies: Kahn and Hampton (1990), Huntington (1998), LeBlanc and Chinn (2004), Cunado and Perez de Gracia (2005), Ewing and Thompson (2007), Farzanegan and Markwardt (2009), De Gregorio et al. (2007), Tang et al. (2010), and Álvarez et al. (2011). These studies reveal that inflation is affected by oil

price. Kahn and Hampton (1990) investigate whether increases in oil price affect the U.S. economy and find that in the short run, higher oil prices can increase inflation and lower real GNP. Huntington (1998) examines the linkages between oil price and inflation from a different perspective and finds that consumer prices appear to respond asymmetrically to energy price increase and decrease in the U.S. LeBlanc and Chinn (2004) show that oil price increases are likely to have only a modest effect on inflation in the U.S., Japan, and Europe. By taking a nonlinear relationship into account, Cunado and Perez de Gracia (2005) report that oil prices have permanent effects on inflation and asymmetric effects on the GNP in European countries. Medina and Soto (2007) use a dynamic stochastic general equilibrium and show that a 13 % increase in the real price of oil leads to an increase in inflation of about 0.4 % in the Chilean economy. Ewing and Thompson (2007) investigate the cyclical co-movements of crude oil price with consumer prices using Hodrick–Prescott methodology. Their findings support that the price of oil is the leading factor in consumer prices in the U.S. Farzanegan and Markwardt (2009) analyze the dynamic relationship between oil price shocks and major macroeconomic variables in Iran by applying a VAR approach. They identify that negative oil price shocks significantly increase inflation. De Gregorio et al. (2007) estimate a Phillips curve equation with lags of inflation, the output gap, and the percentage change in the price of oil for 24 industrial economies and 12 emerging economies. Their study shows the effects of oil shocks on the general level of prices. Van den Noord and André (2007) conclude that the spillover effects of energy prices into core inflation are small in comparison with the effects of the 1970s. Tang et al. (2010) provide statistical support for the adverse economic impacts of oil price shocks for developed economies. Their results show that an oil price increase negatively affects output and investment, but positively affects inflation rate and interest rate. Álvarez et al. (2011) examine the impact of oil price changes on Spanish and euro area consumer price inflation using linear and nonlinear models. They find that crude oil price fluctuations are major drivers of inflation variability.

However, at least some studies in the literature show evidence contradicting the results showing that oil price and inflation are linked. Hooker (2002) examines the effects of oil price shocks on inflation in Phillips curve models that allow structural breaks, and he reports that oil price shocks have caused an increase in inflation in the U.S. before 1981. However, in recent periods this impact was negligible. Also, Olomola and Adejumo (2006) evaluate the effects of oil price changes on output, inflation, real exchange rate, and money supply in Nigeria using VAR models. They find that oil price shocks do not have any substantial effect on output and inflation.

When limited numbers of articles that elaborate the impact of oil price changes on inflation in Turkey are examined, it is seen that nonlinearity has been neglected. Kibritcioglu and Kibritcioglu (1999) analyze the effect of oil price shocks on the general price level. They suggest that a 20 % increase in the price of crude oil has an insignificant effect on the general price level. Berument and Tasci (2002) use an input–output table and conclude that general price level increases for a given increase in the price of oil depend on the behavior of the wages, profits, interest, and rents. Aktas et al. (2010) carry out a VAR model and observe that a positive relationship between oil price and inflation exists. They assert that a response of macroeconomic variables against oil price shocks becomes stationary only after 1 year. Aydin and Acar (2011) analyze the economic effects of oil price shocks by developing a dynamic multi-sectoral general equilibrium model. Their results show that the price of oil has significant effects on consumer price inflation. Nazlioglu and Soytas (2011) examine the interdependence between world oil prices and individual agricultural commodity prices using the Toda–Yamamoto causality approach and generalized impulse-response analysis. Their results reveal the neutrality of agricultural commodity markets to the effects of oil price changes. Oksuzler and Ipek (2011) examine whether negative oil supply shocks can increase inflation. According to the results of Granger causality analysis, they find that there is no causality between oil price and inflation, but impulse-response functions showed that a positive oil price shock increases inflation. The empirical evidence obtained from a bound testing approach in the study of Peker and Mercan (2011) shows that the inflationary effect of oil products price increases is positive and statistically significant in the long term. Celik and Akgul (2011) investigate whether there is a linkage between the consumer price index and the fuel oil price index using the vector error correction model. Their results reveal that a 1 % increase in fuel oil prices can cause the consumer price index to rise by 1.26 %. Yaylali and Lebe (2012) specify the importance of crude oil prices in the general level of prices by employing Vector Autoregressive methodology. By identifying the source of change in inflation, their analysis results show that import crude oil prices are one of the important sources of inflation in Turkey. With a different approach, Catik and Onder (2011) investigate the oil price pass-through inflation for Turkey by considering nonlinearity. They find evidence for asymmetric oil pass-through in the high inflation regime by estimating Markov-switching (MS) models. In contrast to these studies, in this article, the effects of the relationship between crude oil import price and domestic gasoline prices on consumer price inflation and core inflation are examined separately using nonlinear models.

3 Data and econometric methodology

3.1 Data

The aim of this study is to empirically investigate the effects of sudden changes in the crude oil import price on inflation in Turkey. We use the first difference of logarithmic crude oil price (Brent-$/barrel) (OIL) and the first difference of logarithmic unleaded gasoline price ($/barrel) (DGAS) in the analyses. Also, inflation (INF) and core inflation (CINF) are calculated using the first difference of logarithmic consumer price index and logarithmic special comprehensive consumer price index, which excludes energy. Crude oil prices are taken from the Federal Reserve Bank of St. Louis Data Delivery page.[1] Historical domestic gasoline prices (TL/liter), which are available on the OPET company web page,[2] are added to the analysis with the values ($/barrel) obtained through the transformation. Consumer price indexes are taken from the Turkish Statistical Institute corporate web page.[3] We employ monthly data from October 2005 to December 2012. The choice of October 2005 as the starting period is based on the following: Between 2002 and 2006, implied inflation targeting was applied by the Central Bank of the Republic of Turkey; at the beginning of 2006 they adopted explicit inflation targeting, so analysis can be made starting from 2006. But to prevent any loss of data, the analysis period was initiated at October 2005.

3.2 Econometric methodology

Ever since the study by Hamilton (1989), Markov regime-switching models have been utilized by researchers for modeling many macroeconomic time series, which exhibit asymmetries and nonlinear behavior (Hansen 1992; Goodwin 1993; Gray 1996; Cologni and Manera 2009). Therefore, the use of the MS approach has become popular for determining asymmetries. Goldfeld and Quandt (1973) introduced the MS model, in which the latent state variable controls the regime shifts. Hamilton (1989) and Krolzig (1998) made important contributions by developing the MS-VAR model, which is able to characterize macroeconomic fluctuations in the presence of structural breaks or shifts. These approaches allow researchers to overcome the shortcomings of linear models in dealing with the asymmetry between expansions and contractions.

In these models, parameters of the VAR model depend on the regime variable (S_t), which makes the process regime dependent. The general idea behind this class of

---

[1] http://research.stlouisfed.org/fred2/.

[2] http://www.opet.com.tr/tr/PompaFiyatlariArsiv.aspx?cat=4&id=34.

[3] http://www.tuik.gov.tr/UstMenu.do?metod=temelist.

regime-switching models is that the parameters of a K-dimensional time series vector $y_t = (y_{1t}, \ldots, y_{Kt})'$, $t = 1, \ldots, T$,

$$y_t = v + A_1 y_{t-1} + \cdots + A_p y_{t-p} + u_t, \tag{1}$$

where $u_t \sim \mathrm{IID}(0, \sum)$ and $y_0, \ldots, y_{1-p}$ are fixed. The general idea behind the MS-VAR models is that the parameters of the underlying data-generating process of the observed time series vector y_t depend on the unobservable regime variable s_t, which represents the probability of being in a different state. The assumption of the MS model is that the unobservable realization of the regime $s_t \in \{1, \ldots, M\}$ is governed by a discrete time, discrete state Markov stochastic process, which is defined by the transition probabilities,

$$p_{ij} = \Pr(s_{t+1} = j | s_t = i), \sum_{j=1}^{M} p_{ij} = 1 \; \forall_i, \quad j \in \{1, \ldots, M\}. \tag{2}$$

It is assumed that s_t follows an irreducible ergodic M state Markov process with the transition matrix.

$$P = \begin{bmatrix} p_{11} & p_{12} & \cdots & p_{1M} \\ p_{21} & p_{22} & \cdots & p_{1M} \\ \vdots & \vdots & \ddots & \vdots \\ p_{11} & p_{12} & \cdots & p_{1M} \end{bmatrix}, \tag{3}$$

where $p_{iM} = 1 - p_{i1} - \cdots - p_{i,M-1}$ for $i = 1, \ldots, M$. We can write the MS-VAR model of order p;

$$y_t = v(s_t) A_1(s_t) y_{t-1} + \cdots + A_p(s_t) y_{t-p} + u_t, \tag{4}$$

where $u_t \sim \mathrm{NID}(0, \sum(s_t))$. The parameter shift functions and $v(s_t), A_1(s_t), \ldots, A_p(s_t)$, and $\sum(s_t)$ describe the dependence of the parameters on the realized regime s_t,

$$v(s_t) = \begin{cases} v_1 & \text{if } s_t = 1, \\ \vdots & \\ v_M & \text{if } s_t = M \end{cases} \tag{5}$$

The MS-VAR model allows for a variety of specifications. Krolzig (1997) made a representation with the general MS models with regime-dependent parameters in order to establish a common notation for each model, such as MSM-VAR, MSH-VAR, and MSIH-VAR.

Recently, there have been some developments in impulse-response relations in nonlinear models. Koop et al. (1996) offer a general analysis of impulse responses in nonlinear models and introduce the concept of generalized impulse response, which can measure the responses of the system to shocks to the variables in h period as,

$$\mathrm{IR}_{\nabla u}(h) = E[y_{t+h}|\xi_t, u_t + \nabla_u; Y_{t-1}] - E[y_{t+h}|\xi_t, u_t; Y_{t-1}], \tag{6}$$

where ∇_u is the shock at time t and the responses to shocks to the variables as in the case of the linear VAR process,

$$\mathrm{IR}_{uk}(h) = \frac{\partial E[y_{t+h}|\xi_t, u_t; Y_{t-1}]}{\partial u_{kt}}. \tag{7}$$

And the responses to shifts in regime are defined in the spirit of the generalized impulse-response concept:

$$\mathrm{IR}_{\nabla u}(h) = E[y_{t+h}|\xi_t + \nabla \xi, u; Y_{t-1}] - E[y_{t+h}|\xi_t, u_t; Y_{t-1}], \tag{8}$$

where $\nabla \xi$ is the shift in regime at time t.

Estimating MS-VAR models that are based on Hamilton's (1989) algorithm consists of two steps. In the first step, population parameters, including the joint probability density of unobserved states, are estimated, and in the second step probabilistic inferences about the unobserved states are made using a nonlinear filter and smoother. Filtered probabilities are inferences about s_t's conditional on information up to time t, and smoothed probabilities are inferences about s_t using all information available in the sample. However, this method becomes more disadvantageous as the number of parameters to be estimated increases. Accordingly, the expectation maximization algorithm, originally described by Dempster et al. (1977) is used. This technique starts with the initial estimates of the hidden data and iteratively produces a new joint distribution that increases the probability of observed data.

3.3 Empirical analysis

The analysis was initiated by calculating the certain statistics of the series used in the study, and the results are given in Table 1.

According to the Jarque–Bera test statistics in Table 1, INF and CINF series are normally distributed but OIL and DGAS series are not. Also the series are found to be stationary at the 5 % level of significance using Dickey and Fuller (1979) τ-test statistics. However, it is known that DF-type unit root tests are not strong in case of a regime change in the series. Therefore, the MS-ADF test, which is a unit root test appropriate for MS models, is also applied (Hall et al. 1999) and it is confirmed that the series are stationary. In order to reveal the nonlinear structure in the

Table 1 Descriptive statistics of series

	INF	CINF	OIL	DGAS
Mean	0.67	0.65	0.34	0.26
SD	0.84	0.99	9.35	4.94
Skewness	0.32	0.10	−1.16	−1.33
Kurtosis	3.20	2.63	5.62	7.63
Jarque–Bera	1.71	0.62	44.80	103.51
ADF	−7.37	−7.49	−6.21	−6.55
MS-ADF	−6.14	−5.36	−5.19	−5.59

Table 2 Results of linearity test

	$d = 1$	$d = 2$	$d = 3$	$d = 4$	$d = 5$	$d = 6$	$d = 7$	$d = 8$	$d = 9$	$d = 10$
INF	0.50	2.77	0.91	1.14	0.85	1.33	0.69	0.47	0.57	0.51
p value	0.44	0.02*	0.15	0.04	0.15	0.09	0.25	0.42	0.34	0.43
CINF	0.32	0.51	1.13	3.29	0.84	0.74	0.98	0.77	0.63	0.21
p value	0.67	0.53	0.11	0.01*	0.43	0.44	0.22	0.44	0.59	0.75
OIL	0.99	3.08	1.24	1.14	0.75	0.77	0.89	0.60	0.44	0.32
p value	0.64	0.02*	0.09	0.11	0.32	0.33	0.27	0.44	0.86	0.85
DGAS	0.62	1.14	3.01	1.21	1.99	0.87	0.92	0.79	0.51	0.52
p value	0.73	0.25	0.01*	0.19	0.09	0.31	0.47	0.50	0.63	0.66

Significance at 5 % is denoted with asterisks

Table 3 LR test results

	INF-OIL	INF-DGAS	CINF-OIL	CINF-DGAS
Ho: linear VAR Ha: two-regime MS-VAR	61.04*	59.14*	64.18*	62.19*
Ho: two-regime MS-VAR Ha: three-regime MS-VAR	11.02	9.07	12.41	10.77

Significance at 5 % is denoted with asterisks

series, the approach suggested by Tsay (1989) is used, and the linearity test results for different delay lengths are presented in Table 2.

The probability values reported in Table 2 calculated for 10 delay show that linearity is rejected more strongly in the second delay for INF and OIL, in the third delay for DGAS, and in the fourth delay for CINF. Afterward, an LR test is made in order to determine the number of regimes of the models, which is the first stage of model selection. Subsequently, the linear VAR model is tested against the 2-regime MS-VAR model. And later the 2-regime MS-VAR model is tested against the 3-regime MS-VAR model.

According to the results shown in Table 3, it is determined that the 2-regime MS-VAR models are appropriate for the analyses. Using the Schwarz Information Criterion, the delay lengths are selected and it is decided that for INF-OIL, MSI(2)-VAR(8); for INF-DGAS, MSI(2)-VARX(2); for CINF-OIL, MSIAH(2)-VAR(5) and for CINF-DGAS, MSIA(2)-VARX(8) models are appropriate. The MSI(2)-VAR(8) model is estimated for INF-OIL and given in Table 4.

Regime 1 represents low inflation; regime 2 represents high inflation periods in the model. As seen in Table 4, the effect of oil price change on inflation is significant and positive in time $t - 2$ and $t - 3$. Transition probabilities and regime durations are given in Table 5.

According to the regime probabilities shown in Table 5, it is seen that the probability of staying in the low inflation

Table 4 MSI(2)-VAR(8) model for INF-OIL

	INF	OIL
Constant (Reg. 1)	0.82* (4.16)	1.27 (0.65)
Constant (Reg. 2)	1.62* (7.07)	11.57* (5.05)
INF_{t-1}	−0.04 (−0.46)	−1.94* (−1.99)
INF_{t-2}	−0.09 (−0.97)	−0.50 (−0.55)
INF_{t-3}	−0.17* (−2.01)	−0.24 (−1.14)
INF_{t-4}	−0.31* (−3.44)	−1.60* (−1.99)
INF_{t-5}	−0.08 (−0.84)	−0.16 (−0.17)
INF_{t-6}	0.16 (1.68)	−1.04 (−1.05)
INF_{t-7}	−0.18* (−2.01)	−2.09* (−2.19)
INF_{t-8}	−0.17 (−1.81)	−1.26 (−1.31)
OIL_{t-1}	0.01 (0.21)	0.28* (3.23)
OIL_{t-2}	0.02* (2.29)	0.21* (2.32)
OIL_{t-3}	0.01* (2.13)	−0.04 (−0.49)
OIL_{t-4}	−0.03 (−0.38)	0.08 (0.95)
OIL_{t-5}	0.03 (0.02)	−0.19* (−2.16)
OIL_{t-6}	0.00 (1.08)	−0.27* (−3.11)
OIL_{t-7}	0.01 (0.94)	0.15 (1.68)
OIL_{t-8}	0.01 (1.91)	−0.01 (−0.19)

t Statistics are given in parentheses (). Probabilities are given in parentheses []. Significance at 5 % is denoted with asterisks

Log-likelihood: 154.12, AIC criterion: 9.79, LR linearity test: 16.50 [0.03], $\chi^{(2)} = [0.003]$, $\chi^{(4)} = [0.002]$, Davies = [0.004], Vector normality test: $\chi^{(4)} = 4.57$ [0.18], Vector hetero test: $\chi^{(24)} = 14.44$ [0.93], Vector portmanteau(5) $\chi^{(12)} = 13.42$ [0.33]

Table 5 Transition probabilities and regime durations for MSI(2)-VAR(8)

Transition probabilities	Regime 1	Regime 2	Nobs	Durations
Regime 1	0.68	0.32	39.9	3.08
Regime 2	0.33	0.67	39.1	3.00

Table 6 MSI(2)-VARX(2) model for INF-DGAS

	INF	DGAS
Constant (Reg. 1)	−0.01 (−0.11)	−2.43* (−3.34)
Constant (Reg. 2)	1.09* (7.56)	3.01* (4.22)
INF_{t-1}	0.19* (2.26)	−1.07* (−2.33)
INF_{t-2}	0.07 (0.88)	0.18 (0.41)
$DGAS_{t-1}$	0.07* (3.84)	0.11 (1.29)
$DGAS_{t-2}$	0.03* (2.65)	0.11 (1.40)
D1	2.59* (4.27)	−3.34* (−6.06)
D2	1.76* (2.32)	−2.28* (−2.35)

t Statistics are given in parentheses (). Probabilities are given in parentheses []. Significance at 5 % is denoted with asterisks

Log-likelihood: 134.88, AIC criterion: 8.22, LR linearity test: 12.52 [0.04], $\chi^{(2)} = [0.021]$, $\chi^{(4)} = [0.012]$, Davies = [0.031], Vector normality test: $\chi^{(4)} = 3.73$ [0.23], Vector hetero test: $\chi^{(24)} = 12.45$ [0.54], Vector portmanteau(5):$\chi^{(12)} = 14.88$ [0.47]

Table 7 Transition probabilities, regime durations for MSI(2)-VARX(2)

Transition probabilities	Regime 1	Regime 2	Nobs	Durations
Regime1	0.40	0.60	33.7	2.48
Regime2	0.36	0.64	51.3	3.78

is 0.68 and staying in the high inflation regime is 0.67. Also, the first regime tends to last 3.08 months on average, while the second regime is less persistent with 3.00 months. The average period of 3 months reveals that the number of stages is high and the transitions are rapid. Also, it is seen that the low inflation and high inflation probabilities and transition probability from one regime to another are almost the same. Observation numbers of regime 1 and regime 2 and duration of remaining at one regime are approximately the same, as well. These results show that the stages of both regimes would be the same. The results obtained for INF-DGAS are reported in Table 6.

It is seen in Table 6, the effect of gasoline price change on inflation is significant and positive in time $t − 1$ and $t − 2$. Transition probabilities and regime durations for MSI(2)-VARX(2) are given in Table 7.

According to the regime probabilities, it is seen that the probability of staying in the low inflation is 0.40 and staying in the high inflation regime is 0.64. Also durations show that the first regime tends to last 2.48 months on average, while the second regime is more persistent, lasting 3.78 months. Transition probabilities show that there is a high probability of transition from low inflation to high inflation. This finding reveals that a sudden increase in domestic gasoline oil prices has the effect of a crisis, and this should be taken into consideration in adjusting gasoline prices to achieve the inflation target. The model selected for CINF-OIL is a MSIAH(2)-VAR(5) model and the results are given in Table 8.

The coefficient in Table 8 reveals that the effects of oil price change on core inflation is significant and positive in time $t − 1$ but negative in $t − 3$. Transition probabilities and regime durations are given in Table 9.

Table 8 MSIAH(2)-VAR (5) model for CINF-OIL

	Regime 1		Regime 2	
	CINF	OIL	CINF	OIL
Constant	1.13* (3.83)	−1.68 (−2.40)	1.18* (3.31)	2.16* (5.94)
$CINF_{t-1}$	−0.11 (−0.92)	0.90 (0.59)	0.35 (1.40)	1.10 (2.53)*
$CINF_{t-2}$	−0.35 (−2.65)*	2.54 (1.98)*	−0.40 (−2.24)*	−1.86 (−3.24)*
$CINF_{t-3}$	−0.62 (−4.07)*	4.08 (2.07)*	−0.08 (1.24)	0.88 (−0.79)
$CINF_{t-4}$	−0.52 (−3.57)*	−1.26 (−0.70)	−0.06 (−2.90)*	−2.05 (−0.61)
$CINF_{t-5}$	0.23 (1.66)	2.25 (1.22)	−0.33 (3.40)*	2.49 (−2.79)*
OIL_{t-1}	0.02 (2.96)*	0.40 (2.18)*	0.01 (−0.53)	−0.04 (0.86)
OIL_{t-2}	0.01 (0.09)	0.31 (1.48)	−0.02 (−2.70)*	−0.22 (−0.71)
OIL_{t-3}	−0.03 (−2.71)*	0.25 (1.20)	0.02 (1.44)	0.12 (1.53)
OIL_{t-4}	−0.02 (−1.34)	−0.13 (−0.64)	0.03 (1.97)*	0.15 (2.54)*
OIL_{t-5}	0.03 (2.25)*	−0.21 (−1.20)	−0.02 (−1.91)*	−0.13 (−1.41)

t Statistics are given in parentheses (). Probabilities are given in parentheses []. Significance at 5 % is denoted with asterisks

Log-likelihood: 121.93, AIC criterion: 8.14, LR linearity test: 15.33 [0.04], $\chi^{(2)} = [0.024]$, $\chi^{(4)} = [0.019]$, Davies = [0.023], Vector normality test: $\chi^{(4)} = 3.73$ [0.25]; Vector hetero test: $\chi^{(24)} = 10.45$ [0.47], Vector portmanteau(5): $\chi^{(12)} = 13.42$ [0.43]

Table 9 Transition probabilities, regime durations for MSIAH(2)-VAR (5)

Transition probabilities	Regime 1	Regime 2	Nobs	Durations
Regime 1	0.47	0.53	34.9	2.11
Regime 2	0.36	0.64	56.1	2.80

Table 10 MSIAH(2)-VARX (8) model for CINF-DGAS

	Regime 1		Regime 2	
	CINF	DGAS	CINF	DGAS
Constant	0.49 (2.71)*	2.31 (2.74)*	0.90 (4.76)*	0.76 (0.86)
$CINF_{t-1}$	0.22 (2.11)*	0.42 (0.89)	0.26 (2.97)*	0.16 (0.39)
$CINF_{t-2}$	0.25 (2.23)*	2.50 (4.75)*	−0.20 (−2.64)*	0.08 (0.22)
$CINF_{t-3}$	−0.39 (−2.24)*	−0.89 (−4.41)*	−0.14 (−1.47)	0.45 (1.04)
$CINF_{t-4}$	−0.28 (−3.02)*	1.42 (3.31)*	0.14 (1.34)	−1.26 (−2.68)*
$CINF_{t-5}$	0.27 (2.14)*	−3.34 (−5.29)*	−0.23 (−2.62)*	1.45 (3.54)*
$CINF_{t-6}$	0.66 (5.18)*	0.67 (1.06)	0.17 (2.14)*	0.09 (0.22)
$CINF_{t-7}$	−0.37 (−2.48)*	−3.28 (−5.17)*	0.14 (1.83)	0.66 (1.84)
$CINF_{t-8}$	−0.49 (−4.08)	−2.37 (−4.37)*	0.03 (0.35)	0.50 (0.97)
$DGAS_{t-1}$	−0.01 (−0.01)	0.49 (6.23)	0.02 (0.97)	0.16 (1.31)
$DGAS_{t-2}$	−0.02 (−1.26)	−0.36 (−3.72)*	0.01 (0.23)	−0.13 (−1.46)
$DGAS_{t-3}$	−0.05 (−2.45)*	0.89 (8.04)*	0.08 (0.58)	0.07 (1.06)
$DGAS_{t-4}$	0.05 (2.22)	−0.09 (−0.75)	−0.03 (−1.97)*	−0.06 (−0.86)
$DGAS_{t-5}$	0.01 (0.55)	−0.22 (−1.69)	0.05 (−0.43)	0.07 (1.04)
$DGAS_{t-6}$	0.10 (3.00)*	1.19 (7.71)*	0.02 (0.85)	−0.11 (−1.78)
$DGAS_{t-7}$	−0.06 (−2.39)*	0.53 (4.30)*	0.03 (2.10)*	−0.01 (−0.04)
$DGAS_{t-8}$	−0.01 (−0.85)	−0.22 (−2.57)*	−0.08 (−4.15)*	−0.30 (−3.16)*
D1	0.86	1.92	1.29	3.56

t Statistics are given in parentheses (). Probabilities are given in parentheses []. Significance at 5 % is denoted with asterisks

Log-likelihood: 114.023, AIC criterion: 6.17, LR linearity test: 11.24 [0.03], $\chi^{(2)} = [0.029]$, $\chi^{(4)} = [0.034]$, Davies = [0.014], Vector normality test: $\chi^{(4)} = 3.14$ [0.33], Vector hetero test: $\chi^{(24)} = 11.12$ [0.64], Vector portmanteau(5): $\chi^{(12)} = 13.11$ [0.29]

The regime probabilities show that, staying in the low inflation is 0.47 and staying in the high inflation regime is 0.64. Therefore, there is a high probability of transition from low inflation to high inflation, and a sudden increase in oil prices has a crisis period effect on core inflation. Also first regime tends to last 2.11 months on average, while the second regime is more persistent, lasting 2.80 months. The results obtained for CINF-DGAS are reported in Table 10.

According to Table 10, the effect of gasoline price change on core inflation is significant and positive in time $t − 4$. This finding reveals that gasoline price shocks have a delayed effect on core inflation. Transition probabilities and regime durations are shown in Table 11.

MSIAH(2)-VARX (8) is selected and the transition probabilities suggest that the persistence of the high inflation regime is higher than that of the low inflation regime. Regime 1 is determined to last, on average, 2.37 months, and the average duration of the high inflation phase is

Table 11 Transition probabilities, regime durations for MSIAH(2)-VARX (8)

Transition probabilities	Regime 1	Regime 2	Nobs	Durations
Regime 1	0.42	0.58	33.6	2.37
Regime 2	0.30	0.70	45.4	3.27

3.27 months. Also, it is found that the sudden increase in gasoline prices has a pass-through effect to core inflation.

Afterward, the filtered and smoothed probabilities are estimated and the graphics are given in Fig. 1.

The regime graphics show that the number of phases is high, there are rapid transitions, and the time remaining in one regime is short. In addition, graphics indicate that all models stay at high inflation longer than at low inflation. Impulse-response analyses were made after regime transition probabilities were reviewed, and the graphics are given in Fig. 2.

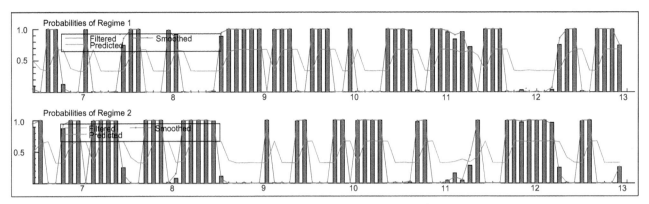

Filtered and smoothed probabilities for INF-OIL

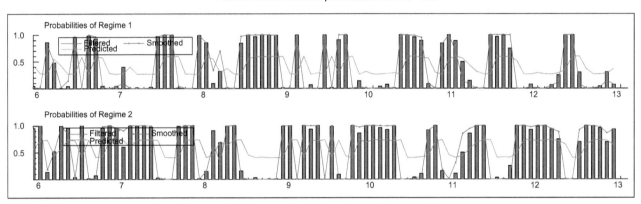

Filtered and smoothed probabilities for INF-DGAS

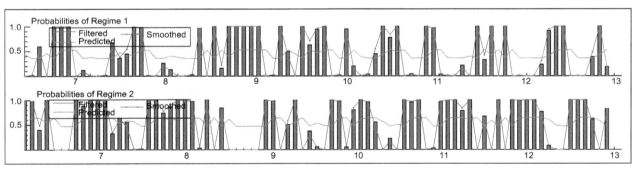

Filtered and smoothed probabilities for CINF-OIL

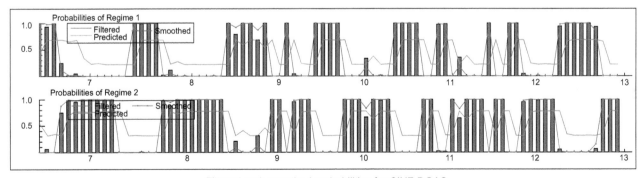

Filtered and smoothed probabilities for CINF-DGAS

Fig. 1 Regime probabilities graphics

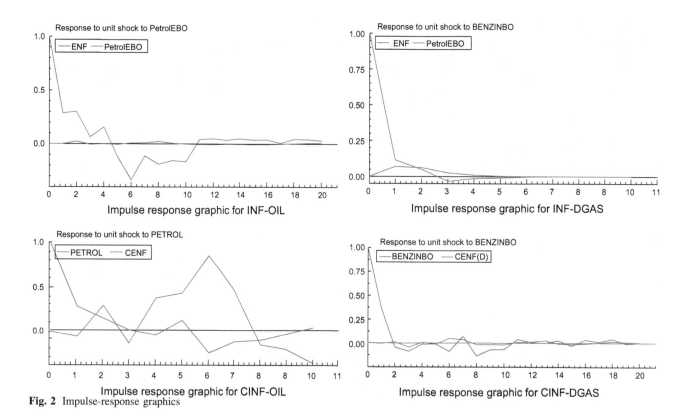

Fig. 2 Impulse-response graphics

When the impulse-response graphics in Fig. 2 are examined, it is seen that, in case of a shock in the price of oil, inflation shows a small and positive response in the second and eighth months, then turns back to its former equilibrium level and becomes stable after the eighth month. This shows that the inflation target level does not deviate because of the increase in crude oil prices. When there is a shock in gasoline prices, it is seen that inflation responds by increasing from the first to the fifth month, with the response disappearing from the sixth month onward. This reveals that increases in gasoline prices cause an inflationary response, and when there is deviation from monthly inflation targets, a significant part of this can be due to the increases in domestic gasoline prices. Core inflation shows response to a shock in crude oil prices with a small decrease in the first period, an increase in the second period, and a high increase between the fourth and seventh periods. After the eighth period, it can be said that the response ends. When there is a sudden increase in domestic gasoline prices, core inflation responds with an increase in the seventh and eleventh months and then loses the impact later on. These findings reveal that the response of core inflation would last for 1 year with shocks in domestic gasoline prices, and therefore there is a long-term pass-through effect.

In this context, the finding of the present study suggests that the effect of crude oil and domestic gasoline price changes on consumer prices and core inflation differs according to regimes, and this indicates that increases in crude oil prices have no inflationary effects, while increases in gasoline prices do. This result is consistent with the finding of Kibritcioglu and Kibritcioglu (1999), where the authors conclude that a 20 % increase in crude oil price has an insignificant effect on general price levels, as well as the finding of Oksuzler and Ipek (2011), who find that there is no causality between oil prices and inflation. However, it is noteworthy to mention that in these studies neither the nonlinear relationship between the variables nor the effect of gasoline prices (which is the main source of inflation) was analyzed.

4 Conclusion

Inflation and the price of oil are seen by academicians and politicians as being connected. The reason for this opinion is that oil is a major factor in the economy. Especially in the non-oil-producing countries, oil price fluctuations can have a great impact on economic variables such as consumer prices and core inflation. Also, oil price and inflation series may exhibit nonlinear behavior due to factors such as policy changes and energy crises.

In this context, the effects of crude oil and domestic gasoline price changes on consumer price inflation and core inflation have been investigated with MS-VAR models in Turkey for the period October 2005–December 2012.

Two regimes have been determined for all variables, and the existence of different regimes has revealed that the series show different behaviors in each regime. This result indicates that the political targets should be changed depending on the change of the inflation rate. Additionally, the findings of the study show that Turkish policy makers should not ignore the pass-through of oil and domestic gasoline prices to macroeconomic variables such as consumer price inflation and core inflation. Also, we find that, unlike the previous studies carried out for Turkey, the impact of sudden increases in gasoline prices on consumer inflation and core inflation is more significant than shocks in crude oil prices. The probabilities of being in low and high inflation regimes and transition from one to another are very close in the inflation/crude oil price relationship. In the inflation/domestic gasoline price relationship, the probability is higher for remaining in high inflation and for transition from low to high inflation. This indicates that increases in domestic gasoline prices affect the inflation rate and make a crisis impact.

References

Aktas E, Ozenc C, Arica F. The impact of oil prices in Turkey on macroeconomics. Pune: Munich Personal RePEc Archive; 2010.

Álvarez LJ, Hurtado S, Sánchez I, et al. The impact of oil price changes on Spanish and euro area consumer price inflation. Econ Model. 2011;28(1):422–31.

Aydin L, Acar M. Economic impact of oil price shocks on the Turkish economy in the coming decades: a dynamic CGE analysis. Energy Policy. 2011;39(3):1722–31.

Berument H, Tasci H. Inflationary effect of crude oil prices in Turkey. Phys A. 2002;316(1):568–80.

Catik AN, Onder AO. Inflationary effects of oil prices in Turkey: a regime-switching approach. Emerg Mark Financ Trade. 2011;47(5):125–40.

CBRT Monetary Policy Report. http://www3.tcmb.gov.tr/yillikrapor/2012/en/m-2-1-2.php. 2012.

Celik T, Akgul B. Changes in fuel oil prices in Turkey: an estimation of the inflation effect using VAR analysis. J Econ Bus. 2011;2:11–21.

Cologni A, Manera M. The asymmetric effects of oil shocks on output growth: a Markov–switching analysis for the G-7 countries. Econ Model. 2009;26(1):1–29.

Cologni A, Manera M. Oil prices, inflation and interest rates in a structural co-integrated VAR model for the G-7 countries. Energy Econ. 2005;30(3):856–88.

Cunado J, Perez de Gracia F. Do oil price shocks matter? Evidence for some European countries. Energy Econ. 2003;25(2):137–54.

Cunado J, Pérez de Gracia F. Oil prices, economic activity and inflation: evidence for some Asian countries. Quart Rev Econ Financ. 2005;45(1):65–83.

Davis S J, Loungani P and Mahidhara R. Regional labor fluctuations: oil shocks, military spending, and other driving forces. Manuscript, Working Paper, 1996.

De Gregorio J, Landerretche O and Neilson C. Another pass-through bites the dust? Oil prices and Inflation. Central Bank of Chile Working Papers. 2007. No. 417.

Dempster AP, Laird NM, Rubin DB. Maximum likelihood from incomplete data via the EM algorithm. J R Stat Soc. 1977;39(1):1–38.

Dickey DA, Fuller WA. Distribution of the estimators for autoregressive time series with a unit root. J Am Stat Assoc. 1979;74(36):427–31.

Dotsey M, Reid M. Oil shocks, monetary policy, and economic activity. Econ Rev. 1992;78(4):14–27.

Du X, Yu CL, Hayes DJ. Speculation and volatility spillover in the crude oil and agricultural commodity markets: a Bayesian analysis. Energy Econ. 2011;33(3):497–503.

Ewing BT, Thompson MA. Dynamic cyclical co-movements of oil prices with industrial production, consumer prices, unemployment, and stock prices. Energy Policy. 2007;35(11):5535–40.

Farzanegan MR, Markwardt G. The effects of oil price shocks on the Iranian economy. Energy Econ. 2009;31(1):134–51.

Federal Reserve Economic Data http://research.stlouisfed.org/fred2/.

Goldfeld SM, Quandt RE. A Markov model for switching regressions. J Econom. 1973;1(1):3–15.

Goodwin TH. Business-cycle analysis with a Markov-switching model. J Bus Econ Stat. 1993;11(3):331–9.

Gray SF. Modeling the conditional distribution of interest rates as a regime-switching process. J Financ Econ. 1996;42(1):27–62.

Hall SG, Psaradakis Z, Sola M. Detecting periodically collapsing bubbles: a Markov-switching unit root test. J Appl Econom. 1999;14(2):143–54.

Hamilton JD. A new approach to the economic analysis of nonstationary time series and the business cycle. Econometrica. 1989;57:357–84.

Hamilton JD. This is what happened to the oil price–macroeconomy relationship. J Monet Econ. 1996;38(2):215–20.

Hansen BE. The likelihood ratio test under nonstandard conditions: testing the Markov switching model of GNP. J Appl Econom. 1992;7(1):61–82.

Hooker MA. Are oil shocks inflationary?: Asymmetric and nonlinear specifications versus changes in regime. J Money Credit Bank. 2002;34(2):540–61.

Huntington HG. Crude oil prices and US economic performance: where does the asymmetry reside? Energy J. 1998;19(4):107–32.

Kahn GA, Hampton JR. Possible monetary policy responses to the Iraqi oil shock. Econ Rev. 1990;2:19–32.

Kibritcioglu A, Kibritcioglu B. Inflationary effects of increases in prices of imported crude oil and oil products in Turkey. T. C. Başbakanlık Hazine Müsteşarlığı Araştırma ve İnceleme Dizisi. 1999;21:77–96.

Kilian L, Vigfusson RJ. Pitfalls in estimating asymmetric effects of energy price shocks. Cent Econ Policy Res. 2009;23(3):27–52.

Kilian L, Murphy DP. The role of inventories and speculative trading in the global market for crude oil. J Appl Econom. 2013;29(3):454–78.

Koop G, Pesaran M, Potter S. Impulse response analysis in nonlinear multivariate models. J Econom. 1996;74(1):119–47.

Krolzig HM. Econometric modeling of Markov-switching vector autoregressions using MSVAR for Ox. 1998. Oxford: Institute of Economics and Statistics and Nuffield College; 1998.

Krolzig HM. Markov-switching vector autoregressions: modelling, statistical inference, and application to business cycle analysis. Berlin: Springer; 1997.

LeBlanc M, Chinn M D. Do high oil prices presage inflation? The evidence from G-5 countries. SCCIE Working Paper. 2004.

Medina J P, Soto C. The Chilean business cycles through the lens of a stochastic general equilibrium model. Central Bank of Chile Working Papers. 2007. No: 457.

Mork KA. Oil and the macro economy. When prices go up and down: an extension of Hamilton's results. J Polit Econ. 1989;97(3): 740–4.

Nazlioglu S, Soytas U. World oil prices and agricultural commodity prices: evidence from an emerging market. Energy Econ. 2011;33(3):488–96.

Oksuzler O, Ipek E. The effects of the world oil price changes on growth and inflation: example of Turkey. Zonguldak Karaelmas University. J Soc Sci. 2011;7(14):16–21.

Olomola PA, Adejumo AV. Oil price shock and macroeconomic activities in Nigeria. Int Res J Financ Econ. 2006;3(1):28–34.

OPET Pump Prices. http://www.opet.com.tr/tr/PompaFiyatlariArsiv. aspx?cat=4&id=34.

Peker O, Mercan M. The inflationary effect of price increases in oil products in Turkey. Ege Acad Rev. 2011;11(4):553–62.

Reboredo JC. How do crude oil prices co-move? A copula approach. Energy Econ. 2011;33(5):948–55.

Robert H, Tatom JA. Energy resources and potential GNP. Fed Reserve Bank St. Louis Rev. 1977;59:10–24.

Santini D. The energy-squeeze model: energy price dynamics in U.S. business cycles. Int J Energy Syst. 1986;5(1):159–94.

Tang W, Wu L, Zhang Z. Oil price shocks and their short- and long-term effects on the Chinese economy. Energy Econ. 2010;32:3–14.

Tsay RS. Testing and modeling multivariate threshold models. J Am Stat Assoc. 1998;93(443):1188–202.

Turkish Statistical Institute, Main Statistics. http://www.tuik.gov.tr/ UstMenu.do?metod=temelist.

Van Den Noord P, André C. Why has core inflation remained so muted in the face of the oil shock?. Mexico: OECD Publishing; 2007.

Yaylali M, Lebe F. Effects of important crude oil prices on macroeconomic activities in Turkey. Marmara Üniv IIBF Derg. 2012;32(1):43–68.

11

Distribution and geochemical significance of phenylphenanthrenes and their isomers in selected oils and rock extracts from the Tarim Basin, NW China

Shao-Ying Huang[1] · Mei-Jun Li[2] · Ke Zhang[1] · T.-G. Wang[2] ·
Zhong-Yao Xiao[1] · Rong-Hui Fang[2] · Bao-Shou Zhang[1] ·
Dao-Wei Wang[2] · Qing Zhao[1] · Fu-Lin Yang[2]

Abstract Twenty-two oil samples and eight source rock samples collected from the Tarim Basin, NW China were geochemically analyzed to investigate the occurrence and distribution of phenylphenanthrene (PhP), phenylanthracene (PhA), and binaphthyl (BiN) isomers and methylphenanthrene (MP) isomers in oils and rock extracts with different depositional environments. Phenylphenanthrenes are present in significant abundance in Mesozoic lacustrine mudstones and related oils. The relative concentrations of PhPs are quite low or below detection limit by routine gas chromatography–mass spectrometry (GC–MS) in Ordovician oils derived from marine carbonates. The ratio of 3-PhP/3-MP was used in this study to describe the relative abundance of phenylphenanthrenes to their alkylated counterparts—methylphenanthrenes. The Ordovician oils in the Tabei Uplift have quite low 3-PhP/3-MP ratios (<0.10), indicating their marine carbonate origin, associating with low Pr/Ph ratios (pristane/phytane), high ADBT/ADBF values (relative abundance of alkylated dibenzothiophenes to alkylated dibenzofurans), low C_{30} diahopane/C_{30} hopane ratios, and low Ts/(Ts + Tm) (18α-22, 29, 30-trisnorneohopane/(18α-22, 29, 30-trisnorneohopane + 17α-22, 29, 30-trisnorhopane)) values. In contrast, the oils from Mesozoic and Paleogene sandstone reservoirs and related Mesozoic lacustrine mudstones have relatively higher 3-PhP/ 3-MP ratios (>0.10), associating with high Pr/Ph, low ADBT/ ADBF, high Ts/(Ts + Tm), and C_{30} diahopane/C_{30} hopane ratios. Therefore, the occurrence of significant amounts of phenylphenanthrenes in oils typically indicates that the organic matter of the source rocks was deposited in a suboxic environment with mudstone deposition. The phenylphenanthrenes may be effective molecular markers, indicating depositional environment and lithology of source rocks.

Keywords Phenylphenanthrene · Methylphenanthrene · Depositional environment · Source rock

1 Introduction

Phenyl-substituted polycyclic aromatic hydrocarbons (PAHs) and their heterocyclic counterparts are important components in aromatic fractions of some crude oils and sedimentary rock extracts (Marynowski et al. 2001, 2002, 2004; Rospondek et al. 2007, 2009; Li et al. 2012a; Grafka et al. 2015). A series of phenylphenanthrene (PhP), phenylanthracene (PhA), and binaphthyl (BiN) isomers have been firmly identified by using authentic standards (Rospondek et al. 2009).

The 9-phenylphenanthrene and other isomers were detected in volatiles formed during pyrolytic carbonization of coal tar pitches (zu Reckendorf 1997, 2000). All PhP, PhA, and BiN isomers have been discovered in marine sedimentary rocks (Rospondek et al. 2009; Grafka et al. 2015), Tertiary and Jurassic lacustrine shales (Li et al. 2012a) and tire fire products (Wang et al. 2007).

The phenyl-substituted PAHs in combustion products may be generated by consecutive reactions of phenyl free radicals with unsubstituted PAHs in the gaseous phase during combustion (zu Reckendorf 2000). Less work has

✉ Mei-Jun Li
  meijunli2008@hotmail.com

[1] Research Institute of Petroleum Exploration and Development, Tarim Oilfield Company, PetroChina, Korla 841000, Xinjiang, China

[2] State Key Laboratory of Petroleum Resources and Prospecting, College of Geosciences, China University of Petroleum, Beijing 102249, China

Edited by Jie Hao

been done on the origin and formation of phenylphenanthrene in crude oils and sedimentary rocks. According to Marynowski et al. (2001), Rospondek et al. (2009), and Grafka et al. (2015), diagenetic/catagenetic oxidation of sedimentary organic matter at the redox interface in buried sedimentary rocks is likely to be the main source of arylated polycyclic aromatic compounds in the geosphere. Laboratory experiments indicate that the reaction of free radical phenylation with phenanthrene or anthracene moieties can account for the distribution of phenylphenanthrenes and phenylanthracenes in oxidized rock samples with Type II and III kerogen (Marynowski et al. 2001; Rospondek et al. 2009; Grafka et al. 2015). Therefore, a significant amount of PhPs and other phenyl-substituted PAHs in ancient sedimentary rocks are commonly associated with oxic to suboxic depositional environments.

The distribution patterns of PhPs and BiNs in mass chromatograms (m/z 254) of aromatic fractions in sedimentary organic matters are relative to the maturation levels. For example, the most stable isomers 2-PhP and 3-PhP predominate, whereas the thermally unstable 9-PhP, 1-PhP, and 4-PhP disappear in highly mature sedimentary organic matter (Rospondek et al. 2009; Li et al. 2012a; Grafka et al. 2015). Among all binaphthyl isomers, 1,1-BiN is the most thermally unstable one and was found only in less mature samples; while 2,2-BiN is more stable and also present above the oil window range (Rospondek et al. 2009; Li et al. 2012a). Therefore, some indices, such as phenylphenanthrene ratio [defined as (2- + 3-PhP)/(2- + 3- + 4- + 1- + 9-PhP)] (Rospondek et al. 2009) and 2,2′-BiN/1,2′-BiN (defined as 2,2′-binaphthyl/1,2′-binaphthyl) (Li et al. 2012a) have been proposed as maturity indicators. In addition, some maturity indicators associated with aromatic compounds including phenylphenanthrene ratios have also been used as frictional stress indicators (Polissar et al. 2011).

Previous studies mainly focused on the formation and application of PhPs in maturation assessment. This paper reported the occurrence of PhPs in Ordovician oils, Mesozoic lacustrine sedimentary rocks, and related oils from the Tarim Basin, NW China. Their potential significance to the depositional environment and lithology of source rocks and application in oil-to-source correlation in oil petroleum system are discussed. The result can further broaden the geochemical significance and application of phenylphenanthrenes in sedimentary rocks and related oils.

2 Samples and geological settings

A total of eight cores and outcrop samples were collected from the Tarim Basin, NW China. Two cores were sampled from Well S5, which is located in the Yakela Faulted Uplift (Fig. 1). One outcrop was sampled in the Kuchehe profile in the Kuqa Depression, which is the prolific hydrocarbon-bearing foreland basin in the Tarim Basin (Zhao et al. 2005). These three rock samples are Upper Triassic mudstones.

Four Jurassic sandy mudstones were collected from the Kuzigongsu profile in the southwest of the Tarim Basin (Fig. 1). The Middle Jurassic in the Tarim Basin is represented by deep lacustrine deposits (Zhang et al. 2000; Liu et al. 2006; Cheng et al. 2008; Wang et al. 2009; Song et al. 2013). In addition, one Jurassic coal sample from the Well YL1 in the eastern Tarim Basin is also investigated. All these samples are good source rocks with total organic carbon (TOC) content of 0.71 %–1.12 % (Table 1), and they underwent moderate to relatively higher thermal maturation with vitrinite reflectance (R_o %) of 0.51 %–1.10 % (Table 1).

A total of 22 oil samples were collected from the Ordovician carbonate reservoirs in the Halahatang Sag, Yakela Faulted Uplift and Akekule Uplift, and the Mesozoic and Paleogene sandstone reservoirs in the Kuqa Depression and Yakela Uplift of the Tarim Basin (Table 1). The Ordovician carbonate oils from the Tabei Uplift were sourced from Paleozoic carbonate source rocks (Zhang and Huang 2005; Wang et al. 2008; Pang et al. 2010; Li et al. 2012b). Oils in wells QL1, Ku1, and S3 were derived from Mesozoic lacustrine mudstones (Xiao et al. 2004; Song et al. 2015).

3 Methods

All rocks were ground into powder in a crusher to <80 mesh. The TOC content was measured on an LECO CS-230 carbon/sulfur analyzer. The vitrinite reflectance values (%) were measured on polished rock blocks using a Leitz MPV-microscopic photometer.

To extract soluble bitumen, the powder was processed for 24 h in a Soxhlet apparatus using 400 mL of dichloromethane and methanol as the solvent (93:7, v:v). Asphaltenes were removed from approximately 20–50 mg oils and bitumen by precipitation using 50 mL of n-hexane and then fractionated by liquid chromatography using alumina/silica gel columns into saturated and aromatic hydrocarbons using 30 mL n-hexane and 20 mL dichloromethane: n-hexane (2:1, v:v) as respective eluents (Fang et al. 2015).

The GC–MS analyses of the aromatic fractions were performed on an Agilent 5975i GC–MS system equipped with an HP-5MS (5 %-phenylmethylpolysiloxane)-fused silica capillary column (60 m × 0.25 mm i.d., with a 0.25-µm film thickness). The GC operating conditions were as follows: the temperature was held initially at 80 °C for

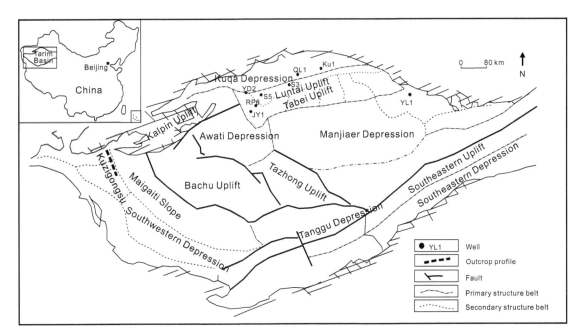

Fig. 1 Map showing the major tectonic terrains in the Tarim Basin (NW China) and the locations of sampled wells and profiles

1 min, increased to 310 °C at a rate of 3 °C/min, and then kept isothermal for 16 min. Helium was used as the carrier gas. The injector temperature was set to 300 °C. The MS was operated in the electron impact (EI) mode with ionization energy of 70 eV, and a scan range of m/z 50–600 Da.

4 Results and discussion

4.1 Identification of phenylphenanthrenes and methylphenanthrenes

The identification and elution order of all isomers of methylphenanthrenes (MPs), phenylphenanthrenes (including their isomers: phenylanthracene and binaphthyl) were determined by the comparison of their mass spectra and standard retention indices (I_{HP-5MS}) with those reported in literature (Lee et al. 1979; Rospondek et al. 2009). Figure 2 shows the chemical structures of MPs and PhPs and the mass chromatograms (m/z 178, 192, 254) of aromatic fractions of selected sediment extracts in this study. The methyl and phenyl substitution pattern on parent rings is indicated on the corresponding peaks (Fig. 2).

Phenanthrene and its alkylated homologues are important polycyclic aromatic hydrocarbons (PAHs) and present in significant concentrations in crude oils and sedimentary rock extracts. Four methylphenanthrene isomers and one methylanthracene isomer (2-methylanthracene: 2-MA) were identified in all rocks and oils in this study. The elution order of the m/z 192 isomers is as follows: 3-MP, 2-MP, 2-MA, 9-MP, and 1-MP. The isomer 2-MA is

present in quite low concentration or below detection limit in oils (Figs. 3, 4). However, it seems abundant in some rocks and coals (Figs. 2c, 4a, b e).

The distribution of arylated homologues of phenanthrenes (phenylphenanthrenes) is shown in Fig. 2. In addition, their binaphthyl (BiN) isomers were also detected in oils and rock extracts. The elution sequence of m/z 254 isomers on an HP-5MS capillary column is as follows: 1,1'-BiN, 4-PhP, 9-PhA, 1,2'-BiN, 9-PhP, 1-PhP, 3-PhP, 2,2'-BiN, 2-PhP, and 2-PhA. The 1-PhA isomer may co-elute with 1,2'-BiN on HP-5MS column, but it is typically absent in geochemical samples (Rospondek et al. 2009). The isomers 3-PhP, 2-PhP, and 2,2'-BiN are typically present in higher abundance relative to other PhP and BiN isomers.

4.2 Depositional environment and lithologies of crude oils and source rocks

Previous studies suggested that oils in wells Ku1 and QL1 of the Kuqa Depression and wells YD2 and S3 of the Tabei Uplift (Fig. 1) are of typical lacustrine mudstone origin (Li et al. 2004; Xiao et al. 2004; Song et al. 2015). Oils from Ordovician carbonate reservoirs were derived from Paleozoic carbonate source rocks (Zhang and Huang 2005; Wang et al. 2008; Chang et al. 2013).

Dibenzothiophene, dibenzofuran, and their alkylated homologues are effective molecular markers in inferring depositional environment, maturation assessment and in tracing oil charging pathways (Bao et al. 1996; Wang et al. 2014; Li et al. 2008; Zhang and Philp. 2010; Li et al. 2011, 2014). A cross-plot of alkyldibenzothiophene/alkyldibenzofuran ratio

Table 1 Bulk properties and selected geochemical parameters for oils and rocks in this study

Sample no.	Fm.	Description	TOC, %	R_o, %	Methylphenanthrene index (MPI1)	3-PhP/ 3-MP	Pr/ Ph	ADBT/ ADBF	C_{30}DiaH/ C_{30}H	Ts/(Ts + Tm)
YD2	Cretaceous	Oil, Tabei Uplift	–	–	0.51	0.12	2.17	0.54	0.48	0.67
S3	Cretaceous	Oil, Yakela Faulted Uplift	–	–	1.11	0.09	1.63	0.33	0.92	0.66
Ku1	Jurassic	Oil, Kuqa Depression	–	–	0.13	0.17	1.84	0.26	0.50	0.55
QL1	Paleogene	Oil, Kuqa Depression	–	–	0.66	0.60	1.79	0.48	0.64	0.63
RP10	Ordovician	Oil, Halahatang Sag	–	–	0.78	0.03	0.81	7.52	0.09	0.37
RP4	Ordovician	Oil, Halahatang Sag	–	–	0.38	0.01	0.76	8.06	0.08	0.42
RP3013	Ordovician	Oil, Halahatang Sag	–	–	0.96	0.03	0.92	9.06	0.12	0.53
RP11	Ordovician	Oil, Halahatang Sag	–	–	0.78	0.02	0.97	6.71	0.12	0.45
RP8	Ordovician	Oil, Halahatang Sag	–	–	0.87	0.03	0.75	10.5	0.12	0.45
JY1	Ordovician	Oil, Halahatang Sag	–	–	0.96	0.02	0.83	7.16	0.12	0.58
JY7	Ordovician	Oil, Halahatang Sag	–	–	0.92	0.02	1.04	9.06	0.09	0.49
JY3	Ordovician	Oil, Halahatang Sag	–	–	0.70	0.02	1.03	8.61	0.09	0.45
XK5	Ordovician	Oil, Halahatang Sag	–	–	0.85	0.02	0.82	6.81	0.10	0.42
XK4	Ordovician	Oil, Halahatang Sag	–	–	0.71	0.01	0.97	5.39	0.08	0.40
XK7	Ordovician	Oil, Halahatang Sag	–	–	0.79	0.03	0.81	7.69	0.09	0.29
XK9005	Ordovician	Oil, Halahatang Sag	–	–	0.94	0.01	1.05	9.83	0.08	0.46
Ha601	Ordovician	Oil, Halahatang Sag	–	–	0.72	0.02	0.88	5.63	0.17	0.44
Ha7-1	Ordovician	Oil, Halahatang Sag	–	–	0.80	0.01	1.00	9.08	0.17	0.43
Ha8	Ordovician	Oil, Halahatang Sag	–	–	0.70	0.02	0.98	6.52	0.12	0.37
Ha13-6	Ordovician	Oil, Halahatang Sag	–	–	0.74	0.01	1.01	6.84	0.11	0.46
TP12-8	Ordovician	Oil, Akekule	–	–	0.89	0.03	0.91	2.72	0.08	0.49
TP14	Ordovician	Oil, Akekule	–	–	0.65	0.02	0.87	10.10	0.06	0.45
S5	Triassic	Mudstone, core, 5400.8 m	0.71	0.74	0.48	0.11	1.26	0.17	0.65	0.85
S5	Triassic	Mudstone, core, 5405.4 m	0.73	0.75	0.42	0.10	1.21	0.75	n.d.	0.87
Ku-13	Jurassic	Sandy mudstone, outcrop, Kuzigongsu	1.64	1.10	0.72	0.17	1.56	0.47	1.55	0.76
Ku-18	Jurassic	Sandy mudstone, outcrop, Kuzigongsu	2.02	0.99	0.64	0.16	2.10	0.22	2.97	0.86
Ku-26	Jurassic	Sandy mudstone, outcrop, Kuzigongsu	3.53	0.94	0.19	0.98	1.59	0.12	6.71	0.96
Ku-30	Jurassic	Sandy mudstone, outcrop, Kuzigongsu	3.27	0.93	0.22	0.90	2.00	0.10	4.48	0.92
YL1	Jurassic	Coal, core, 2878.0 m	38.4	0.51	0.35	0.33	3.77	0.50	n.d.	n.d.
KCH-01	Triassic	Mudstone, outcrop, Kuchehe	1.12		0.26	1.44	2.08	0.31	n.d.	0.88

n.d.: no data

(ADBT/ADBF) versus pristane/phytane ratio (Pr/Ph) provides a powerful and convenient way to infer crude oil source rock depositional environments and lithologies (Radke et al. 2000). The oils from Ordovician reservoirs in the Tabei Uplift were characterized by lower Pr/Ph ratio and higher ADBT/ADBF ratio (Table 1), and the data points were plotted in Zone 1A of the cross-plot of ADBT/ADBF versus Pr/Ph ratios, indicating their marine carbonate origin (Fig. 5).

Oils from Jurassic and Paleogene reservoirs in the Kuqa Depression and oils from Cretaceous reservoirs in the Tabei Uplift have relatively higher Pr/Ph values and very low ADBT/ADBF values. All these data points fall into Zone 3 (Fig. 5), suggesting the mudstone lithology of their source rocks. On the basis of oil-to-source correlation results, all these oils were derived from Mesozoic lacustrine mudstone source rocks (e.g., Song et al. 2015).

Selected Mesozoic mudstones from the Kuqa Depression and the Tabei Uplift were also analyzed. All rock samples have relatively higher Pr/Ph ratios and quite low values of ADBT/ADBF. The points also fall into Zone 3, indicating their lacustrine mudstone lithology (Fig. 5). The coal sample from Well YL1 in the eastern Tarim Basin has very high Pr/Ph ratio, which falls into Zone 4 (fluvial/deltaic carbonaceous shale and coal zone).

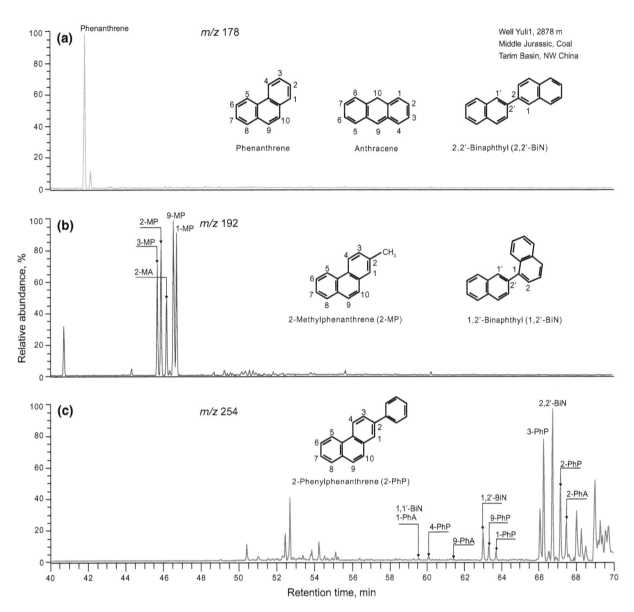

Fig. 2 Identification of phenanthrene (*m/z* 178), methylphenanthrene and methylanthracene (*m/z* 192), and phenylphenanthrene, phenylanthracene, and binaphthyl isomers (*m/z* 254) in sedimentary rocks and their chemical structures

Therefore, the relative abundances of alkylated dibenzothiophenes to alkylated dibenzofurans and pristane to phytane confirm that the oils in the Ordovician reservoirs are of marine carbonate origin and the oils from Mesozoic and Paleogene sandstone reservoirs in the Kuqa Depression and Tabei Uplift of the Tarim Basin are of lacustrine mudstone origin.

4.3 Distribution of PhPs and MPs in oils and source rocks

Most of the phenylphenanthrene isomers were detected in oils from wells Ku1, QL1, YD2, and S3 (Fig. 3a–d). They are present in differing abundance in these oils. The isomer of 2,2′-biphenyl is typically present in quite low

concentrations, and 1,2′-biphenyl is commonly absent or under detection limit. The 3-PhP is the dominant compound among all PhP, PhA, and BiN isomers in *m/z* 254 mass chromatograms. The concentrations of PhPs are generally lower than those of their methylated counterparts—MPs in all oils. Here we defined 3-PhP/3-MP to indicate the relative abundance of PhPs to MPs. The oils from wells Ku1, QL1, YD2, and S3 have 3-PhP/3-MP ratios higher than 0.10. However, oils from Ordovician reservoirs in the Tabei Uplift, including Ha6, Repu, Xinken, Jinyue, and Tuofutai blocks, have extremely low concentrations of phenylphenanthrenes with 3-PhP/3-MP ratios lower than 0.10 (Fig. 4e, f; Table 1).

Selected Mesozoic lacustrine mudstones were also investigated to analyze the distribution patterns of PhPs and

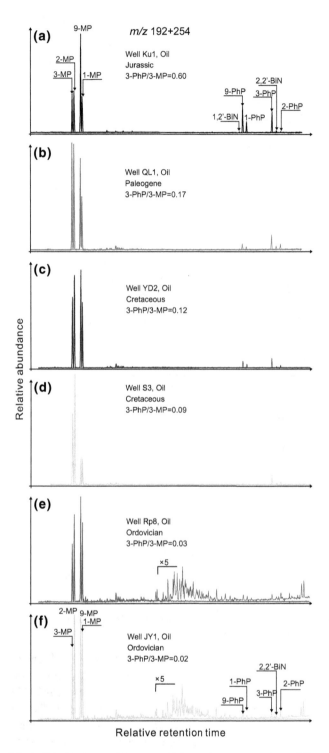

Fig. 3 Distribution of methylphenanthrenes (*m/z* 192), phenylphenanthrenes (*m/z* 254), and their isomers in selected oils from the Tarim Basin

MPs in sedimentary rock extracts. Most of the PhP isomers and 2,2'-BiN were detected in all rocks in this study. For the lower thermodynamic stability, PhAs and 1,1'-BiN are generally below detection limit in this study. The ratio of 3-PhP/3-MP in Mesozoic lacustrine mudstones is from 0.11 to 1.44, which is consistent with oils from wells Ku1, QL1,

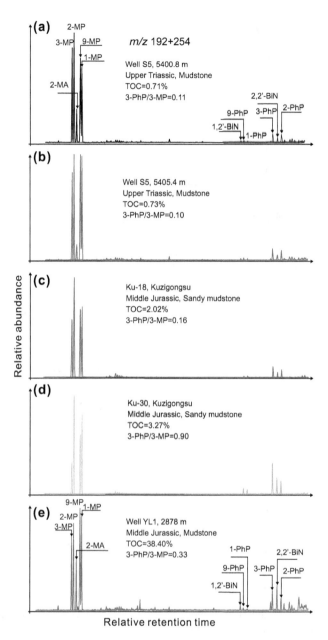

Fig. 4 Distribution of methylphenanthrenes (*m/z* 192), phenylphenanthrenes (*m/z* 254), and their isomers in Mesozoic source rocks in the Tarim Basin

YD2, and S3. Therefore, the occurrence and distribution of phenylphenanthrene and methylphenanthrenes in Mesozoic oils and source rocks further confirmed their genetic affinity.

4.4 Effect of environment and lithology on the distribution of MPs and PhPs

Much work has been done on the occurrence and distribution of methylated phenanthrenes. For example, Alexander et al. (1995) demonstrated that the sedimentary methylation process can form some alkylphenanthrene isomers. Due to the ubiquitous occurrence, the methylphenanthrenes appear to

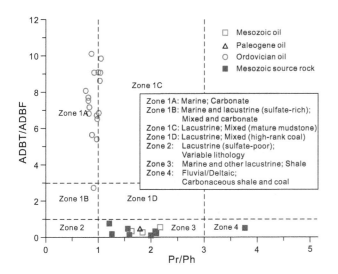

Fig. 5 Cross-plot of alkyldibenzothiophene/alkyldibenzofuran ratio (ADBT/ADBF) versus pristane/phytane ratio (Pr/Ph) showing crude oil source rock depositional environments and lithologies (after Radke et al. 2000)

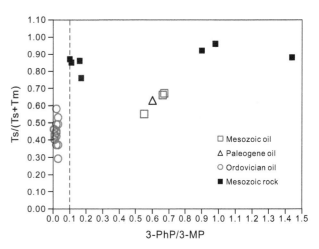

Fig. 6 The cross-plot of Ts/(Ts + Tm) versus 3-PhP/3-MP ratio provides a useful way to infer the depositional environment and lithology of oils and related source rocks. Ts—C_{27} 18α-22,29,30-trisnorneohopane; Tm—C_{27} 17α-22,29,30-trisnorhopane

have limited depositional environment and lithology significance. They are effective molecular markers for thermal maturity assessment (e.g., Radke et al. 1982; Boreham et al. 1988; Voigtmann et al. 1994; Szczerba and Rospondek 2010).

Here we use 3-PhP/3-MP to show the abundance of phenylphenanthrenes relative to methylphenanthrenes in oils and sedimentary organic matter. The abundance of parent phenanthrene may be different for the differences in the input of organic matter in source rocks and/or thermal maturation level of sedimentary organic matter. As proposed by previous studies (Marynowski et al. 2001; Rospondek et al. 2009), the phenylation of phenanthrene is mainly associated with an oxidizing depositional environment. Therefore, an oxic to suboxic environment may favor the formation of phenylphenanthrenes. The cross-plot of Ts/(Ts + Tm) versus 3-PhP/3-MP is used here to investigate the effect of the depositional environment and lithology on the distribution patterns of MPs and PhPs. The Ts/(Ts + Tm) ratio (18α-22,29,30-trisnorneohopane/(18α-22,29,30-trisnorneohopane + 17α-22,29,30-trisnorhopane)) depends on both source and maturity (Moldowan et al. 1986). It is a reliable maturity indictor when assessing oils from a common source of consistent organic facies (Peters et al. 2005). This ratio increases with the increasing maturity. It is sensitive to clay-catalyzed reactions. For example, oils from carbonate source rocks appear to have low Ts/(Ts + Tm) ratios compared with those from shales (e.g., Rullkötter et al. 1985).

Here we found that oils from Ordovician reservoirs have relatively lower Ts/(Ts + Tm) values (<0.50) and very low 3-PhP/3-MP ratios (<0.10). In contrast, the Mesozoic and Paleogene oils from wells Ku1, QL1, S3, and YD2 and Mesozoic lacustrine mudstones have higher Ts/(Ts + Tm)

(>0.55) and 3-PhP/3-MP ratios (>0.10) (Fig. 6). The higher Ts/(Ts + Tm) ratios for Mesozoic oils and source rocks are mainly attributed to a suboxic depositional environment and clay-rich lithology, because Ordovician oils sourced from Paleozoic marine source rocks have generally higher maturation levels than those from the Mesozoic lacustrine source rocks.

The C_{30} diahopane in sedimentary rock extracts and oils may be derived from bacterial hopanoid precursors that have experienced oxidation and rearrangement by clay-mediated acidic catalysis. Therefore, the presence of a significant amount of C_{30} diahopane indicates bacterial input to sediments containing clays deposited under oxic or suboxic environments (Peters et al. 2005). In this study, the C_{30} diahopane is generally present in very low concentration in Ordovician oils (with C_{30} diahopane/C_{30} hopane ratios lower than 0.20). While Mesozoic and Paleogene oils from wells Ku1, QL1, S3, and YD2 and Mesozoic lacustrine mudstones have higher C_{30} diahopane/C_{30} hopane ratios (Table 1). Therefore, our study suggests that the presence of a significant amount of phenylphenanthrenes is generally related to clay-enriched sediments deposited under suboxic conditions.

5 Conclusions

Phenyl phenanthrenes and their isomers have been detected in oils and source rocks from the Tarim Basin, NW China. 3-phenylphenanthrene (3-PhP), 2-PhP, and 1,2′-binaphthyls (1,2′-BiN) are typically predominant compounds among all isomers. The abundances of phenylphenanthrenes are extremely low in Ordovician oils in the Tabei Uplift of the Tarim Basin. The Mesozoic oils from wells QL1 and

Ku1 from the Kuqa Depression and wells S3 and YD2 from the Tabei Uplift have relatively higher concentrations of phenylphenanthrenes.

The ratio of 3-PhP/3-MP (3-phenylphenanthrene/3-methylphenanthrene) is used to indicate the relative abundances of phenylphenanthrenes to methylphenanthrenes. We discovered that oils from marine carbonate source rocks have a very low ratio (<0.10). However, oils of lacustrine mudstone origin and related source rocks have relatively high 3-PhP/3-MP ratios, associating with higher Pr/Ph, Ts/(Ts + Tm) and C_{30} diahopane/C_{30} hopane values.

The occurrence and distribution of phenylphenanthrenes in oils and sedimentary rock extracts in the Tarim Basin clearly show an environment dependence. Relatively higher abundance of phenylphenanthrenes (3-PhP/3-MP higher than 0.10) generally suggests clay-enriched sediments under suboxic depositional environment. Certainly, further work is needed to investigate whether this is valid in other basins.

Acknowledgments The work was financially supported by the National Natural Science Foundation of China (Grant No. 41272158) and the State Key Laboratory of Petroleum Resources and Prospecting (PRP/indep-2-1402). The authors thank the assistance of Zhu Lei in the GC–MS analysis. We are grateful to three anonymous reviewers for their constructive comments and suggestions. We thank the Tarim Oilfield Company of PetroChina for providing samples and data, and for permission to publish this work.

References

Alexander R, Bastow TP, Fisher SJ, et al. Geosynthesis of organic compounds: II. Methylation of phenanthrene and alkylphenanthrenes. Geochim Cosmochim Acta. 1995;59(20):4259–66.

Bao JP, Wang TG, Chen FJ. Relative abundance of alkyl dibenzothiophenes in the source rocks and their geochemical significances. J China Univ Pet. 1996;20:19–23 (**in Chinese**).

Boreham CJ, Crick IH, Powell TG. Alternative calibration of the Methylphenanthrene index against vitrinite reflectance: application to maturity measurements on soils and sediments. Org Geochem. 1988;12:289–94.

Chang X, Wang T-G, Li Q, et al. Geochemistry and possible origin of petroleum in Palaeozoic reservoirs from Halahatang Depression. J Asian Earth Sci. 2013;74:129–41.

Cheng X, Liao L, Chen X, et al. Jurassic sedimentary facies and paleoenvironmental reconstruction of southeastern Tarim Basin, Northwestern China. J China Univ Min Technol. 2008;37:519–25 (**in Chinese**).

Fang R, Li M, Wang T-G, et al. Identification and distribution of pyrene, methylpyrenes and their isomers in rock extracts and crude oils. Org Geochem. 2015;83–84:65–76.

Grafka O, Marynowski L, Simoneit BRT. Phenyl derivatives of polycyclic aromatic compounds as indicators of hydrothermal activity in the Silurian black siliceous shales of the Bardzkie Mountains, Poland. Int J Coal Geol. 2015;139(1):142–51.

Lee ML, Vassilaros DL, White CM, et al. Retention indices for programmed-temperature capillary-column gas chromatography of polycyclic aromatic hydrocarbons. J Chromatogr A. 1979;51:768–74.

Li JG, Liu WH, Zheng JJ, et al. Dibenzofuran series of terrestrial source rocks and crude oils in Kuqa Depression. Acta Pet Sin. 2004;25(1):40–43, 47 (**in Chinese**).

Li M, Wang T, Liu J, et al. Total alkyl dibenzothiophenes content tracing the filling pathway of condensate reservoir in the Fushan depression, South China Sea. Sci China (Ser D). 2008;51:138–45.

Li M, Shi S, Wang T-G, et al. The occurrence and distribution of phenylphenanthrenes, phenylanthracenes and binaphthyls in Palaeozoic to Cenozoic shales from China. Appl Geochem. 2012a;27(12):2560–9.

Li M, Wang T, Shi S, et al. The oil maturity assessment by maturity indicators based on methylated dibenzothiophenes. Pet Sci. 2014;11(2):234–46.

Li M, Wang T-G, Lillis PG, et al. The significance of 24-norcholestanes, triaromatic steroids and dinosteroids in oils and Cambrian-Ordovician source rocks from the cratonic region of the Tarim Basin, NW China. Appl Geochem. 2012b;27(8):1643–54.

Li S, Pang X, Shi Q, et al. Origin of the unusually high dibenzothiophene concentrations in Lower Ordovician oils from the Tazhong Uplift, Tarim Basin, China. Pet Sci. 2011;8(4):382–91.

Liu S, Qiu M, Chen X, et al. Sedimentary setting of Mesozoic and its petroleum geologic features in western Tarim Basin. Xinjiang Pet Geol. 2006;27:10–4 (**in Chinese**).

Marynowski L, Czechowski F, Simoneit BRT. Phenylnaphthalenes and polyphenyls in Palaeozoic source rocks of the Holy Cross Mountains, Poland. Org Geochem. 2001;32:69–85.

Marynowski L, Rospondek MJ, zu Reckendorf RM, et al. Phenyldibenzofurans and phenyldibenzothiophenes in marine sedimentary rocks and hydrothermal petroleum. Org Geochem. 2002;33:701–14.

Marynowski L, Piet M, Janeczek J. Composition and source of polycyclic aromatic compounds in deposited dust from selected sites around the Upper Silesia, Poland. J Geol Q. 2004;48:169–80.

Moldowan JM, Sundararaman P, Schoell M. Sensitivity of biomarker properties to depositional environment and/or source input in the Lower Toarcian of SW-Germany. Org Geochem. 1986;10(4–6):915–26.

Pang X, Tian J, Pang H, et al. Main progress and problems in research on Ordovician hydrocarbon accumulation in the Tarim Basin. Pet Sci. 2010;7(2):147–63.

Peters KE, Walters CC, Moldowan JM. The biomarker guide. 2nd ed. New York: Cambridge University Press; 2005.

Polissar PJ, Savage HM, Brodsky EE. Extractable organic material in fault zones as a tool to investigate frictional stress. Earth Planet Sci Lett. 2011;311:439–47.

Radke M, Vriend SP, Ramanampisoa LR. Alkyldibenzofurans in terrestrial rocks: influence of organic facies and maturation. Geochim Cosmochim Acta. 2000;64(2):275–86.

Radke M, Welte DH, Willsch H. Geochemical study on a well in the Western Canada Basin: relation of the aromatic distribution pattern to maturity of organic matter. Geochem Cosmochim Acta. 1982;46:1–10.

Rospondek MJ, Marynowski L, Góra M. Novel arylated polyaromatic thiophenes: phenylnaphtho[b]thiophenes and naphthylbenzo[b]thiophenes as markers of organic matter diagenesis buffered by oxidising solutions. Org Geochem. 2007;38:1729–56.

Rospondek MJ, Marynowski L, Chachaj A, et al. Novel aryl polycyclic aromatic hydrocarbons: phenylphenanthrene and phenylanthracene identification, occurrence and distribution in sedimentary rocks. Org Geochem. 2009;40:986–1004.

Rullkötter J, Spiro B, Nissenbaum A. Biological marker characteristics of oils and asphalts from carbonate source rocks in a rapidly subsiding graben, Dead Sea, Israel. Geochim Cosmochim Acta. 1985;49(6):1357–70.

Song D, Li M, Wang T. Geochemical studies of the Silurian oil reservoir in the Well Shun-9 prospect area, Tarim Basin, NW China. Pet Sci. 2013;10(4):432–41.

Song D, Wang T, Li H. Geochemical characteristics and origin of the crude oils and condensates from Yakela Faulted-Uplift, Tarim Basin. J Pet Sci Eng. 2015;133:602–11.

Szczerba M, Rospondek MJ. Controls on distributions of methylphenanthrenes in sedimentary rock extracts: critical evaluation of existing geochemical data from molecular modelling. Org Geochem. 2010;41:1297–311.

Voigtmann MF, Yang K, Batts BD, et al. Evidence for synthetic generation of methylphenanthrenes in sediments. Fuel. 1994;73(12):1899–1903.

Wang Z, Li K, Lambert P, et al. Identification, characterization and quantitation of pyrogenic polycyclic aromatic hydrocarbons and other organic compounds in tire fire products. J Chromatogr A. 2007;1139:14–26.

Wang B, Ma H, Liu J, et al. Triassic source rock characteristics in Yakela of Shaya Uplift, Tarim Basin. Spec Oil Gas Reserv. 2009;16(2):43–6, 9 (in Chinese).

Wang T-G, He F, Li M, et al. Alkyldibenzothiophenes: molecular tracers for filling pathway in oil reservoirs. Chin Sci Bull. 2014;49(22):2399–404.

Wang T-G, He F, Wang C, et al. Oil filling history of the Ordovician oil reservoir in the major part of the Tahe Oilfield, Tarim Basin, NW China. Org Geochem. 2008;39:1637–46.

Xiao ZY, Huang GH, Lu YH, et al. Rearranged hopanes in oils from the Quele 1 Well, Tarim Basin, and the significance for oil correlation. Pet Explor Dev. 2004;31(2):35–7 (in Chinese).

Zhang C, Xiao C, Li J, et al. Depositional feature of the Jurassic fault basin in southwest Tarim depression. J Mineral Petrol. 2000;20:41–5 (in Chinese).

Zhang M, Philp P. Geochemical characterization of aromatic hydrocarbons in crude oils from the Tarim. Pet Sci. 2010;7(4):448–57.

Zhang S, Huang H. Geochemistry of Palaeozoic marine petroleum from the Tarim Basin, NW China: Part 1. Oil family classification. Org Geochem. 2005;36:1204–14.

Zhao W, Zhang S, Wang F, et al. Gas systems in the Kuche Depression of the Tarim Basin: source rock distributions, generation kinetics and gas accumulation history. Org Geochem. 2005;36(12):1583–601.

zu Reckendorf RM. Identification of phenyl-substituted polycyclic aromatic compounds in ring furnace gases using GC–MS and GC–AED. Chromatographia. 1997;45:173–82.

zu Reckendorf RM. Phenyl-substituted polycyclic aromatic compounds as intermediate products during pyrolytic reactions involving coal tars, pitches and related materials. Chromatographia. 2000;52:67–76.

12

Economic appraisal of shale gas resources, an example from the Horn River shale gas play, Canada

Zhuoheng Chen[1] · Kirk G. Osadetz[2] · Xuansha Chen[3]

Abstract Development of unconventional shale gas resources involves intensive capital investment accompanying large commercial production uncertainties. Economic appraisal, bringing together multidisciplinary project data and information and providing likely economic outcomes for various development scenarios, forms the core of business decision-making. This paper uses a discounted cash flow (DCF) model to evaluate the economic outcome of shale gas development in the Horn River Basin, northeastern British Columbia, Canada. Through numerical examples, this study demonstrates that the use of a single average decline curve for the whole shale gas play is the equivalent of the results from a random drilling process. Business decision based on a DCF model using a single decline curve could be vulnerable to drastic changes of shale gas productivity across the play region. A random drilling model takes those drastic changes in well estimated ultimate recovery (EUR) and decline rates into account in the economic appraisal, providing more information useful for business decisions. Assuming a natural gas well-head price of $4/MCF and using a 10 % discount rate, the results from this study suggest that a random drilling strategy (e.g., one that does not regard well EURs), could lead to a negative net present value (NPV); whereas a drilling sequence that gives priority to developing those wells with larger EURs earlier in the drilling history could result in a positive NPV with various payback time and internal rate of return (IRR). Under a random drilling assumption, the breakeven price is $4.2/MCF with more than 10 years of payout time. In contrast, if the drilling order is strictly proportional to well EURs, the result is a much better economic outcome with a breakeven price below the assumed well-head price accompanied by a higher IRR.

Keywords Drilling order · EUR · Risk aversion · Shale petroleum resource

Abbreviations

BC	British Columbia
BCF	One billion cubic feet
BCOGC	British Columbia Oil and Gas Commission
DCF	Discount cash flow
EIA	Energy Information Administration (US)
EUR	Estimated ultimate recovery
IRR	Internal rate of return
MBTU	Thousand British thermal units
MCF	Thousand cubic feet
MMCF	Million cubic feet
NPV	Net present value
PVs	Present values
TCF	Trillion cubic feet

✉ Zhuoheng Chen
zchen@nrcan.gc.ca

[1] Geological Survey of Canada, Calgary, AB, Canada

[2] University of Calgary, Calgary, AB, Canada

[3] ZLR Valeon, Beijing, China

Edited by Xiu-Qin Zhu

1 Introduction

Recent advances in horizontal drilling coupled with multistage hydraulic fracturing have extended our ability to produce commercial oil and natural gas from low porosity–permeability fine-grained reservoirs. These technical advances have resulted in a fundamental shift in North American energy markets, as illustrated by the fast and large-scale development of shale gas and tight oil (EIA 2014). Although shale gas development projects face various challenges due to the uncertainties of environmental impacts and the long-term sustainability of economic development of this energy resource (Cueto-Felguerosoa and Juanesa 2013; Hughes 2013; Inman 2014), the potential economic benefits in North America have inspired shale exploration activities in other continents (Aguilera and Radetzki 2013; Huang et al. 2012; Jia et al. 2012; Zou et al. 2012), such as China with the recent discovery of the Fuling field, a giant shale gas field in the Sichuan Basin (Zuo et al. 2015). It is likely that many other countries, particularly in Europe where natural gas supply is tight and politically sensitive, will follow North American shale gas development trends (Aguilera and Radetzki 2013; Monaghan 2014; Slingerland et al. 2014; Weijermars 2013).

Bringing shale gas to market involves all aspects of upstream petroleum industry activities, from exploration for, pilot tests of, and commercial production of gas from shale reservoirs in a basin. Many business aspects of conventional petroleum exploration and development have been discussed in the "Business of Petroleum Exploration" edited by Steinmetz (1992). For example, Megill (1992) defined the economics of conventional petroleum exploration to evaluate the economic viability of transforming potential resources to producible reserves for commercial development and he discussed various economic measuring-sticks for supporting decision-making. Development economics refers to the determination of the investment opportunities that occur after the discovery of the resource, with specific regard to the techniques used and the economic yardsticks available for investment decisions (Roebuck 1992). In principle, the economic analysis of unconventional shale gas resources is no different than the economic analysis of conventional petroleum resources. However, because of the very nature of a pervasive regionally occurring, but spatially variable resource density and reservoir characteristics, the development of shale gas is characterized by lower geologic discovery risk, but higher commercial production risk (Gray et al. 2007). The challenge in shale gas plays is not to find where they contain natural gas, but to find the areas that are economically viable, which will lead to the best production and recovery volumes at a rate of return that satisfies investors (Kaiser and Yu 2012). Thus, shale gas economic appraisal has its primary emphasis on development and production evaluation.

Natural gas in shale formations has been known for over a century in North America (Zagorski et al. 2012). Economic evaluation of the feasibility of commercial production of natural gas from shale formation can be traced back many years prior to the shale gas evolution in the early 2000s. For example, TRW (1977) in a scientific report documented the economic side of the Devonian shale gas drilling ventures involving fracturing stimulation of the Appalachian Basin in the United States. In recent years since the start of the shale gas boom, methods and procedures of economic feasibility evaluation of shale gas development have been proposed and examples of economic appraisals of North American shale plays are publically available in literature as well as from industry reports.

Agrawal (2009) conducted an economic analysis on five emerging shale gas plays in the United States for two types of completions—vertical and horizontal to examine which completion method is the best under different economic environments. A discounted cash flow (DCF) model was used to generate economic yardsticks, such as net present value (NPV), internal rate of return (IRR) and payout time for comparing the economic outcomes in those five emerging shale plays. Hammond (2013) also used a DCF model to evaluate shale gas production well economics. He breaks down the breakeven natural gas price by year and shale play to illustrate the variability of economic outcomes in the major North American shale plays. Penner (2013) discussed different input parameters and factors in single well production economic appraisal and illustrated the importance of understanding the range of resource potential for commercial realization. Projection of future production is believed to be one important element in shale gas development economic analysis. Weijermars (2014) presented the uncertainty range in the future natural gas production output from US shale plays up to 2025 using a bottom up model as opposes to the top–down model of the US National Energy Modeling System (NEMS) by EIA (2014).

Kaiser and Yu (2013, 2014) published a series of articles discussing natural gas well production characteristics, operation costs, and economic analysis of the Haynesville Shale in Louisiana. The Haynesville Shale play has been one of the major shale gas plays in the United States. Although technical advances have lowered the development cost constantly, depressed well-head natural gas price is still the main constraint responsible for decline in the number of drilling rigs and production in the region.

Weijermars (2013) discussed economic feasibility of five emergent shale gas plays on the European continent. The study assumes that natural gas production is from 100 wells, drilled over a decade at a rate of 10 wells/year for each shale gas play and that production performance follows an exponential decline similar to the model used by EIA (Cook and Wagener 2014). The specific input parameters for shale gas productivity, such as initial production (IP), decline rate and well EURs in each play are generated by analogy to the production data of well-known shale plays in the United States. By applying the same production decline model to all wells in each shale gas play, the natural gas production from the realized wells over a 25-year life cycle were analyzed using DCF models to generate financial criteria as economic indicators for ranking the relative prospectivity of the five studied shale gas plays. The risks due to uncertainties in resource potential estimates, production performance and fiscal regimes are translated into the uncertainty in cash flow models under different scenarios. The resulting economic outcomes of gas productions in P10, P50, and P90 were presented to represent the uncertainty.

This paper presents a probabilistic model for shale gas development economic evaluation by considering the differences in well EUR and natural gas productivity due to variable geological/reservoir characteristics across a basin. The Horn River shale gas play in western Canada is used as an example to demonstrate the differences in economic outcomes under various financial terms for the same shale gas play.

2 Methods

2.1 Discount cash flow model

DCF analysis values a shale gas development project using the concept of the time value of money (such as inflation and bank interest). All future cash flows are estimated and discounted to give their present values (PVs). The sum of all future cash flows is the NPV, which is taken as the value of the cash flows. NPV, breakeven price, IRR and payout time provide the economic yardsticks for supporting investment decisions. Megill (1992) described the use of cash flow models and NPV to generate economic yardsticks in exploration economic analysis in detail. Weijermars (2013) provides a quantitative description of formulating a DCF model.

The DCF approach involves many aspects in petroleum business, some of which are technical and the others are fiscal terms. All those have to be analyzed and quantified as input parameters. Iledare (2014) discussed different fiscal terms in upstream petroleum economic analysis in the United States.

2.2 Gas price model

The future natural gas price is an external uncertainty factor affecting the revenue and economic margin of natural gas sales. In the last 10 years, well-head prices have been extremely volatile, varying from $2/MCF to over $10/MCF

in the North American natural gas market, and is reflective of the regional demand and supply dynamics. In North America, the spot natural gas prices continue to depress the well-head prices of all natural gas producers due to rapid growth of natural gas production from shale gas reservoirs. The cash flow models in this study adopt an initial natural gas price set at $4/MCF, with a forward correction for inflation modeled by an annual inflation factor of 1.5 % using the following equation (Weijermars 2013).

$$P_n = P_1(1 + r_{\text{inf}})^n, \qquad (1)$$

where P_n is the well-head natural gas price in year n, P_1 the well-head natural gas price in the 1 year, and r_{inf} the annual inflation rate affecting the natural gas price, and n the number of years of natural gas production. In our models 1000 cubic feet (1 MCF) of natural gas is equivalent to a calorific value of 1 million British thermal units (1 M Mbtu) in market pricing. Alternative functions for modeling natural gas price trends and background on what drives regional natural gas prices are highlighted in EIA (2014) and Weijermars (2013).

2.3 Production decline models

Forecasting long-term natural gas production from shale gas reservoirs is an essential element in economic evaluation. Single well estimated ultimate recovery (EUR) can be estimated directly from historical production data by fitting a decline model (Lee 2012). Among the available models, the Arps model is a well-known model proposed for various conventional reservoirs that has been applied to estimate well EUR for several decades (Arps 1945, 1956; Lee 2012). It has the following form.

$$q = q_i \frac{1}{(1 + bD_i t)^{(1/b)}}, \qquad (2)$$

where q is the production rate, t the time, q_i the IP rate, b and D_i are model parameters.

The application of the Arps model (i.e., the exponential and hyperbolic decline relations) is restricted to boundary-dominated flow regimes and may lead to significant overestimation of reserves if it is applied to transient flow (Ilk et al. 2010). Another problem is when in the hyperbolic decline form, the Arps model may lead to a physically impossible result in some cases (Lee 2010).

For a better empirical fit to production from tight reservoirs, like shale gas reservoirs, many new decline models have been proposed in recent years. The Valko model, also called the stretched exponential model, was proposed by Valko and Lee (2010) for unconventional shale gas and tight reservoirs. It is applied commonly to unconventional resource assessment (e.g., Valko and Lee

2010; Lee 2012; Chen and Osadetz 2013). The model has the following form:

$$q = q_i \exp\left[-\left(\frac{t}{\tau}\right)^n\right], \tag{3}$$

where q is the production rate, t the time, q_i the IP rate, n and τ are model parameters.

A simple exponential decline function is often used for modeling production decline in many academic and government energy supply analyses (EIA 2014; Cook and Wagener 2014; Weijermars 2013, 2014) with the following form:

$$Q_n = Q_1(1 - r)^{n-1}, \tag{4}$$

where Q_1 is the production in year 1 and decline factor r. The production in year n is given by Q_n.

However, the fundamental mechanisms controlling shale gas production remain poorly understood, and the classic theories and simulation techniques used by the oil and natural gas industry have proven inadequate for shale gas reservoirs (Cipolla et al. 2010; Monteiro et al. 2012). Under these circumstances, there is an obvious need for the theoretically based rate-decline models that are applicable for all flow regimes. Ilk et al. (2010) proposed a hybrid model for performance forecasting. The proposed model consists of three main rate-time relations and five supplementary rate-time relations, which utilize power-law, stretched exponential, hyperbolic, and exponential components to properly model the behavior of a given set of rate-time data from unconventional reservoirs.

In a recent study, Patzek et al. (2013) proposed a "diffusion-type" equation to model production decline from shale reservoirs. By introducing physical mechanisms behind natural gas recovery for shale formations to the production rate, this model represents a contribution toward reducing uncertainty (Cueto-Felguerosoa and Juanesa 2013). However, it is conceivable that the conceptual model of a linear, single-phase flow of gas into parallel, equidistant fractures may not be universally applicable. The complexity of induced hydraulically fractured geometries, the stimulation of networks of pre-existing (natural) fractures, adsorption and desorption processes, and non-Darcy multiphase flow through the chemically heterogeneous shale are all phenomena that could potentially result in a departure from the scaling behavior of the Patzek model (Cueto-Felguerosoa and Juanesa 2013).

For this study, the Valko model is used to estimate well EURs and generate production profiles (decline curves) from historical monthly natural gas production records in all the production wells. This is used to represent the anticipated variations in ultimate recoverable resource and production performance throughout the Horn River Basin shale gas play.

3 Horn River example

3.1 Geological background and data

The Horn River Basin, located in northeastern British Columbia (BC) in Canada (Fig. 1), covers about 11,000 km^2 (BCOGC 2014). IP wells of 30 MMCF/day (close to 1.0 million m^3/day) suggest that the Horn River Basin is one of most proliferous dry gas shale plays in North America. Three organic-rich stratigraphic units of Devonian age are the current targets for shale gas development in the basin (Fig. 2). The thickness of the three units varies from 100 to 300 m with burial depth of around 2500 m. Natural gas production started in 2005 (BCOGC 2014) and historical monthly production data (up to the end of October 2014) from 242 stimulated wells are collected from a commercial dataset. Natural gas composition analysis indicates that the methane content of the natural gas accounts for only 82 %–92 % depending on the stratigraphic units. Other natural gas contents include CO_2 and small amount of H_2S. Table 1 lists some major parameters of the reservoir, horizontal well and completion in this basin. Table 2 lists the production and financial parameters of the Horn River shale gas play for economic appraisal.

Figure 3 shows a histogram and a cumulative probability distribution of well EURs derived from fitting the historical production records of 206 production wells with production longer than 12 months by the Valko model. The EURs vary considerably from less than 3 BCF/well to over 40 BCF/well with a mean of 11.9 BCF/well and a median of 9.9 BCF/well, indicating a drastic variation in productivity across the basin, similar to other North American shale basins (Maugeri 2015).

The natural gas resource evaluation of the Horn River Basin was conducted by a joint effort from the British Columbia Ministry of Energy and Mines and the National Energy Board (2011) using a volumetric approach. Chen and Osadetz (2015) conducted a natural gas resource assessment of the Horn River Shale Basin using historical production data. The estimated recoverable natural gas resource potential varies from 32 TCF to 235 TCF (90 %–10 % confidence interval) with a median 118 TCF. The estimated number of production wells required to drain all the recoverable resource was estimated from a few thousand to over twenty thousand.

The North America natural gas price has been fluctuated between $2.5/MCF and $10/MCF in the last 20 years. We assume a start well-head price of US$4/MCF with an annual inflation adjustment of 1.5 % (Fig. 4) in this study and a comparison of the natural gas price used in this study with the projected future North America natural gas price by EIA (2013) is provided in Fig. 4.

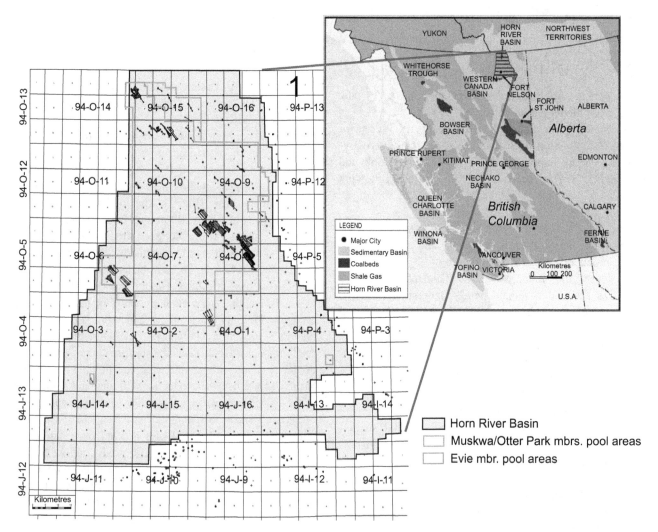

Fig. 1 Map showing the location of the study area and the production wells used in this study (compiled from BC MEM and NEB 2011 and BCOGC 2014)

3.2 Single well development economics

To illustrate the workflow of the economic analysis and method, the economic evaluation of a single production well with fixed financial and fiscal terms is conducted using the DCF method. The capital expenditures (CAPEX), operational expenses (OPEX) and other expenses and fiscal terms are listed in Table 2. Figure 5 is a typical production curve fitted to a decline model using the Valko model (200/C-096-H 094-O-08/00 well), showing general characteristics of well performance of the Horn River shale gas play. A projection of the fitted decline curve from the Valko model for 40 years of production lifetime gives a well EUR of 10.9 BCF/well, which is close to the mean value of 11.9 BCF/well for all the 206 production wells. Figure 6a shows monthly and cumulative natural gas sales and Fig. 6b shows the cumulative present value of cash flows under different discount rates with a payback period of about

11 years. With current fiscal terms and economic condition, any scenarios with a discount rate >7 % will not generate positive NPV (Fig. 6c). Given a discount rate of 10 %, the NPVs with different natural gas well-head prices is shown in Fig. 6d, suggesting that under this discount rate, only with a well-head price greater than $4.2/MCF does the NPV become positive (breakeven price). When the well EUR is greater than 17 BCF/well, the model indicates a positive NPV (Fig. 6e). Any production well with EUR < 10 BCF needs more than 10 years to pay back the investment (Fig. 6f) under the scenario of well-head price at a given discount rate of 10 %.

3.3 Shale gas play development economics

Chen and Osadetz (2015) estimated that it may need at least eight thousand production wells to exhaust the natural gas resource in the Horn River shale gas play. To discuss

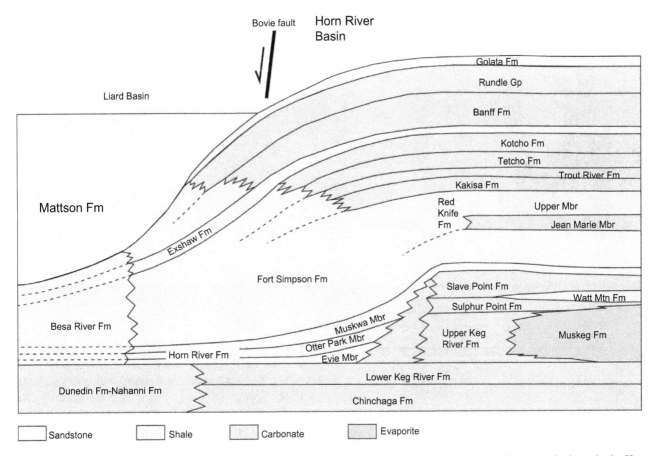

Fig. 2 Geological cross section demonstrating the major structural and depositional features of the basin and the target horizons in the Horn River Basin (cross section from Fiess et al. 2013)

Table 1 General geological information and reservoir parameters of the Horn River shale gas play, British Columbia, Canada (BCOGC 2014)

Reservoir	Parameters
Depth range	1900–3100 m
Gross thickness	140–280 m
TOC range	1 %–5 %
Porosity	3 %–6 %
Water saturation	25 %
Pressure	20–53 MPa
Pressure regime	Normal-over pressure
Temperature	80–160 °C
Drilling	
Wells/pad up to	16
Well spacing	100–600 m, avg 300 m
HZ length up to	3100 m, avg ∼1500 m
Wellbore	Cased
Completion data	
Fractype	Perf and plug
Fracfluid	Slick water
Fracstages	Up to 31, avg 18
Pump rate	8–16 m³/min
Water/well avg.	64,000 m³

Table 2 Fiscal and economic parameters used for this study, Horn River shale gas play, British Columbia, Canada

Item	Minimum	Mean	Maximum
EUR/well/metre, MMCF	1	6	23
Well CAPEX, Million$/well	12	15	20
OPEX, $/mcf	0.4	0.6	0.8
Royalty rate, %	15	15	15
Corporate tax, %	25	25	25
Discount rate, %	10	10	10

the shale gas development economics, we assume that a total of 1440 wells are to be drilled in the next 20 years with a drilling rate of six wells/month with a production lifetime of 40 years for each production well. The same economic and fiscal terms used for the single production well case study are applied. In the shale gas development economic analysis, we present two scenarios of utilizing a simple mean production decline model and a random drilling process to illustrate the advantages of including full range of variations in well EUR and production decline on economic outcomes of the same shale gas play.

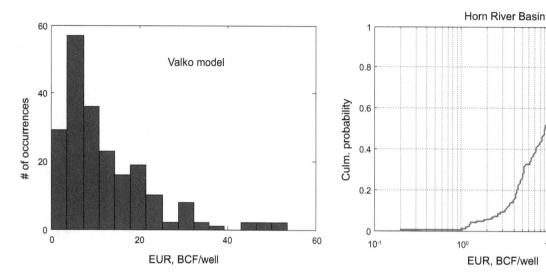

Fig. 3 *Histogram* (left) and *cumulative distribution curve* (right) of EURs from monthly production records in 206 production wells available for this study

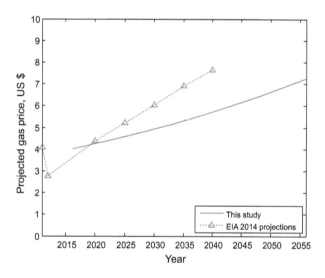

Fig. 4 Comparison of the EIA forecast natural gas price (EIA 2013) and the model used in this study

3.3.1 Average production decline model

A single decline curve derived by averaging the 206 decline models derived from the production wells is used to represent the average natural gas productivity of the Horn River play. The cumulative natural gas production of continuous drilling shows rapid increase of natural gas production with time and reaches the maximum production capacity at the end of 20 years when the last production well completed starts producing gas (Fig. 7a). A rapid decline of gas production follows because of the fast depletion of gas in the vicinity of the hydraulic fracture networks in the fractured horizontal wells such that the decline rate gradually becomes flatter with time (Fig. 7a).

The production rate drops again at 40 years as the oldest production wells come to the end of their production lifetime. The discounted cash flow model with a discount rate of 10 % suggests no positive NPV even though the model generated large amounts of cash, as most of the positive cash flow occurs after 20 years of production (Fig. 7b).

3.3.2 Random drilling scenario

It is interesting to compare the above-mentioned discount cash flow model using a single average production decline curve with a model using a random drilling procedure. A random drilling procedure can represent the full variation in the well EUR and production decline on the economic outcome. For the random sampling model, the following assumptions are made:

1. The historical production data from the 206 production wells represent a sample from a parent population of all possible natural gas production outcomes.
2. The 206 production decline curves derived from fitting the Valko model to the historical production data are extended to 40-year production lifetime to represent the production performance of the wells in the Horn River shale.
3. A minimum of 8000 wells are required to extract the shale gas resource in the Horn River Basin.

A random drilling procedure means that a well with any given value of EUR could appear at any order in a drilling sequence. Thus, there are many equally probable combinations that can form a random drilling sequence. To capture all possible outcomes, Monte Carlo simulation is

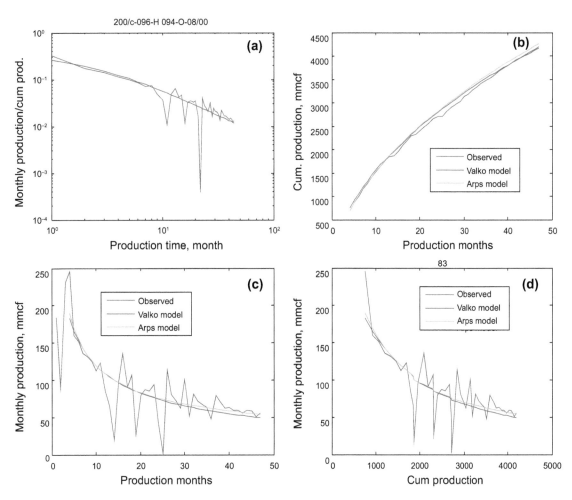

Fig. 5 Typical production curves (200/c-096-H 094-O-08/00 well) in the Horn River Basin showing the decline characteristics and fitted models

used to generate drilling sequences randomly. A total of 2000 equal probable drilling sequences (also called random realizations) were generated and each of the drilling sequences has 1440 production wells. The effect of many equally possible combinations of drilling sequences on economic outcomes can be captured by analyzing the results of the 2000 repeated random realizations. Figure 7c shows the average gas production from the 2000 random drilling sequences. Except for a rougher gas production curve (Fig. 6a), the rest of the curves looks similar (Fig. 7a, c). Figure 7d shows the mean cash flows with and without discount by averaging 2000 modeled outcomes from the random drilling development plan and the results suggest no positive NPV with a 10 % discount rate. It is no surprise that the two discount models in Fig. 7b, d have the same outcomes. This is because the average of a random sample is equivalent to the mean of the population. However, the random sampling procedure produces more than a mean, and other statistical measures provide the uncertainty or variation of the possible outcome, yielding more information for business decisions (Fig. 8a), from which it

is clear that under the assumed economic and fiscal terms, a random drilling development strategy does not result in a positive NPV unless the well-head gas price is higher than $4.3/MCF (Fig. 8b), or the average well EUR is greater than 15 BCF/well. In this case, only if the discount rate is lower than 7 %, the NPV is expected to be positive (Fig. 8e).

3.3.3 When someone does better than the average

In fact, one may do things better than others. Depending on the technology applied, the understanding of the rock or early entry of the play, some companies can do better than average, which means that a better gas recovery from the same shale formation can be achieved. Owning the best acreage or the worst makes a huge difference in economic outcomes (Maugeri 2015). If information on potential size of well EURs for selecting drilling sites is available, an optimal development strategy can be planned. By drilling sweet spots with higher productivity than an average EUR in the early stage of shale gas development, better

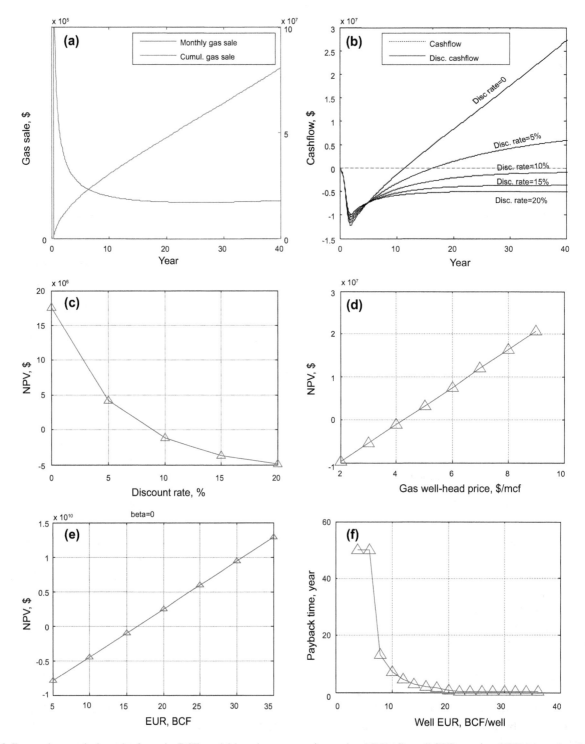

Fig. 6 Economic appraisal results from the DCF model based on a single well shale gas development project. Monthly production and cumulative production curve (**a**), PV of cumulative cash flow with different discount rates (**b**), discount rate against NPV (**c**), gas initial price against NPV (**d**), well EURs against NPV (**e**), and well EURs against payout time with a discount rate of 10 % (**f**)

economic outcomes can be achieved. To illustrate this effect, we generated drilling sequences better than average statistically based on a sampling scheme adopted from

"discovery process" models (Andreatta and Kaufman 1996; Chen et al. 2013), commonly used in conventional petroleum resource assessment. Such a drilling sequence,

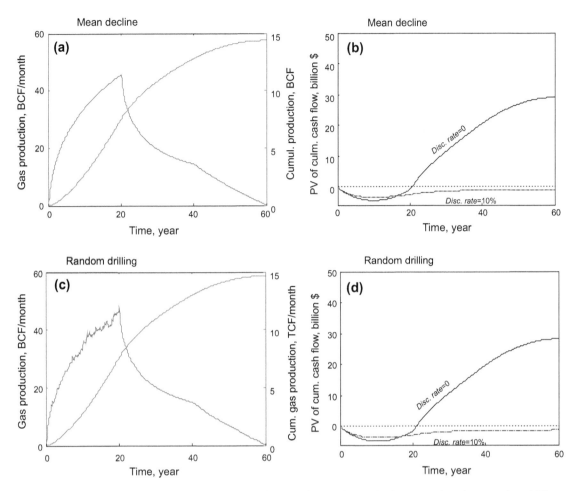

Fig. 7 Economic evaluation results at a play level based on a total of 1440 wells with a drilling rate of 6 wells/month. Production (monthly and cumulative) curves with time under the scenario of a single average decline production model (**a**), PV of cumulative discount cash flow model (discount rate = 0: *solid*; discount rate = 10 %: *dashed*) (**b**), production curves based on a random drilling model (**c**), PV of cumulative discount cash flows of the random drilling model (discount rate = 0: *solid*; discount rate = 10 %: *dashed*) (**d**)

which is affected by, but not exactly ranked by well EUR, is of probabilistic character in nature. The possible drilling sequence is formulated using the probabilistic model as a result of sampling from the parent population with a probability proportional to well EUR without replacement (Andreatta and Kaufman 1996). This probabilistic model and its application have been presented by Chen and Osadetz (2015) and will be discussed elsewhere in a separate paper.

Following the same development plan as applied to the random drilling scenario, a total of 1440 wells are drilled in 20 years and 2000 equally probable realizations of drilling sequences were generated to represent the possible combinations wells with different EURs and decline rates in the drilling sequence. After 60 years drilling and production (20 years drilling program and 40-year production), 22 TCF of natural gas was sold (Fig. 9a). This scenario resulted in an average drilled well EURs of 17.5 BCF/well,

much higher than the mean of 11.7 BCF/well of all 206 production wells. Under the same fiscal and economic terms, it gives a mean NPV of $2.6 billion Canadian dollars at the end of 60 years (Fig. 9b, c). Compared to the previous random or average production scenarios, this approach results in better economic outcomes given the lower breakeven price and higher IRR. The internal return rate is 14 % (Fig. 9d), the breakeven price is only $3.2/ MCF (Fig. 9g) and the payback time is about 13 years (Fig. 9b).

4 Conclusions

The current practice of shale gas economic evaluation commonly uses a mean EUR and a single production decline model for the entire lease or play (e.g., Penner 2013; Weijermars 2013; Cook and Wegener 2014).

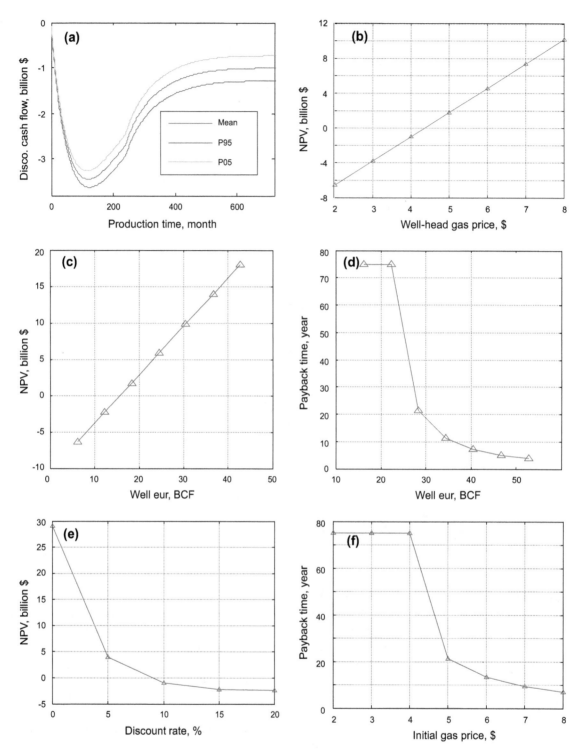

Fig. 8 Economic outcomes of the discounted cash flow models plotted as mean, p05 and p95 from Monte Carlo simulation of 2000 equally probable realizations under a random drilling scenario (**a**) and other *diagrams* showing the economic measures for the Horn River Basin shale gas development strategies. Initial natural gas price versus NPV (**b**), well EUR versus NPV (**c**), well EUR versus payback time under a discount rate of 10 % (**d**), discount rate versus NPV; and initial gas price versus payback time under a discount rate of 10 % (**f**)

Through numerical examples, this study shows that the use of a single average decline curve for the whole shale gas play is equivalent to a random drilling process. Business decisions based on economic forecasting derived from a single decline curve could be vulnerable to drastic changes of shale gas productivity across the region. While the

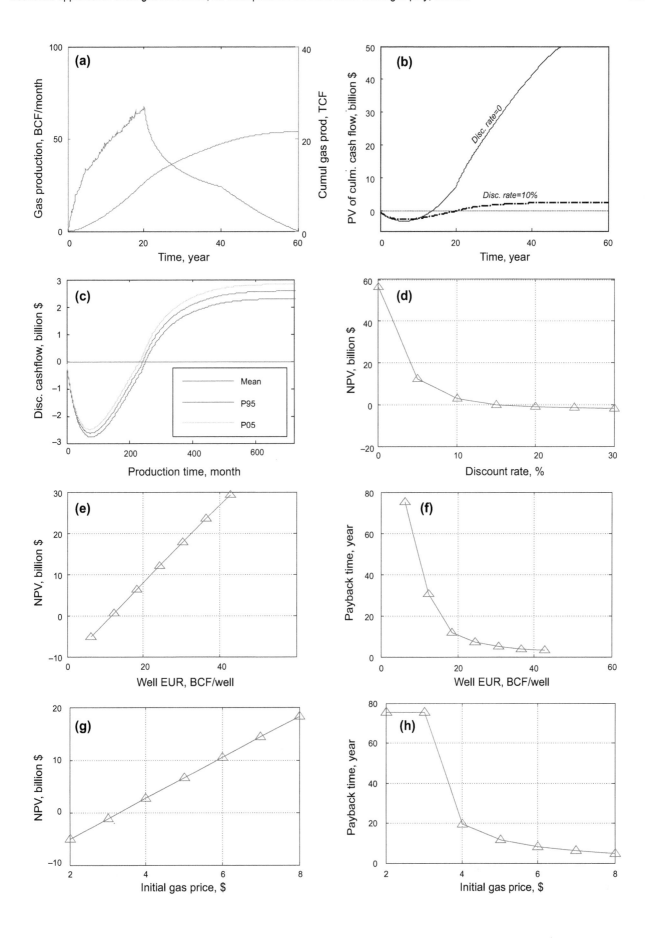

◄**Fig. 9** Economic outcome forecasting of the discounted cash flow models from 2000 Monte Carlo simulations for the selective drilling strategy with probability proportional to well EUR in Horn River Basin. Natural gas production with time (**a**), PV of cumulative cash flow plot with two different discount rates (disc. rate = 0: *solid line*; disc. rate = 10 %: *dashed line*) (**b**), mean, p05 and p95 of PV of cumulative discount cash flow (disc. rate = 10 %) from 2000 equal probable Monte Carlo simulations (**c**), discount rate versus NPV (**d**), well EUR versus NPV (**e**), and initial natural gas price versus payback time under 10 % discount rate (**f**)

random drilling model, taking account of variable well EURs and decline rates in economic appraisal, can result in large variation in economic outcomes and more statistical measures, it is useful for various business decisions, risk aversion and alternative development strategy planning.

Shale gas development drilling is based on pre-screening criteria and appears to be selective for potentially high productivity. A random drilling model may not fully capture the dynamic resource development process. To illustrate the impact of a selective drilling program on the economic outcome, a better than average drilling scenario is used in the DCF model. Under the assumed fiscal and economic terms, this study indicates that a shale gas resource development strategy with a random drilling strategy most likely results in, a negative NPV with a breakeven price of $4.2/MCF. However, under a selected drilling strategy, where the order of drilling is affected by well EUR, the economic outcome is improved with the breakeven price dropping below the well-head gas price and resulted in more than a 10 % IRR. This may imply that early identification of and drilling on "sweet spots" is critical to success in shale gas development.

Acknowledgments The study is funded partly by the ecoEII program and supported by Geoscience for New Energy Supply Program of Natural Resources Canada. The authors thank internal reviewer Dr. Stephen Grasby of the Geological Survey of Canada and journal reviewer Dr. Keyu Liu for their helpful comments and suggestions. This is ESS contribution #20150145.

References

Agrawal A. A technical and economic study of completion techniques in five emerging US gas shale plays. Master of Science thesis, Texas A&M University; 2009.

Aguilera RF, Radetzki M. Shale gas and oil: fundamentally changing global energy markets. Oil Gas J. 2013;111(12):54–61.

Andreatta G, Kaufman G. Estimation of finite population properties when sampling is without replacement and proportional to magnitude. J Am Stat Assoc. 1996;81(395):657–66.

Arps JJ. Analysis of decline curves. Trans AIME. 1945;160:228–47.

Arps JJ. Estimation of primary oil reserves: transactions AIME. J Petrol Technol. 1956;207:182–91.

British Columbia Ministry of Energy and Mines and National Energy Board, 2011. Ultimate potential for unconventional natural gas in northeastern British Columbia's Horn River Basin; Oil and Gas Reports 2011-1, 39 p. https://www.neb-one.gc.ca/nrg/sttstc/ntrlgs/rprt/archive/ncnvntnlntrlgshrnrvrbsnhrnrvr2011/ncnvntnln trlgshrnrvrbsnhrnrvr2011-eng.pdf

British Columbia Oil and Gas Commission (BCOGC). Horn River Basin unconventional shale gas play Atlas. 2014. http://www.google.ca/url?url=http://www.bcogc.ca/node/11238/download&rct=j&frm=1&q=&esrc=s&sa=U&ei=N-ihVOyZM4WhyASky4CoBg&ved=0CCMQFjAB&sig2=xX6DE2ZvElbdjpTXe2bTLw&usg=AFQjCNH_CEuFPdrCKO6jtHtD0aADdyFDsA.

Chen Z, Osadetz KG. Characteristics and resource potential evaluation of tight oil resources in Upper Cretaceous Cardium Formation, Western Canada Sedimentary Basin. Pet Explor Dev. 2013;40(3):344–53.

Chen Z, Osadetz KG. Economic implications of drilling sequence in shale gas development, example from Horn River Basin, Canada. GeoConvention, May 4–8, 2015, Calgary, Alberta.

Chen Z, Osadetz K, Chen G. Application of a least square nonparametric discovery process model to Colorado group mixed conventional and unconventional oil plays, Western Canada sedimentary basin. In: Pardo-Iguzqiza E, Guardiola-Albert C, Heredia J, Moreno-Merino L, Duran JJ, Vargas-Guzman JA, editors. Proceedings of the fifteenth annual conference of the international association for mathematic geosciences. New York: Springer; 2013. p. 617–20.

Cipolla CL, Lolon EP, Erdle JC, et al. Reservoir modeling in shale-gas reservoirs. SPE Reserv Eval Eng. 2010;13:638–53.

Cook T, Wagener DV. Improving well productivity based modeling with the incorporation of geologic dependencies. EIA, working paper series; 2014.

Cueto-Felguerosoa L, Juanesa R. Forecasting long-term gas production from shale. Proc Natl Acad Sci USA. 2013;110(49):19660–1.

EIA. Annual energy outlook 2013. DOE/EIA-0383(2013), June 2011. US Department of Energy, Energy Information Administration, Washington, DC; 2013. http://www.eia.gov/forecasts/aeo/pdf/0383(2013).pdf.

EIA. Annual projections to 2040. Updated data series. 2014. http://www.eia.gov/forecasts/aeo/pdf/tbla13.pdf.

Fiess K, Ferri F, Pyle L, et al. Liard Basin hydrocarbon project: shale gas potential of Devonian-Carboniferous strata in the Northwest territories, Yukon and northeastern British Columbia.GeoConvention 2013, extended abstract, Canadian Society of Petroleum Geologists. 2013. http://cseg.ca/assets/files/resources/abstracts/2013/263_GC2013_Liard_Basin_Hydrocarbon_Project.pdf.

Gray WM, Hoefer TA, ChiappeA, et al. A Probabilistic approach to shale gas economics, SPE-108053-MS. 2007. doi:10.2118/108053-MS

Hammond CD. Economic analysis of shale gas wells in the United States batchelor's thesis. Massachusetts Institute of Technology. 2013. http://dspace.mit.edu/handle/1721.1/83718#files-area.

Huang J, Zou C, Li J, et al. Shale gas generation and potential of the lower Cambrian Qiongzhusi Formation in Southern Sichuan basin. China Pet Explor Dev. 2012;39(1):69–75.

Hughes JD. Energy: a reality check on the shale revolution. Nature. 2013;494:307–8. doi:10.1038/494307a.

Iledare OO. Upstream petroleum economic analysis: balancing geologic prospectivity with progressive, stable fiscal terms and instruments. Way Ahead. 2014;10(1):28–30.

Ilk D, Currie, SM, Symmons D, et al. Hybrid rate-decline models for the analysis of production performance in unconventional reservoirs, SPE 135616. Prepared for presentation at the SPE annual technical conference and exhibition held in Florence, Italy, September 19–22, 2010.

Inman M. The fracking fallacy. Nature. 2014;516(4):28–30.

Jia C, Zheng M, Zhang Y. Unconventional hydrocarbon resources in China and the prospect of exploration and development. Pet Explor Dev. 2012;39(2):129–36.

Kaiser MJ, Yu Y. Louisiana Haynesville shale—2: economic operating envelopes characterized for Haynesville shale. Oil Gas J. 2012; 110(1a):70–74, 87.

Kaiser MJ, Yu Y. Haynesville update—1: north Louisiana gas shale's drilling decline precipitous. Oil Gas J. 2013; p. 62.

Lee WJ. Gas reserves estimation in resource plays. SPE paper 130102, presented at the unconventional gas conference, Pittsburgh, Pennsylvania, USA, February 23–25, 2010.

Lee WJ. Production forecast, reserves estimations and reporting rules for unconventional resources. Course material of SPE training series, hold in conjunction with SPE Canadian unconventional resources conference, Calgary AB, Canada, October 28–29, 2012.

Maugeri L. Comment: beware of break-even and marginal-cost analyses. Oil Gas J. 02/10/2015. http://www.ogj.com/articles/uogr/print/volume-3/issue-1/comment-beware-of-break-even-and-marginal-cost-analyses.html.

Megill RE. Exploration economics: chapter 10: part III. Economic aspects of the business. In: Steinmetz R, editor. The business of petroleum exploration, AAPG treatise of petroleum geology handbook of petroleum geology, vol 2. 1992. p. 107–16.

Monaghan AA. The carboniferous shales of the midland valley of Scotland: geology and resource estimation. British Geological Survey for Department of Energy and Climate Change, London, UK. 2014.

Monteiro PJM, Rycroft CH, Barenblatt GI. A mathematical model of fluid and gas flow in nanoporous media. Proc Natl Acad Sci USA. 2012;109(50):20309–13.

Patzek TW, Male F, Marder M. Gas production in the Barnett shale obeys a simple scaling theory. Proc Natl Acad Sci USA. 2013;110:19731–6.

Penner E. Shale gas production economics spreadsheet model and inputs. Oil Gas Financ J. September 20, 2013. http://www.ogfj.com/articles/2013/09/shale-gas-production-economics-spreadsheet-model-and-inputs.html.

Roebuck F. Development economics. In: Steinmetz R, editor. 382p. ISBN 0-89181-601-1. p. 117–24. American Association of Petroleum Geologists, Tulsa; 1992

Slingerland S, Rothengatter N, van der Veen R, et al. Economic impacts of shale gas in The Netherlands. http://www.tripleeconsulting.com/sites/default/files/Economic%20Impacts%20of%20Shale%20Gas%20in%20the%20Netherlands%20-%20FINAL%20REPORT.pdf.2014.

Steinmetz R. The business of petroleum exploration, AAPG treatise of petroleum geology handbook of petroleum geology, vol 2. American Association of Petroleum Geologists: USA; 1992.

TRW. Economic analysis of Devonian gas shale drilling ventures involving fracture stimulation, a report prepared for the United States Department of Energy, under contract EY-77-C-2 1-8085.1977. http://www.netl.doe.gov/kmd/cds/disk7/disk1/EGS%5CEconomic%20Analysis%20of%20Devonian%20Gas%20Shale%20Drilling%20Ventures%20In.pdf.

Valko PP, Lee WJ. A better way to forecast production from unconventional gas wells. SPE J. 2010;2(3):134–231.

Weijermars R. Economic appraisal of shale gas plays in continental Europe. Appl Energy. 2013;106:100–15. http://www.alboran.com/files/2013/07/SR-7.pdf.

Weijermars R. US shale gas production outlook based on well roll-out rate scenarios. Appl Energy. 2014;124:283–97.

Zagorski WA, Wrightsone GR, Bowman DC. The Appalachian Basin Marcellus gas play: its history of development, geological controls on production, and future potential as a world-class reservoir. In: Breyer JA, editor. Shale reservoirs—giant resources for the twenty-first century AAPG Memoir. 2012;97:172–200.

Zou C, Yang Z, Tao S, et al. Nano-hydro-carbon and the accumulation in coexisting source and reservoir. Pet Explor Dev. 2012;39(1):15–32.

Zuo C, Zhai G, Zhang G, et al. Formation, distribution, potential and prediction of global conventional and unconventional hydrocarbon resources. Pet Explor Dev. 2015;42(2):13–25.

13

Model inference using the Akaike information criterion for turbulent flow of non-Newtonian crude oils in pipelines

Ahmed H. Kamel[1] · Ali S. Shaqlaih[2] · Essam A. Ibrahim[1]

Abstract The friction factor is a crucial parameter in calculating frictional pressure losses. However, it is a decisive challenge to estimate, especially for turbulent flow of non-Newtonian fluids in pipes. The objective of this paper is to examine the validity of friction factor correlations adopting a new informative-based approach, the Akaike information criterion (AIC) along with the coefficient of determination (R^2). Over a wide range of measured data, the results show that each model is accurate when it is examined against a specific dataset while the El-Emam et al. (Oil Gas J 101:74–83, 2003) model proves its superiority. In addition to its simple and explicit form, it covers a wide range of flow behavior indices and generalized Reynolds numbers. It is also shown that the traditional belief that a high R^2 means a better model may be misleading. AIC overcomes the shortcomings of R^2 as a trade between the complexity of the model and its accuracy not only to find a best approximating model but also to develop statistical inference based on the data. The authors present AIC to initiate an innovative strategy to help alleviate several challenges faced by the professionals in the oil and gas industry. Finally, a detailed discussion and models' ranking according to AIC and R^2 is presented showing the numerous advantages of AIC.

Keywords Friction factor · Pipeline · Information theory · Non-Newtonian · Turbulent

List of symbols

f	Fanning friction factor, dimensionless
i	Any models in the set
m	Number of models in the set
n	Flow behavior index, dimensionless
N_{Reg}	Generalized Reynolds number, dimensionless
R^2	Coefficient of determination
SSE	Summation of squared residuals
SSY	Summation of squared errors
ω	Akaike weight, dimensionless
Δ	Information lost compared with the best model
K	Number of the estimated parameters in the model
k	Consistency index of power law fluid, Pa s^n
v	Average fluid velocity, m/s
ρ	Fluid density, g/cc
K_p	Pipe consistency index, Pa s^n
E, φ, α	Parameters, in Eq. (7), function of flow behavior index
y_i	Any predicted data point
$\left(l\left(\hat{\theta}\|y\right)\right)$	Nnumerical value of the likelihood at its maximum

✉ Ahmed H. Kamel
kamel_a@utpb.edu

[1] College of Business & Engineering, University of Texas of the Permian Basin, 4901 E University, Odessa, TX 79762-0001, USA

[2] Department of Mathematics & Information Sciences, University of North Texas at Dallas, 7400 University Hills Blvd, Dallas, TX 75241, USA

Edited by Yan-Hua Sun

1 Introduction

Throughout the world, large numbers of pipelines transport non-Newtonian pseudoplastic fluids including crude oils and petroleum products under turbulent flow conditions.

Pipeline design involves defining pipe size, thickness, and pumping head requirements where adequate provision for flow resistance is essential. The pumping head is required to overcome the inertia, gravity, and friction of the liquid inside the pipeline. The largest resistance force is the friction which is generally expressed in terms of pressure per unit length ($\mathrm{d}p/\mathrm{d}l$) required to overcome this resistance, psi/mile. The Darcy–Weisbach model is generally utilized to calculate the frictional pressure losses. Although it is a simple model, it involves a very essential factor; the friction factor. Numerous studies (Bogue 1962; Trinh 1969; Yoo 1974; Hanks and Ricks 1975; Govier 2008) have indicated that the friction factor is proportional to the kinetic energy of the fluid per unit volume and the area of the solid surface in contact with the fluid. This is the basis of the definition of the friction factor (Streeter and Wylie 1985). This friction factor or flow coefficient is not constant. Instead, it is defined in terms of the pipe specifications and fluid properties, but it is known to a high accuracy within certain flow regimes. For example, it was indicated that turbulent friction factors for non-Newtonian fluids can be obtained from the curves used for Newtonian fluids after the proper viscosity is inserted into the generalized Reynolds number (Govier 2008). However, there have been a number of advances in understanding the flow resistance of non-Newtonian fluids. As a result, many implicit and explicit equations; empirical, semi-empirical, and analytical have been proposed in the literature to accurately predict its value. Yet, they all seem to suffer from some drawbacks, either they are simple but not accurate or they are accurate but not simple.

So, the question is "which equation should be used?". To answer this, a detailed comparative study among the published correlations is indispensable to select the best model while taking into consideration its simplicity. Data published by several authors including Dodge and Metzner (1959), Shaver and Merrill (1959), Yoo (1974), and Szilas et al. (1981) represent the basis of this comparison. The models involved in the comparison are selected based upon their accuracy, precision, simplicity, and range of applicability as indicated in the literature.

Previously, similar comparisons were based upon model selection methods; most commonly R^2. However, it is well documented that these methods still have some shortcomings (Anderson 2008; Shaqlaih 2010; Shaqlaih et al. 2013). For example, the coefficient of determination, R^2 is interpreted as an indication of the "goodness of fit" of the model. However, it may be misleading if data are associated with noise. Another interpretation of the R^2 coefficient is that the higher the coefficient of determination, the better the variance that the dependent variable is explained by the independent variable (Larson and Marx 2007). Yet, R^2 can be potentially increased by adding more independent variables to the model which makes it appear to be better while it is not. A third problem with this coefficient is that it does not give a clear indication on what value of R^2 should be used to categorize the good model versus the weaker model. There are many examples of models with relatively high R^2 but they do not represent good models (Burnham and Anderson 2002).

In this paper, a newly adopted technique in the oil and gas industry, based on the information theory approach (Akaike information criterion, AIC), is used. There are many reasons that make the AIC information theory a much better approach for model selection than many other well-known approaches. First, AIC is derived from the principles of information theory. Therefore, it models the information in the data rather than the data itself which are essential as data have noise (Claeskens and Hjort 2009; Shaqlaih et al. 2013). Second, AIC is theoretically sound as it is a mathematically derived formula not just a definition. The best model in this approach is the model that minimizes the information lost when the model is used to approximate the truth model (the perfect model to represent the data with the highest possible accuracy). Third, AIC penalizes the number of parameters in the model which means applying the parsimony principle in the model selection process and hence preventing over fitting (Burnham and Anderson 2002). Moreover, AIC gives a clear-cut way to distinguish between the poor models and the good models. In other words, AIC excludes the models that have a poor information-based representation of the truth model. Furthermore, it has been proven that the AIC approach is more stable in ranking the models than many other approaches (Shaqlaih 2010).

The analysis presented in this paper allows us to select the most precise model while not neglecting its simplicity. The authors believe that the application of the information theory approach and AIC will resolve various issues faced by oil and gas professionals related to model selection and will initiate an innovative strategy that has been demonstrated in other disciplines. Yet, it has not been used extensively in the oil and gas industry. The models, data, and analysis techniques are discussed thoroughly within the context of this paper.

2 Friction factor equations

The calculation of frictional pressure losses using the Darcy–Weisbach equation requires knowledge of the friction factor. It is worth recalling that the friction factor originally defined by Weisbach and Darcy friction factor is four times the Fanning friction factor (Moody 1944). It was shown, using dimensional analysis, that the friction factor and generalized Reynolds number are the two dimensionless groups obtainable from flow tests, and therefore, they are used to characterize the flow-resistance relationship. Many explicit and implicit equations have been proposed in the literature for the determination of the friction factor. The most common and accurate equations included in this study are explained in detail in the following paragraphs and then examined to check their validity and applicability. In this paper, the generalized Reynolds number can be given mathematically as:

$$N_{Reg} = \frac{d_i^n \rho v^{2-n}}{8^{n-1} K_p}.$$ (1)

2.1 Dodge and Metzner equation

Dodge and Metzner (1959) carried out a semi-theoretical analysis of the turbulent flow of time-independent fluids, pseudoplastic power law fluids, and applied techniques of dimensional analysis to derive their first semi-empirical formula for friction factors in circular pipes:

$$1 \big/ \sqrt{f} = \frac{2}{n^{0.75}} \log\left(N_{Reg} f^{1-\frac{n}{2}}\right) - \frac{0.2}{n^{1.2}}.$$ (2)

They reported an excellent agreement between the calculated and the experimentally determined friction factors over a range of n from 0.36 to 1.0 and for generalized Reynolds numbers between 2900 and 100,000. The validity of this equation has been established for polymeric solutions, solid–liquid suspensions, power law, and non-power law fluids.

2.2 Shaver and Merrill equation

Shaver and Merrill (1959) developed a friction factor equation based on flow of aqueous plastic dispersions in smooth pipes:

$$f = \frac{0.79}{n^5 N_{Reg}^{\frac{2.62}{10.5n}}}.$$ (3)

It was reported that this empirical equation succeeded in correlating all the experimental data for n between 0.4 and 1.0. In fact, this correlation should not be used with n values lower than 0.4 due to its minimal accuracy (Shaver and Merrill 1959).

2.3 Tomita equation

Tomita (1959) developed his friction factor formula applying the Prandtle mixing length concept. The approximate validity of this equation was confirmed by 40 data points taken with starch pastes and lime slurries for flow behavior indices between 0.178 and 0.952 with generalized Reynolds numbers between 2000 and 100,000, as given below:

$$1 \big/ \sqrt{f} = \sqrt{2} \log\left(N_{Reg} \sqrt{\frac{f}{4}}\right) - \sqrt{0.2}.$$ (4)

2.4 Thomas equation

Thomas (1960) modified the Dodge and Metzner relationship to be given as:

$$1 \big/ \sqrt{f} = \frac{\sqrt{2}}{n} \log\left(N_{Reg} \left(\frac{f}{4}\right)^{1-\frac{n}{2}}\right) - \frac{\sqrt{0.2}}{n}.$$ (5)

2.5 Clapp equation

Clapp (1961) applied the Prandtle and Von-Karman approach to derive a universal velocity profile and friction factor correlation for turbulent flow of power law fluids in smooth pipes. This equation reduces to an equation similar to the Nikuradse equation for $n = 1.0$, as following:

$$1 \big/ \sqrt{f} = \frac{1.16}{n} - 1.22 + \frac{1.51}{n} \log\left(N_{Reg} \left(\frac{f}{4}\right)^{1-\frac{n}{2}}\right)$$
$$+ \frac{0.58}{n}(5n - 8).$$ (6)

This equation was validated employing experimentally gathered data with a maximum deviation of $\pm 4\,\%$ for $0.698 < n < 0.813$ and $548 < N_{Reg} < 42,800$.

2.6 Kemblowski and Kolodziejski equation

An alternative expression has been developed for the friction factor of power law fluids in turbulent flow (Kemblowski and Kolodziejski 1973). It is an empirical equation based on experimental data of aqueous suspensions with flow behavior indices ranging from 0.14 to 0.83 and generalized Reynolds numbers from 2680 to 98,600. E, φ, and m are defined elsewhere (Kemblowski and Kolodziejski 1973):

$$f = E \times \frac{\varphi^{\frac{1}{N_{Reg}}}}{N_{Reg}^{\alpha}},$$ (7)

with

$$E = 0.0089 e^{3.57n^2},$$ (7a)

$$\phi = e^{\frac{0.572(1-n^{4.2})}{n^{0.0435}}},$$ (7b)

$$\alpha = 0.314 n^{2.3} - 0.064.$$ (7c)

2.7 Garica and Steffe equation

Another equation was developed for the determination of the friction factor for pseudoplastic power law fluids (Garica and Steffe 1986) as follows:

$$1/\sqrt{f} = 1.318 \ln\left(N_{Reg}\sqrt{f} - 0.398\right). \tag{8}$$

2.8 Szilas et al. equation

A friction factor equation was developed by Szilas et al. (1981) as the first analytical relationship for flow of Non-Newtonian power law crude oils:

$$1/\sqrt{f} = \frac{\sqrt{2}}{n}\log\left(N_{Reg}f^{1-\frac{n}{2}}\right) + 1.23^{\frac{1}{n}}\left(\frac{0.707}{n} + 2.12\right) - \frac{2}{n} - 1.028.$$

$$\tag{9}$$

This equation was experimentally verified using data from the Hungarian Algyo crude oil pipeline for generalized Reynolds numbers varying between 10,000 and 100,000. This equation proved its accuracy when compared with several other equations (Szilas et al. 1981).

2.9 Desouky and El-Emam equation

Desouky and El-Emam (1990) derived an equation for designing a pipeline handling any type of pseudoplastic fluids under turbulent flow conditions by integrating the velocity distribution over the cross sectional area of the pipeline:

$$f = 0.71n^n\left(0.0112 + N_{Reg}^{-0.3185}\right). \tag{10}$$

A comparison with experimental data for pseudoplastic fluids measured by Yoo (1974) showed an excellent agreement with an average error of 2.6 % for all values of n (0.241 to 0.893).

2.10 Hawase et al. equation

Hawase et al. (1994) proposed an explicit expression for friction factor for hydraulically smooth pipes:

$$1/\sqrt{f} = 1.89 \log\left(\frac{N_{Reg}^{\frac{1}{n^{0.615}}}}{6.5n^{\frac{1}{1+0.75n}}}\right). \tag{11}$$

The values of f were within an error bound of ± 2.4 % for $0.3 < n < 1$ and $4000 < N_{Reg} < 100,0000$ when compared with the predictions from the implicit expression of Dodge and Metzner.

2.11 El-Emam et al. equation

El-Emam et al. (2003) employed the data measured by several authors (Dodge and Metzner 1959; Shaver and Merrill 1959; Yoo 1974; Szilas et al. 1981) and developed a new empirical equation to calculate the friction factor for turbulent flow of non-Newtonian fluids. Their equation was

statistically examined versus several other equations and experimental data and proved its accuracy:

$$f = \frac{n}{3.072 - 0.143n}N_{Reg}^{\frac{n}{0.282-4.211n}} - 0.00065. \tag{12}$$

Furthermore, the El-Emam et al. equation, in addition to several other equations, were evaluated using field data from an Egyptian pipeline; the Melieha-Al-Hamrah pipeline (101 miles long and 16-in. in diameter) which confirmed their proposed equation as a more realistic and simple approach (El-Emam et al. 2003).

Other equations are available in the literature as well (Torrance 1963; Trinh 1969; Hanks and Dadia 1971: Hanks and Ricks 1975; Derby and Melson 1981; Shenoy and Saini 1982; Shenoy 1988; Irvine 1988; Tam and Tiu 1988; Hemeida 1993; Trinh 2005). However, they are not included in the analysis due to either complexity, for example, they incorporate other dimensionless numbers such as the Hedstrom number, Deborah number, etc., or their limited validity when evaluated statistically or experimentally (Bogue 1962; Garica and Steffe 1986; Hartnet and Kostic 1990; Khaled 1994; El-Emam et al. 2003; Gao and Zhang 2007).

Table 1 lists the equations used in the present study along with their application ranges for flow behavior indices and generalized Reynolds numbers.

3 Measured data

The measured friction factors at different values of flow behavior indices and generalized Reynolds numbers incorporated in this analysis were gathered and published by several authors. Dodge and Metzner (1959) published friction factor values at flow behavior indices of 0.617, 0.726, and 1.0, while Shaver and Merrill (1959) published friction factor values at flow behavior indices of 0.6, 0.7, and 0.9. Other sets of data are published by Yoo (1974) at different values of flow behavior indices covering a wide range from 0.241 to 0.893 as well as Szilas et al. for $n = 0.5287, 0.6991, 0.7169, 0.8311,$ and 0.948 (1981). The four sets of data are included in the analysis individually and collectively to cover a wide range of both generalized Reynolds number and flow behavior indices.

4 Model selection methods

Model selection methods refer to the criteria or strategy by which one can identify the most accurate model among a set of candidate models. However, there are many different strategies to select the best model from a set of candidate models. In this study, the widely used statistical procedure (R^2) method and the information theory approach, Akaike

Table 1 Fanning friction factor equations and application ranges

Model	Formula	Notes	Year
D & M	$1/\sqrt{f} = \frac{2}{n^{0.75}} \log\left(N_{Reg} f^{1-\frac{n}{2}}\right) - \frac{0.2}{n^{1.2}}$	$0.36 < n < 1.0$ $2900 < N_{Reg} < 100{,}000$	1959
S & M	$f = \frac{0.79}{n^5 N_{Reg}^{\frac{2.63}{10.5n}}}$	$0.4 < n < 1.0$ $4000 < N_{Reg} < 100{,}000$	1959
Tomita	$1/\sqrt{f} = \sqrt{2}\,\log\left(N_{Reg}\sqrt{\frac{f}{4}}\right) - \sqrt{0.2}$	$0.178 < n < 0.952$ $2000 < N_{Reg} < 100{,}000$	1959
Thomas	$1/\sqrt{f} = \frac{\sqrt{2}}{n}\log\left(N_{Reg}\left(\frac{f}{4}\right)^{1-\frac{n}{2}}\right) - \frac{\sqrt{0.2}}{n}$	$0.36 < n < 1.0$ $2900 < N_{Reg} < 100{,}000$	1960
Clapp	$1/\sqrt{f} = \frac{1.16}{n} - 1.22 + \frac{1.51}{n}\log\left(N_{Reg}\left(\frac{f}{4}\right)^{1-\frac{n}{2}}\right) + \frac{0.58}{n}(5n - 8)$	$0.698 < n < 0.813$ $548 < N_{Reg} < 42{,}800$	1961
K & K	$f = E \times \frac{\phi^{\frac{1}{N_{Reg}}}}{N_{Reg}^{\alpha}}$ $E = 0.0089 e^{3.57n^2}$ $\phi = e^{\frac{0.572(1-n^{4.2})}{n^{0.0435}}}$ $\alpha = 0.314 n^{2.3} - 0.064$	$0.14 < n < 0.83$ $2{,}680 < N_{Reg} < 98{,}600$	1973
SBN	$1/\sqrt{f} = \frac{\sqrt{2}}{n}\log\left(N_{Reg} f^{1-\frac{n}{2}}\right) + 1.23^{\frac{1}{n}}\left(\frac{0.707}{n} + 2.12\right) - \frac{2}{n} - 1.028$	$0.24 < n < 1.0$ $10{,}000 < N_{Reg} < 100{,}000$	1981
G & S	$1/\sqrt{f} = 1.318 \ln\left(N_{Reg}\sqrt{f} - 0.398\right)$	$0.4 < n < 0.82$ $3000 < N_{Reg} < 50{,}000$	1986
D & E	$f = 0.71 n^n \left(0.0112 + N_{Reg}^{-0.3185}\right)$	$0.241 < n < 0.893$ $4000 < N_{Reg} < 100{,}000$	1990
HSW	$1/\sqrt{f} = 1.89 \log\left(\frac{N_{Reg}^{\frac{1}{n^{0.615}}}}{6.5 n^{\frac{1}{1+0.75n}}}\right)$	$0.3 < n < 1.0$ $4000 < N_{Reg} < 100{,}000$	1994
El-Emam et al.	$f = \frac{n}{3.072 - 0.143n} N_{Reg}^{\frac{n}{0.282 - 4.211n}} - 0.00065$	$0.178 < n < 1.0$ $4000 < N_{Reg} < 150{,}000$	2003

information criterion (AIC) are used. The coefficient of multiple determinations, R^2 for a model is defined as:

$$R^2 = 1 - \frac{SSE}{SSY} \tag{13}$$

where SSE and SSY are given as:

$$SSE = \sum_{i=1}^{n}(y_i - \hat{y}_i)^2, \tag{13a}$$

$$SSY = \sum_{i=1}^{n}(y_i - \bar{y})^2, \tag{13b}$$

where $\bar{y}$ is the average value of the observed values y_i, and $\hat{y}_i$ is predicted value of y_i under the model (Mendenhall and Sinich 2003). With the R^2 method, the larger the R^2, the more accurate the model (Mendenhall and Sinich 2003; Larson and Marx 2007). Even though R^2 is widely used as an indication of the goodness of fit of models, it should not be used with nonlinear models. However, it is used in this paper to prove that it is not a good measure for the models fit (Anderson 2008; Shaqlaih et al. 2013).

On the other hand, in the information theory approach, it is thought of the full reality as a model to be approximated (Burnham and Anderson 2002), and the objective is to find the model that best approximates the unknown truth model. Akaike (1973) showed that the model that best approximates the truth model is the one with smallest value of AIC:

$$AIC = -2\log\left(l\left(\hat{\theta}|y\right)\right) + 2K, \tag{14}$$

where K is the number of the estimated parameters in the model and $\left(l\left(\hat{\theta}|y\right)\right)$ is the numerical value of the likelihood at its maximum (Akaike 1973). The value of AIC gives the information lost if the chosen model is used to approximate the truth model. In other words in the information theory approach, the smaller the AIC, the more accurate the model. It is useful to define the AIC difference as: $\Delta_i = AIC_i - AIC_{\min}$, where $AIC_{\min}$ is the smallest value of the AIC values for all the set of candidate models. The best model has a Δ value of zero. A candidate model with Δ value higher than 10 should not be considered as a useful model (Anderson 2008; Shaqlaih et al. 2013). Another parameter is the Akaike's weight, ω_i:

$$\omega_i = \frac{\exp\left(-\frac{\Delta_i}{2}\right)}{\sum_{r=1}^{m} \exp\left(-\frac{\Delta_i}{2}\right)} \pi r^2, \qquad (15)$$

where Δ_i is the AIC difference of the model i and m is the number of candidate models. ω_i gives the weight of evidence in favor of model i being the best model in the set of m models. One of the approaches to create a 95 % confidence set of models in the information theory approach is based on Akaike weights. In this approach, we sum the Akaike weights from largest to smallest until the sum is just ≥ 0.95. In the information theory approach, it is essential to find the Akaike weight for each model to be able to see the probability of the model being the best model. Akaike weight, AIC differences, and the confidence set of models are all essential tools in the model selection process in the information theory approach.

5 Results and discussion

In this study, both R^2 and the AIC are used to check for the best model among a set of the 11 candidate models discussed previously. For better understanding of the best model that approximates the friction factor, the published four sets of data are used individually. Later, these four sets are combined and used collectively as one set to examine the same models. The detailed results are discussed in the following paragraphs.

The first set of data was published by Dodge and Metzner (1959) for three different values of flow behavior index (0.617, 0.726, and 1.0) and a wide range of generalized Reynolds numbers. Table 2 shows the results for both R^2 and AIC methods. This table shows that according to R^2 values, Clapp (1961) and Desouky and El-Emam (1990) models are the best fit for the data with R^2 values of 0.92. Dodge and Metzner (1959) and El-Emam et al. (2003) models still have reasonable fit as their R^2 values are 0.88 and 0.85, respectively. Other models have reasonable R^2 values as well. Tomita (1959), Thomas (1960), and Garica and Steffe (1986) models have poor fits as each has R^2 value less than 0.50. As stated earlier, the R^2 method does not give a clear-cut evaluation of which models should be considered. For example, the Shaver and Merrill (1959) model has R^2 value of 0.66 which may be considered reasonably large. However, on the other hand, it is considerably less than 0.92, the largest R^2 value. The same conclusion applies to other models.

Regarding the AIC, as we can see in table, only two models can be used to accurately predict the friction factor; namely the Clapp (1961) and the Desouky and El-Emam

Table 2 Ranking of the correlations using Dodge and Metzner (1959) data

Model	R^2	R^2 ranking	Δ	ω	AIC ranking
D & M	0.88	2	14.3	0.00	Poor
S & M	0.66	6	51.8	0.00	Poor
Tomita	0.37	Poor	101.7	0.00	Poor
Thomas	0.17	Poor	83.5	0.00	Poor
Clapp	0.92	1	0.0	0.59	1
K & K	0.76	5	38.7	0.00	Poor
SBN	0.81	4	31.2	0.00	Poor
G & S	0.08	Poor	87.6	0.00	Poor
D & E	0.92	1	0.8	0.41	2
HSW	0.56	7	61.3	0.00	Poor
El-Emam et al.	0.85	3	21.6	0.00	Poor

models (1990). In fact the best model to use is the Clapp (1961) model with an Akaike weight, ω of 59.0 %. The Desouky and El-Emam (1990) model still has a recognizable Akaike weight of 41.0 %. We recall here that the Akaike weight provides evidence for which model is the best. The other models have no chance of being good models. Even though the Dodge and Metzner (1959) model was developed using this data, its performance is very unsatisfactory. The same results can be attained by looking at the Δ values. Indeed, we can see that the best model is the Clapp (1961) model with a Δ value of zero (or equivalently the smallest AIC value and hence the best model). The Desouky and El-Emam (1990) model is second best with a Δ-value of 0.75. We can clearly see that all other models should not be considered as their Δ-values are higher than 10 and hence their Akaike weights are negligible.

However, AIC states that the Clapp model is better than Desouky and El-Emam (1990) model as the ratio between their Akaike weights is 0.59/0.41 = 1.4 which means that the Clapp model is 1.4 times better than the Desouky and El-Emam (1990) model, as inferred from the weight factor according to AIC definitions. The advantages of the AIC method over R^2 are clear as it gives the set of models that can be considered. Moreover, the AIC method not only ranks the models but also separates the models that should not be considered. Furthermore, the Akaike ratio clarifies how the selected models should be preferred (Burnham and Anderson 2002; Shaqlaih 2010).

Table 3 shows the results using Shaver and Merrill data for flow behavior indices of 0.6, 0.7, and 0.9. According to the R^2, Shaver and Merrill (1959) is the best model (highest R^2 value) which is logically true since this is the data used to develop the model. Also, same conclusion can be inferred for AIC.

Table 3 Ranking of the correlations using Shaver and Merrill (1959) data

Model	R^2	R^2 ranking	Δ	ω	AIC ranking
D & M	0.74	6	133.1	0.00	Poor
S & M	0.99	1	0.0	1.00	1
Tomita	0.12	11	183.7	0.00	Poor
Thomas	0.13	10	102.9	0.00	Poor
Clapp	0.81	4	119.7	0.00	Poor
K & K	0.44	9	164.6	0.00	Poor
SBN	0.89	2	94.9	0.00	Poor
G & S	0.71	8	174.5	0.00	Poor
D & E	0.76	5	129.9	0.00	Poor
HSW	0.74	7	133.2	0.00	Poor
El-Emam et al.	0.89	3	97.2	0.00	Poor

However, according to AIC all other models are weak (zero values for Akaike weights) and should not be considered, a result that could not be attained using R^2 values alone.

Similar conclusions can be generated using Yoo (1974) data and Szilas et al. (1981) data for other ranges of flow behavior indices and generalized Reynolds numbers.

Table 4 shows the results when using Yoo data while Table 5 shows similar results for Szilas et al. (1981) data. For the Yoo (1974) data in Table 4, both R^2 and AIC indicate that the Desouky and El-Emam (1990) model is the best in the set. Again, this is reasonably accepted since the Desouky and El-Emam (1990) showed that their model had an excellent agreement with an average error of 2.6 % for all the values of n when compared with the data measured by Yoo (1974). However, AIC analysis suggests that all other models should not be considered as they all have Akaike weights of 0.0.

Similarly, Table 5 shows that the Szilas et al. (1981) model is the best model when using Szilas data. Again, this

Table 4 Ranking of the correlations using Yoo (1974) data

Model	R^2	R^2 ranking	Δ	ω	AIC ranking
D & M	0.80	2	44.0	0.00	Poor
S & M	0.08	Poor	106.8	0.00	Poor
Tomita	0.28	Poor	153.1	0.00	Poor
Thomas	0.74	5	53.7	0.00	Poor
Clapp	0.79	3	44.5	0.00	Poor
K & K	0.21	Poor	95.1	0.00	Poor
SBN	0.39	Poor	85.2	0.00	Poor
G & S	0.11	Poor	129.5	0.00	Poor
D & E	0.94	1	0.0	1.00	1
HSW	0.78	4	48.4	0.00	Poor
El-Emam et al.	0.63	6	67.1	0.00	Poor

Table 5 Ranking of the correlations using Szilas et al. (1981) data

Model	R^2	R^2 ranking	Δ	ω	AIC ranking
D & M	0.70	7	532.2	0.00	Poor
S & M	0.84	4	487.2	0.00	Poor
Tomita	0.59	9	648.3	0.00	Poor
Thomas	0.97	3	376.9	0.00	Poor
Clapp	0.71	6	529.2	0.00	Poor
K & K	0.45	Poor	574.1	0.00	Poor
SBN	0.99	1	0.0	1.00	1
G & S	0.15	Poor	625.7	0.00	Poor
D & E	0.74	5	521.2	0.00	Poor
HSW	0.67	8	538.9	0.00	Poor
El-Emam et al.	0.99	1	245.9	0.00	Poor

conclusion is reasonably accepted. Even though El-Emam et al. (2003) and Thomas (1960) models have very high value of R^2 (0.99 and 0.97, respectively), they seem to be poor models according to AIC ranking. Recall that this is one of the disadvantages of the R^2 method (Burnham and Anderson 2002; Shaqlaih 2010).

In general, with each set of data, a specific model is believed to be the best either because it was developed using this set of data or because, when developed, it was compared and examined with this set of data to show its accuracy. Now, all four sets of data are combined and the same models are examined using R^2 and AIC. It is worth mentioning that combining all sets of data covers a very wide range of flow behavior indices n and generalized Reynolds numbers N_{Reg}. The analysis in this case is believed to be more realistic and the results should be statistically valid. The results are summarized in Table 6.

From Table 6, it can be seen that none of the previously selected models, for example, the Desouky and El-Emam (1990) model based on Yoo data and the Szilas et al. model based on Szilas data can predict accurate values of the friction factor for this wide range of n and N_{Reg} values. This may be due their application range. Most of these equations were empirically derived and experimentally verified with measured data covering a certain range of n and N_{Reg} values. Extending their application beyond this range is not normally possible and can lead to erroneous results. Using all data collectively showed that a different model seems to be reasonably good and should be used. It is the El-Emam et al. model. This could be reasonably accepted as the model was developed using the four sets of data and was evaluated using pipeline field data. Its R^2 value is the highest, 0.92, and it is ranked first. Also, the same conclusion can be drawn from the AIC results as the El-Emam et al. (2003) model is still ranked number one with an Akaike weight factor of 99.9 % and no information loss, i.e., $\Delta = 0.0$. Furthermore, since the El-Emam et al.

Table 6 Ranking of the correlations using all data

Model	R^2	R^2 ranking	Δ	Ω	AIC ranking
D & M	0.74	7	219.321	0.000	Poor
S & M	0.86	3	99.852	0.000	Poor
Tomita	0.29	Poor	515.978	0.000	Poor
Thomas	0.82	4	147.375	0.000	Poor
Clapp	0.78	5	186.358	0.000	Poor
K & K	0.48	Poor	347.898	0.000	Poor
SBN	0.91	2	14.447	0.001	Poor
G & S	0.06	Poor	457.298	0.000	Poor
D & E	0.78	6	193.122	0.000	Poor
HSW	0.69	8	252.957	0.000	Poor
El-Emam et al.	0.92	1	0.0	0.999	1

(2003) model turned out to be the best model for a wide range of generalized Reynolds numbers and flow behavior indices, it is recommended to be used for the prediction of the friction factor.

From Table 6, we can also notice that even though the Szilas et al. model has a large R^2 value, it is a poor model from the AIC point of view. In fact with the exception of the El-Emam et al. model, all models should not be considered for the prediction.

6 Conclusions and recommendations

The present paper shows that a large number of equations exist to calculate the friction factor for pseudoplastic power law fluids. Yet, selecting the equation represents an immense challenge facing the pipeline engineer. A wrong selection may lead to an error of up to 83.4 % (El-Emam et al. 2003). Eleven equations are discussed and examined using four sets of friction factor measured data. Traditionally, R^2 along with the AIC approach are used throughout the comparative study to select the best model to predict the Fanning friction factor. Both AIC and the R^2 methods suggest that the El-Emam et al. model is reasonably good in predicting friction factors. The suggested model has the highest R^2 (0.92) as well as the highest Akaike weight factor ($\omega = 99.9$ %) with no formation loss ($\Delta = 0.0$). Moreover, this model, unlike other models, covers a wide range of both flow behavior indices and generalized Reynolds number. Nevertheless, other models showed excellent performance when compared with their original data.

The shortcomings of using R^2 are discussed where certain models can have high R^2 values, yet the Akaike weight factors are very low as an indication of their poor performance. A good example is the Szilas et al. model when examined using all the data. The advantages of using AIC

over R^2 are presented which makes it a viable alternative for model selection. It employs the parsimonious principle to trade between the complexity of the model and its accuracy, not only to find a best approximating model, but also to develop statistical inference based on the data.

It is therefore recommended that the El-Emam et al. model is used to predict the Fanning friction factor employing the AIC approach rather than the conventional R^2 approach for model selection.

Finally, the authors introduce AIC to the oil and gas industry as an innovative tool for model selection. We believe this AIC can alleviate the dilemma of model selection encountered by professionals in the oil and gas industry.

References

Akaike H. Information theory as an extension of the maximum likelihood principal. Second international symposium on information theory. Budapest: Akademiai Kiado; 1973. p. 267–81.

Anderson D. Model based inference in the life sciences: a primer on evidence. New York: Springer; 2008.

Bogue D. Velocity profiles in turbulent non-newtonian pipe flow. PhD dissertation. University of Delaware, Newark; 1962.

Burnham D, Anderson D. Model selection and multimode inference, a practical information-theoretic approach. 2nd ed. New York: Springer; 2002.

Claeskens G, Hjort N. Model selection and model averaging. 2nd ed. Cambridge: Cambridge University Press; 2009.

Clapp R. Turbulent heat transfer in pseudoplastic non-newtonian fluids. Int. Developments in Heat Transfer, ASME. 1961; Part III, Sec. A, 652.

Derby R, Melson J. How to predict the friction factor for flow of bingham plastics? Chem Eng. 1981;28:59–61.

Desouky S, EL-Emam N. A generalized pipeline design correlation for pseudoplastic fluids. J Can Pet Technol. 1990;29(5):48–54.

Dodge D, Metzner A. Turbulent flow of non-newtonian systems. AIChE J. 1959;5(2):189–203.

El-Emam N, Kamel AH, Al-Shafie M, et al. New equation calculates friction factor for turbulent flow of non-Newtonian fluids. Oil Gas J. 2003;101(36):74–83.

Gao P, Zhang J. New assessment of friction factor correlations for power law fluids in turbulent pipe flow: a statistical approach. J Central South Univ Technol. 2007;14:77–81.

Garica E, Steffe J. Comparison of the friction factor equations for non-Newtonian fluids in pipe flow. J Food Process Eng. 1986;9(2):93–120.

Govier G. The flow of complex mixtures in pipes. 2nd ed. New Orleans: Society of Petroleum Engineers; 2008.

Hanks R, Dadia BH. Theoretical analysis of the turbulent flow of non-Newtonian slurries in pipes. AIChE J. 1971;17(3):554–7.

Hanks R, Ricks B. Transitional and turbulent pipe flow of pseudoplastic fluids. J Hydronaut. 1975;9:39–44.

Hartnet J, Kostic M. Turbulent friction factor correlations for power law fluids in circular and non-circular channels. Int Commun Heat Mass Trans. 1990;17:59–65.

Hawase Y, Shenoy A, Wakabayashi K. Friction and heat and mass transfer for turbulent pseudoplastic non-Newtonian fluids flowing in rough pipes. Can J Chem Eng. 1994;72:798–804.

Hemeida A. Friction factor for yieldless fluids in turbulent pipe flow. J Can Pet Technol. 1993;32(1):32–5.

Irvine TF Jr. A generalized Blasius equation for power law fluids. Chem Eng Commun. 1988;65:39–47.

Kemblowski Z, Kolodzie J. Flow resistances of non-Newtonian fluids in transitional and turbulent-flow. Int Chem Eng. 1973;13(2): 265–79.

Khaled S. A comparative study among the different methods used for calculating the pressure losses in pipelines transporting non-Newtonian oils. MSc thesis. AL-Azhar University, Egypt; 1994.

Larson R, Marx M. An introduction to mathematical statistics and its applications. 3rd ed. New Jersey: Pearson; 2007.

Mendenhall W, Sincich T. A second course in statistics: regression analysis. 6th ed. New Jersey: Pearson Education Inc; 2003.

Moody L. Friction factors for pipe flow. Trans ASME. 1944;66: 671–84.

Shaqlaih A. Model selection by an information theory approach. PhD dissertation. University of Oklahoma, Norman; 2010.

Shaqlaih A, White L, Zaman M. Resilient modulus modeling with information theory approach. Int J Geomech. 2013;13(4):384–9.

Shaver R, Merrill E. Turbulent flow of pseudoplastic polymer solutions in straight cylindrical tubes. AIChE J. 1959;5(2): 181–8.

Shenoy A. Rheology of highly filled polymers melt systems. In: Cheremisinoff NP, editor. Encyclopedia of fluid mechanics, vol. 7. Houston: Gulf Publishing; 1988. pp. 667–71.

Shenoy A, Saini D. A new velocity profile model for turbulent pipe-flow of power-law fluids. Can J Chem Eng. 1982;60(5):694–6.

Streeter V, Wylie B. Fluid mechanics. 7th ed. New York: McGraw-Hill College; 1985.

Szilas A, Bobok E, Navratil L. Determination of turbulent pressure losses of non-Newtonian oil flow in rough pipes. Rheol Acta. 1981;20(5):487–96.

Tam K, Tiu C. A general correlation for purely viscous non-Newtonian fluids flowing in ducts of arbitrary cross sections. Can J Chem Eng. 1988;66:542–9.

Thomas DG. Heat and momentum transport characteristics of non-Newtonian aqueous thorium oxide suspensions. AIChE J. 1960;6(4):631–9.

Tomita Y. A study of non-Newtonian flow in pipelines. Bull JSME. 1959;5:10–16.

Torrance B. Friction factors for turbulent non-Newtonian fluid flow in circular pipes. S Afr Mech Eng. 1963;13:89–91.

Trinh K. A boundary layer theory for turbulent transport phenomena, MSc thesis. University of Canterbury New Zealand; 1969.

Trinh K. A zonal similarity analysis of wall-bounded turbulent shear flows. Paper presented at the 7th world congress of chemical engineering; 2005.

Yoo SS. Heat transfer and friction factors for non-newtonian fluids in turbulent pipe flow. PhD dissertation. Graduate College, University of Illinois, Chicago; 1974.

14

The use of grass as an environmentally friendly additive in water-based drilling fluids

M. Enamul Hossain[1] · Mohammed Wajheeuddin[1]

Abstract Excellent drilling fluid techniques are one of the significant guaranteed measures to insure safety, quality, efficiency, and speediness of drilling operations. Drilling fluids are generally discarded after the completion of drilling operations and become waste, which can have a large negative impact on the environment. Drilling materials and additives together with drill cuttings, oil, and water constitute waste drilling fluids, which ultimately are dumped onto soil, surface water, groundwater, and air. Environmental pollution is found to be a serious threat while drilling complex wells or high-temperature deep wells as these types of wells involve the use of oil-based drilling fluid systems and high-performance water-based drilling fluid systems. The preservation of the environment on a global level is now important as various organizations have set up initiatives to drive the usage of toxic chemicals as drilling fluid additives. This paper presents an approach where grass is introduced as a sustainable drilling fluid additive with no environmental problems. Simple water-based drilling fluids were formulated using bentonite, powdered grass, and water to analyze the rheological and filtration characteristics of the new drilling fluid. A particle size distribution test was conducted to determine the particle size of the grass sample by the sieve analysis method. Experiments were conducted on grass samples of 300, 90, and 35 µm to study the characteristics and behavior of the newly developed drilling fluid at room temperature. The results show that grass samples with varying particle sizes and concentrations may improve the viscosity, gel strength, and filtration of the bentonite drilling fluid. These observations recommend the use of grass as a rheological modifier, filtration control agent, and pH control agent to substitute toxic materials from drilling fluids.

Keywords Rheology · Filtration · Filter cake · Apparent viscosity · Plastic viscosity · Gel strength

1 Introduction

The use of drilling fluids (DFs, also called drilling mud) is an essential part of a rotary drilling process. Different types of chemicals and polymers are used in designing a drilling fluid to meet functional requirements such as appropriate mud rheology, density, mud activity, fluid loss control property, etc. (Amanullah et al. 1997). Today, the choice of drilling fluids and their additives has become complex (Caenn et al. 2011), considering both the technical and environmental factors (Amanullah 1993).

The preservation of the environment on a global level is now important as various organizations have set up initiatives to drive the usage of toxic chemicals as DF additives. Environmental pollution has been considered a serious threat while drilling complex wells or high-temperature deep wells, which are now managed by using oil-based DF systems and high-performance water-based DF systems. As environmental protection has become a consideration before any oil and gas resources exploration, people have paid more and more attention to the DF for environmental safety. Advances in recent technologies led to the development of novel environmentally friendly DF

✉ M. Enamul Hossain
menamul@kfupm.edu.sa

[1] Department of Petroleum Engineering, College of Petroleum Engineering and Geosciences, King Fahd University of Petroleum & Minerals (KFUPM), Box: 2020, Dhahran 31261, Kingdom of Saudi Arabia

Edited by Yan-Hua Sun

systems (Kok and Alikaya 2003; Zhao et al. 2009; Lan et al. 2010). However, problems such as complicated treating chemical synthesis technology, the lack of raw material for treatment agents, and high initial cost have limited the development of the DF (Li et al. 2014). The application of DF for environmental protection is limited in oil resources exploration as the treating chemicals from natural macromolecular materials are often of poor quality. Synthetic and mineral oils are used in oil-based DF systems to reduce the environmental impact on the surrounding localities and the habitats. Earlier, little attention was paid to conserving the initial environmental conditions at less environmentally sensitive areas for onshore operations. However, later this delay brought the realization of negative environmental impact from DF additives such as chemicals, polymers, salt water, and oil-based fluids. Minimization of the environmental impact as well as safety considerations of a drilling operation directly affects the choice of DF additive systems. Due to the environmental regulatory agencies, products that have been used in the past may no longer be acceptable. As more environmental laws are enacted and new safety rules are applied, the choices of additives and fluid systems must also be reevaluated. To meet the challenges of a changing environment, product knowledge and product testing become essential tools for selecting suitable additives and DF systems.

There are many factors that are to be weighed when choosing a DF. However, the key considerations are well design, anticipated formation pressures and rock mechanics, formation chemistry, the degree of damage the DF imparts to the formation, temperature, environmental effects and regulations, logistics, and economics. To meet these key design factors, DFs offer a complex array of interrelated properties. Five basic properties are usually defined by the well program and monitored during drilling. These properties are listed as viscosity, density, filter cake or filtration of water loss, solids content, and quality of water make up. Once the properties and their parameters are determined, the DF can be controlled and adjusted accordingly.

2 Natural elements as additives

2.1 The need for natural substitutes

Working with DFs can be dangerous as some DF ingredients emit noxious or hazardous vapors that may reach levels that exceed the maximum recommended short-term or long-term safe exposure limits. Some shale and corrosion inhibitors and some emulsifiers in oil-based drilling fluids tend to produce ammonia or other lethal volatile

amines, particularly in hot areas on a rig. Other products are flammable or combustible (flash point <140 °F) and must be handled with caution. Various mud products such as brines, cleaning agents, solvents, and base oils commonly found on drilling rigs are irritating or even hazardous to body tissues. Perilous effects of additives such as defoamers, descalers, thinners, viscosifiers, lubricants, stabilizers, surfactants, and corrosion inhibitors on marine life and human life have been reported by several authors (Becket et al. 1976; Miller and Pesaran 1980; Younkin and Johnson 1980; Murphy and Kehew 1984; Candler et al. 1992; Ameille et al. 1995; Greaves et al. 1997). These effects range from minor physiological changes to reduced fertility and higher mortality rates. Therefore, it is very important to replace toxic ingredients from conventional DFs by a truly nontoxic natural substitute. In addition, the current trend in the DF development is to come up with novel environmentally friendly DFs that will rival the present day DFs in terms of reduced toxicity levels, performance, efficiency, and cost (Apaleke et al. 2012). Several researchers proposed substitutes which give better or at least the same level of results as their toxic counterparts (see Table 1). As a result, these materials have become vital ingredients for the DF. Table 1 shows a list of these natural elements used as additives during the formulation of a DF.

2.2 Grass

Grass is the principal fodder for cattle across the globe, and its use is known to humankind for centuries. The preamble of this research is to introduce grass as an environment friendly additive in the DF.

As stated earlier, in its quest to explore hydrocarbons, the drilling industry today uses a lot of chemically toxic additives for the formulation of DFs. This leads the Environmental Protection Agencies (EPA) to closely monitor the operations of the oil and gas industry for the usage of such fluids (with high toxicity) subjecting the industry to strict environmental legislations. The objective of this research is to introduce a naturally available material (powdered grass) with low or no cost as a suitable substitute to the toxic additives used to formulate a DF. This initiative of using such a material could help in reducing the environmental concerns and improving the work environment of people involved daily in this business.

3 Particle size distribution and compositional analysis

3.1 Particle size distribution

Particle size distribution is extensively used by geologists in geomorphological studies to evaluate sedimentation and

Table 1 Use of natural elements as DF additives

Inventor	Material	Function
Morris (1962)	Ground peach seeds	Filtration control agent
Lummus and Ryals (1971)	Ground nut shells and nut flour	Filtration control agent
Burts and Boyce (2001)	Corn cob outers	Filtration control agent
Nestle (1952)	Tree bark (douglas fir)	Filtration control agent
Sampey (2006)	Sugar cane ash	Filtration control agent
Green (1987)	Ground cocoa bean shells	Lost circulation material
Burts (1997)	Rice fractions (rice hulls, rice tips, rice straw and rice bran)	Lost circulation material
Ghassemzadeh (2011)	Fibers	Lost circulation material
Cremeans (2003)	Cotton seed hulls	Lost circulation material
Macquiod and Skodack (2004)	Coconut coir	Lost circulation material
Sharma and Mehto (2004)	Tamarind gum	Viscosifier

alluvial processes and by civil engineers to evaluate materials used for foundations, road fill, and other construction purposes. In the oil and gas industry, analysis of particle size distribution is used to determine filtration loss properties, and the amount of solids retained in the DF after the fluid is pumped into the system. A DF containing particles of sizes ranging up to the requisite maximum should be able to effectively bridge the formation and form a filter cake (in the case of a water-based drilling fluid). Above 10 Darcys or in fractures, larger particles are required, and most likely the amounts needed to minimize spurt losses increase with the size of the openings. In general, with the increasing concentration of bridging particles, bridging occurs faster, and spurt loss declines (Barrett et al. 2005; Growcock and Harvey 2005). Filtrate invasion into the formation can substantially reduce the permeability of the near wellbore region either by particle plugging, clay swelling, or water blocking. Permeability of the filter cake is dependent on the particle size distribution as an increase in the particle size decreases the permeability due to the fact that colloidal particles get packed very tightly. For non-reservoir applications, enough particles of the required size range are usually present in most DFs after cutting just a few feet of rock. These particles impact the choice of various drilling equipment (i.e., shale shakers, desanders, desilters, etc.) at the surface and thus can be effectively designed by having a prior knowledge of the particle sizing in the drilling fluid (Wajheeuddin and Hossain 2014).

The literature shows that DF properties (plastic viscosity (PV), yield point (YP), and gel strength (GS)) affect the rate of penetration (ROP) drastically because the presence of unremoved drill solids can cause a phenomenon known as the chip hold down effect, which increases both the equivalent circulation density (ECD) and the differential pressure causing a decrease in the ROP. For instance, it is

an established fact that the PV is influenced by the amount of colloidal particles present in the DF. Colloids present in the drilling fluid increase the fluid viscosity, which reduces the mobility of the cuttings as these cuttings stick to the bottom, requiring a re-drilling operation which severely affects the bit life. However, it is inferred that although DF properties affect the ROP, their net effect may not be as significant as it is thought to be except for the annular pressure losses in the laminar flow regime.

3.2 Use of X-ray fluorescence (XRF) in the petroleum industry

Commercial clays such as bentonite or other chemically treated clays are added to the DF for controlling rheological and filtration properties. The combined mix of commercial clays and drilled solids is called the "low-gravity solids" (LGS). Weighting materials (e.g. barite, barium sulfate, hematite etc.) are added to the fluid to make up the required density. This additive is necessary to densify mud and keeps the desired the hydrostatic pressure exerted by the DF in the drillpipe column and annulus. The concentration of these weighting materials is known as "high-gravity solids" (HGS). It is important for effective control of the properties of the fluid to know the individual concentrations of all types of solids (i.e., LGS and HGS). These entities are either measured directly or calculated from the density and volume fraction of solids in the DF, both of which can be measured but this is laborious. Traditionally, the LGS–HGS volume ratio is measured using a retort, a technique that requires good operator skills, takes at least 45 min, and has an error margin of 15 %.

The X-ray fluorescence (XRF), introduced into the oil and gas industry for analyzing core samples, is now deployed to monitor the concentrations and differentiate various solids types (LGS and HGS) in the DFs (Gilmour et al. 1996). XRF has the advantage of more frequent measurement, greater precision, and less dependence on operator skills. It is extensively used for the characterization of bentonite and other clay types for different clay applications. XRF is used to determine the elemental composition of additives to limit the usage of toxic chemicals in environmentally sensitive areas. For this purpose and due to the unavailability of the elemental composition of grass in the literature, the authors have taken the initiative to conduct XRF studies on the three said specimens.

4 Experimental methods and calculation procedure

4.1 Sample collection and preparation

Grass was collected from the Eastern Province of Saudi Arabia. The sample was dried in a sunny area for about a week, and then in a moisture extraction oven. The obtained grass was then pulverized in a grinding machine to obtain the desired grass samples.

4.2 Particle size distribution of the grass sample

Figure 1 shows a normal distribution curve of the particle size of the grass sample. Sieve sizes of 300, 180, 125, 90, 75 μm, and a no-sieve pan were used. The highest percentage of weight retained was on the 180-μm sieve, which indicates that the maximum of particles of the grass sample belonged to the medium category of particle size classification. The frequency distribution curve (Fig. 2) of the grass sample shows that at and above 50 % cumulative

weight, the sample consisted of fine particles with 6 % of the sample retained on the pan (finest fraction). In order to determine the average particle size of the finest fraction, a laser particle size analyzer (PSA) was used with three attempts of measurements. The particle size is plotted on the X-axis of Fig. 3, while the normal and frequency distributions are plotted on the Y-axis (right and left of the Y-axis, respectively). The test reveals the average particle size of the finest fraction of grass at 50 % net weight as 35 μm, thus implying that this grass sample (comprising various particle sizes) is a suitable candidate to be tested for use as an additive in the DF.

XRF analysis of the finest grass sample reveals calcium, potassium, and chlorine as the highest contributors by net normal weight percentage. Sulfur, silicon, iron, phosphorous, and manganese are also found in this specimen as small traces. Table 2 illustrates the elements present in the grass sample. Compounds of calcium are used as bridging and weighing agents in the DF. Calcium carbonate ($CaCO_3$) is used as an inhibitor to control active shale, and calcium chloride ($CaCl_2$) is used as a clay dispersion additive. Potassium compounds are used in the DF as alkalinity control agents (potassium chloride, KCl), acidity control agent (potassium hydroxide, KOH), and weighing agents (potassium formate, $CHKO_2$). Compounds of chlorine are used as disinfectants to clean surface pipes as it is used with source materials in the form of sodium hypochlorite and calcium hypochlorite. It is also used as a polymer oxidizer for drilling, completion, and work-over clean up in the form of chlorine bleach. Silica is used to exhibit various functions in the DF: it is added to a drilling

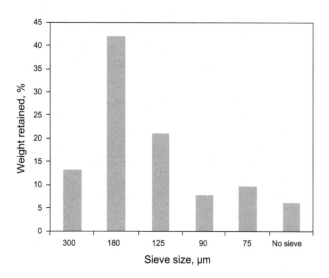

Fig. 1 Normal distribution curve of the grass sample on various sieves

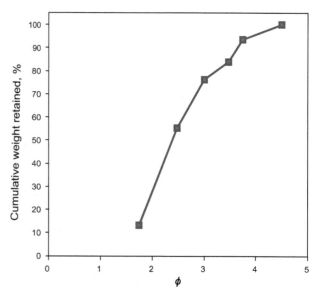

Fig. 2 Frequency distribution curve of the grass sample (where ϕ is a dimensionless unit of the sieve sizes and is defined as $\phi = -\log_{10}d/\log_{10}2$

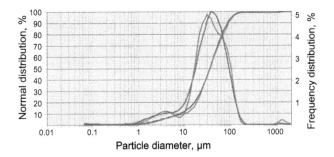

Fig. 3 Particle size distribution of the finest grass fraction obtained using a particle size analyzer

Table 2 XRF analysis of grass

Element	Net normal weight, %
Calcium (Ca)	53.80
Potassium (K)	19.83
Chlorine (Cl)	15.54
Sulfur (S)	3.89
Silicon (Si)	3.13
Iron (Fe)	2.46
Phosphorous (P)	1.24
Manganese (Mn)	0.12

fluid to change density, ionic strength, charges, etc. that are needed for special functions of DF such as drill-bit cooling, bit cleaning, effective cuttings removal to surface, downhole pressure control, and shale stabilization. Similarly, the use of silicate drilling fluids offers the advantages of prevention of bit-balling, differential sticking, and lost circulation and, in addition, promotes corrosion inhibition. Phosphoric acid is used to reduce the pH of the drilling fluid which is done conventionally.

The intention of mentioning the said compounds is to highlight the principal elements (K, Ca, Cl, Si, etc.). It is expected that the presence of these elements may contribute to mimic the performance of their toxic counterparts in an eco-friendly manner as grass is organic in nature. Moreover, readers may argue that grass is composed of lignin which itself is structured with C, H, and O. A separate SEM–EDX study conducted revealed that grass comprises of 95 % of these elements combined, and hence, were ignored from analysis (as an option present within the software of the XRF machine) as the authors focused on the applicability of other elements (as discussed in the previous paragraph) found in the grass sample.

4.3 Composition of the developed drilling fluid systems

Table 3 shows the compositions of the DF systems developed. The use of grass as an additive for DFs is

unknown to the industry. Hence, the formulations are kept simple with water, bentonite, and grass (in varying concentrations) to study the effect of grass in the DF. The bentonite formulation was kept under agitation for 24 h so as to achieve a homogenized suspension and allow bentonite to swell to its capacity. Also, it is noteworthy to mention that the viscosities and yield point obtained can be normalized using barite.

5 Results and discussion

5.1 Rheological properties of bentonite drilling fluids added with 300-μm grass particles

This section presented here shows the rheological profile of grass of a particle size of 300 μm. All DF systems show good dial readings with values increasing progressively from 3 rpm dial speed to 600 rpm. Figure 4a shows the consistency curves for all concentrations of grass. All these curves are in good agreement with the Bingham plastic model, and it is observed that the shear stress increased with the concentration of grass at a given shear rate. It is seen that the apparent viscosity gradually increased as the concentration of grass increased in the DF system, whereas the PV remained constant after the initial concentration of 0.25 ppb (Fig. 4b). This is practically good as a DF with higher PV increases the ECD, surge, and swab pressures and also reduces the ROP with chances of differential sticking. Figure 4c indicates that the yield point remained constant at lower concentrations, and increased gradually as the concentration of grass increased in the DF system. It is a known fact that a high yield point fluid has more significance as it indicates better cutting carrying capacity. As observed in Fig. 4d, the initial and final gel strengths are found to be increasing gradually, which indicates better suspension of cuttings in the DF. Moreover, it has been observed from experience that high gel strength values are not sought as this requires high pumping pressure once drilling is resumed after a period of shut down.

5.2 Filtration properties of bentonite drilling fluid added with 300-μm grass particles

Filtration is an important phenomenon seen in the wellbore due to pressure exerted by the hydrostatic column of the drilling fluid. Due to a pressure differential, a mud cake or filter cake with very low permeability is formed on the walls of the borehole which acts as a barrier between the formation and the drilled bore. The amount of filtrate loss to the formation is also essential as a DF with greater filtrate loss will exhibit higher density due to reduction in the

Table 3 DF types used in this research

Sample	Size of grass particles, μm	Additives	Amount	Fluid weight, ppg
Sample 1	–	Water + bentonite	Water: 350 mL; bentonite: 22.5 g	8.6
Sample 2	300	Water + bentonite + grass	Water: 350 mL; bentonite: 22.5 g; grass: 0.25, 0.50, 0.75, or 1.0 g	8.6
Sample 3	90	Water + bentonite + grass	Water: 350 mL; bentonite: 22.5 g; grass: 0.25, 0.50, 0.75, or 1.0 g	8.6
Sample 4	35	Water + bentonite + grass	Water: 350 mL; bentonite: 22.5 g; grass: 0.25, 0.50, 0.75, or 1.0 g	8.6

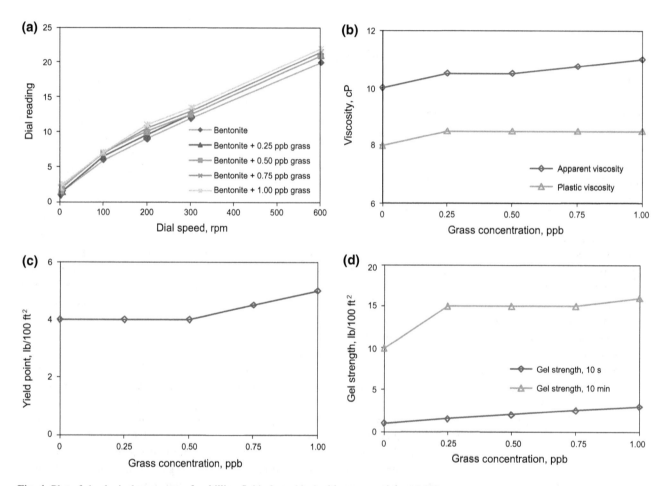

Fig. 4 Plot of rheological parameters for drilling fluids formulated with grass particles of 300 μm

water content of the fluid. Moreover, this creates a zone of damage near the well bore region and is one of the factors considered for formation damage. Figure 5 shows the trend of the filtrate loss of the drilling fluid formulated using the 300-μm sample. The filtration properties exhibit a decrease in the filtrate loss to a maximum of 24.7 % as the concentration of grass increased in the drilling fluid. This ensures that a firm filter cake is formed and a lesser amount of filtrate invades the formation which is an important property of a drilling fluid.

5.3 Selection of optimal concentration for 300-μm grass particles

Figure 6 shows all rheological properties and filtration characteristics for drilling fluids formulated with 300-μm grass particles. This is done in order to find out a concentration where all rheological properties (PV, YP, and GS) as well as the filtration characteristics are reasonable. It is concluded from Fig. 6 that the optimal concentration of grass particles of 300 μm is 0.75 ppb.

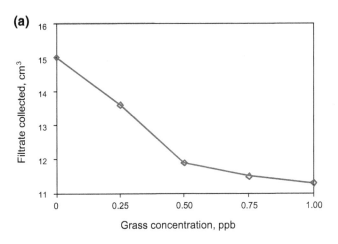
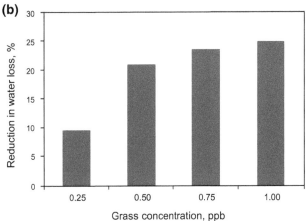

Fig. 5 Filtration characteristics of drilling fluids formulated with grass particles of 300 μm

5.4 Rheological properties of bentonite drilling fluid added with 90-μm grass particles

This section here shows the rheological profiles of drilling fluids containing grass particles of 90 μm. All DF systems show good dial readings with values increasing progressively from 3 rpm dial speed to 600 rpm. Again it is seen in Fig. 7a that the consistency curves confirm to the Bingham plastic model where shear stresses increased as a function of the shear rate. As observed in Fig. 7b, the viscosities increased as the concentration of grass increased in the DF system. As stated earlier, a DF with higher PV increases the ECD, surge, and swab pressures, and also reduces the ROP with chances of differential sticking. From Fig. 7c, it is clearly evident that the grass particles (at this particular particle size) did not contribute enough to impart high yield points as this defines the cutting carrying

ability of the DF. The gel strengths (Fig. 7d) are found to be increasing progressively which indicates that this drilling fluid had good cuttings suspension quality.

5.5 Filtration properties of bentonite drilling fluid added with 90-μm grass particles

Figure 8 shows filtration characteristics of the drilling fluid formulated using 90-μm grass particles. Figure 8a illustrates that as the grass particles were added to the bentonite DF system, the filtration characteristics of the drilling fluid improved as evident. The reduction in water loss observed was 23.3 % at a concentration of 1.0 ppb (Fig. 8b).

5.6 Selection of optimal concentration for 90-μm grass particles

Figure 9 shows all rheological properties and filtration characteristics of the bentonite drilling fluid containing 90-μm grass particles. The optimal concentration of grass particles was selected where all rheological properties (PV, YP, and GS) as well as the filtration characteristics are rational. It is concluded that the optimal concentration of grass particles of 90 μm is 1.0 ppb (Fig. 9).

5.7 Rheological properties of bentonite drilling fluid added with 35-μm grass particles

This section here presents the rheological profile of bentonite drilling fluid containing grass particles of 35 μm. All DF systems show good dial readings with values increasing gradually from 3 rpm dial speed to 600 rpm. Figure 10a shows the consistency curves for all concentrations of grass. All these curves are in good agreement with the Bingham plastic model, and it is observed that the shear stress increased with the concentration of grass at a given

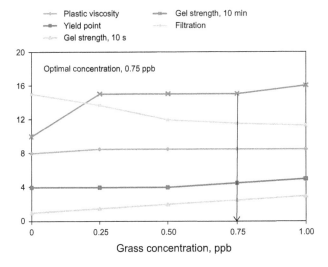

Fig. 6 Selecting the optimal concentration for 300-μm grass particles

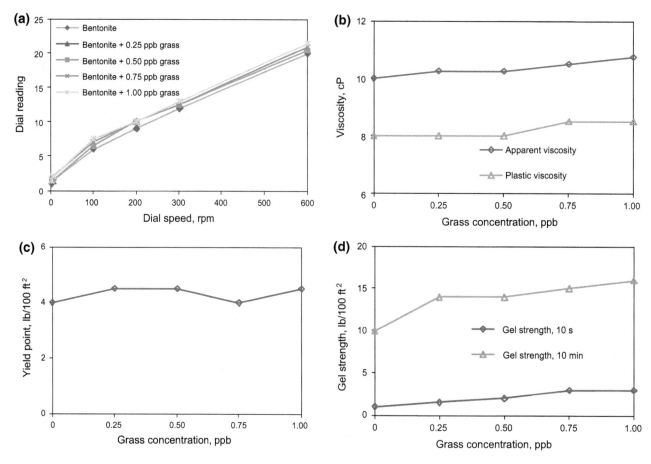

Fig. 7 Plot of rheological parameters for drilling fluids formulated with grass particles of 90 μm

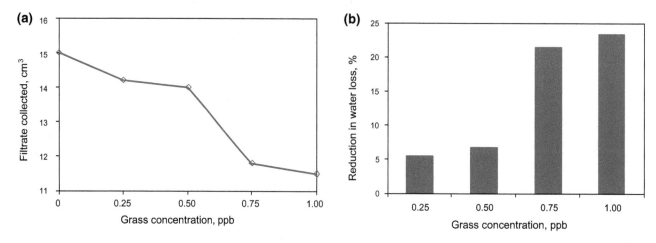

Fig. 8 Filtration characteristics of drilling fluids formulated with grass particles of 90 μm

shear rate. It is seen from Fig. 10b, c that the viscosities and the yield point gradually increased as the concentration of grass increased in the DF system. A DF with higher PV increases the ECD, surge, and swab pressures and also reduces the ROP with chances of differential sticking. It is

known that a high yield point fluid has more practical significance as it indicates better cutting carrying capacity. As observed in Fig. 10d, the initial and final gel strengths are found to be increasing gradually, which indicates better suspension of cuttings in the DF.

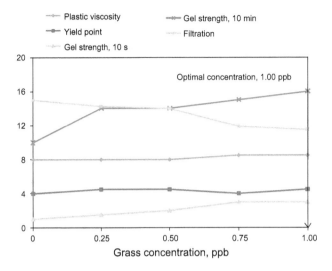

Fig. 9 Selecting the optimal concentration for 90-μm grass particles

5.8 Filtration properties of bentonite drilling fluids added with 35-μm grass particles

Fluid loss is a common occurrence in drilling operations. DFs are designed to seal porous formations intentionally

while drilling, by the creation of a mud cake. However, some part of the fluid is lost through the mud cake, and thus, fluid loss control additives are required. In this section, the fluid loss characteristics of grass drilling fluids are depicted. The filtration properties for the drilling fluid added with different concentrations of grass are shown in Fig. 11. It is seen in Fig. 11a that as the grass was introduced into the drilling fluid, filtration was controlled as evident by the decreasing trend. A reduction in water loss of 19.3 % was observed at a concentration of 1.0 ppb which is the least at this particle size (35 μm).

5.9 Selection of optimal concentration for 35-μm grass particles

Figure 12 shows all rheological properties and filtration characteristics for the drilling fluids formulated with 35-μm grass particles. The optimal value of grass concentration is selected where all rheological properties (PV, GS, and GS) as well as the filtration characteristics are reasonable. It is concluded Fig. 12 that the optimal concentration of grass particles of 35 μm is 0.75 ppb. Here, 1.0 ppb is not selected as the optimal concentration as in previous cases because of the gel

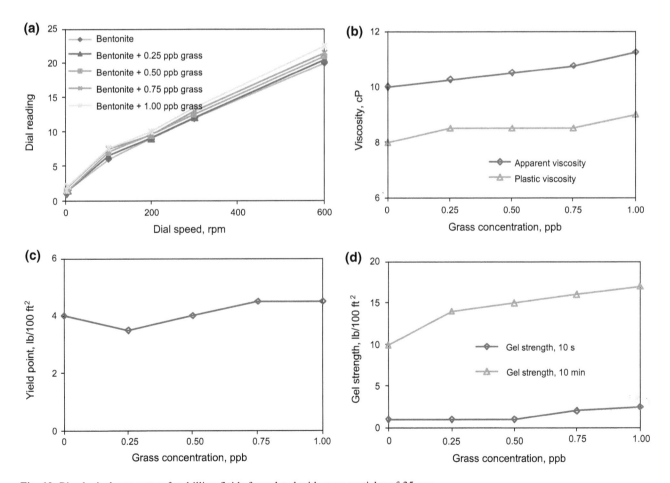

Fig. 10 Rheological parameters for drilling fluids formulated with grass particles of 35 μm

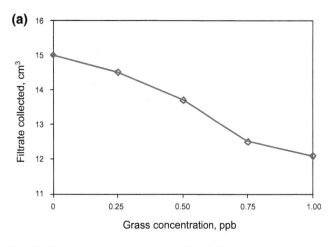
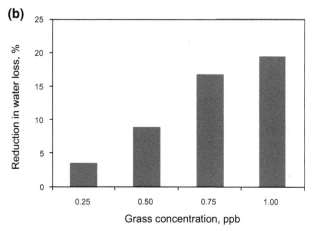

Fig. 11 Filtration characteristics of drilling fluids formulated with grass of 35 μm

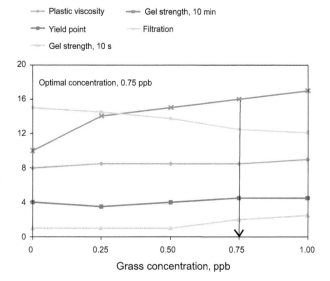

Fig. 12 Selecting the optimal concentration for 35-μm grass particles

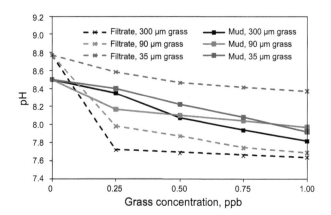

Fig. 13 Effect of the concentration of grass particles on pH of bentonite drilling fluids

strength which is significantly high at the highest concentration as this would require high pump pressures for recirculation in case of pump shut down during fishing operations.

5.10 Effect of grass on the pH of the drilling fluid

pH is a measure of the hydrogen ion concentration of a solution. Figure 13 illustrates the trend of pH followed by the drilling fluid with added grass particles. It can be inferred from the plot that as the grass particles were added to the bentonite drilling fluid, the pH of the filtrate, as well as the drilling fluid decreased (solution tends to become acidic). It is well-known that the drilling fluid gets contaminated through its various trips into the borehole which either increases or decreases its pH. Owing to the experiments conducted, we proposed to use grass as a greener alterative to lower the pH of a contaminated drilling fluid whose pH has been raised to an unacceptable level.

Potassium hydroxide, sodium hydroxide, calcium hydroxide, and magnesium hydroxide are commonly used as alkalinity and acidity control agents in DFs. These chemicals are declared as very hazardous in case of skin contact, eye contact, ingestion, and inhalation. An alternative solution would be to use grass as it modifies the pH of the drilling fluid and is environmentally friendly imparting no ill effects on the health of personnel who are daily involved in this trade.

6 Comparison of the rheology of grass drilling fluids with different water-based drilling fluids

A comparison is made between the existing water-based DF systems and the newly formulated grass drilling fluid using data from Amoco Production Company available in an open source web link (Drilling Fluids Manual, Amoco Production Company. Accessed on February 24, 2015 at 8:15 PM. http://www.academia.edu/6348534/Drilling_Fluids_Manual). Table 4 is prepared based on the data available on different water-based DF types to compare the proposed grass drilling fluid. The parameters such as PV,

Table 4 Comparison between proposed grass drilling fluid and various water-based drilling fluid systems

Drilling fluid type	Density, ppg	Plastic viscosity, cP	Yield point, lb/100 ft^2	Gel strength, lb/100 ft^2		Filtrate, cm^3/30 min	Cost
				10 s	10 min		
Lignite/lignosulfonate muds* (deflocculated)	9	8–12	6–10	2–4	4–10	8–12	Moderate
Lime muds*	10	15–18	6–10	0–2	0–4	6–12	Low
Lime muds* (deflocculated)	9	9–12	2–20	0–5	1–20	8–12	Low
Gypsum muds*	9	12–15	6–10	2–4	8–12	8–12	Moderate
Brackish-water muds*	9	16	10–18	2–4	5–10	6–10	Moderate
KCl-polymer muds*	9–10	12–25	10–20	6–8	8–20	10–12	Moderate
KOH-lignite muds*	9	12–24	9–12	2–4	4–8	10–12	Moderate
KOH-lime muds*	9	10–12	8–12	4–6	6–10	6–9	Moderate
New grass mud	8.6	8–9	3.5–5	1–3	10–78	11–14.5	N/A

* Source-Amoco Production Company

N/A not available

YP, GS, and the filtrate loss are included for the comparison. Table 4 shows that different DF systems have different properties and show a clear contrast. The grass drilling fluid seems quite comparable with these drilling fluids. All drilling fluids are formulated using additives which include a viscosifier, a weighting agent, a filtration control agent, and an alkalinity control agent, whereas the proposed system comprises only a viscosifier and powdered grass. It is expected that the cost of formulating grass drilling fluid is very low solely based on experience as well as owing to the abundance of the source material, grass. However, no formal cost analysis is conducted.

7 Conclusions and recommendation

Grass was used as an additive for the formulation of an environmentally friendly DF with different particle sizes and concentrations. The results obtained show that grass added to the bentonite DF (all concentrations at various particle sizes) improved the rheological properties such as apparent and plastic viscosities and gel strength. The filtration characteristics of the bentonite drilling fluid also enhanced because lower filtration losses were observed for all the samples. Tests carried out on the pH indicated that the addition of grass decreased the pH of the drilling fluid. The obtained results can be summarized as below:

(1) Tests carried out on drilling fluids formulated using 300-μm grass particles exhibited a control in filtration loss of about 25 %. Significant increases in the viscosities, yield point, and gel strengths were also observed. The optimal concentration of grass particles was 0.75 ppb in the bentonite drilling fluid (at 300 μm).

(2) The formulation containing 90-μm grass particles revealed a 23 % decrease in the filtration loss.

Increases in the viscosities, yield point, and gel strengths were also significant. An optimal concentration of 1 ppb grass particles was suggested at this particle size.

(3) Formulations containing fine-sized grass particles, i.e., 35 μm, helped decrease the filtration loss to 19 %. The viscosities, yield point, and gel strengths also show some increments. The optimal concentration witnessed was at 0.75 ppb of grass material.

Grass is proposed as a rheology modifier, filtration control agent as well as an alkalinity control agent for a DF. We further recommend carrying out investigations with this additive at elevated temperatures to analyze its performance so that a strong decision can be made in favor of the proposed grass which can be a better choice to replace the current toxic chemicals. Also, it is highly encouraged to develop a cost analysis model so as to study the applicability of grass as an additive for a DF system.

Acknowledgments The authors would like to acknowledge the support provided by the Deanship of Scientific Research (DSR) at King Fahd University of Petroleum & Minerals (KFUPM) for funding this work through Project No. IN141008. The authors are also grateful for the support and guidance received from Saudi Aramco during the completion of the work.

References

Amanullah M. Shale drilling mud interactions. Ph.D. Thesis. University of London; 1993.

Amanullah M, Marsden JR, Shaw HF. An experimental study of the swelling behavior of mud rocks in the presence of drilling mud systems. Can J of Pet Technol. 1997;36:45–50. doi:10.2118/97-03-04.

Ameille J, Wild P, Choudart D, et al. Respiratory symptoms, ventilatory impairment, and bronchial reactivity in oil mist-exposed automobile workers. Am J Ind Med. 1995;27:247–56. doi:10.1002/ajim.4700270209.

Apaleke AS, Al-Majed AA, Hossain ME. State of the art and future trend of drilling fluid: an experimental study. In: SPE Latin America and Caribbean petroleum engineering conference, 16–18 April, Mexico City; 2012. doi:10.2118/153676-MS.

Barrett B, Bouse E, Carr B, DeWolfe B, et al. Drilling fluids processing handbook. Houston: Gulf Professional Publishing; 2005.

Becket A, Moore B, Weir RH. Acute toxicity of drilling fluid components to rainbow trout. In: Industry/government working group in disposal waste fluids form exploratory drilling in the Canadian North, Yellowknife, N.W.T. Canada. Department of the Environment, Environmental Protection Service. 1976; vol. 9: p. 88.

Burts Jr BD. Lost circulation material with rice fraction. US Patent 5599776. Assigned to M & D Inds Louisiana Inc.; 1997.

Burts Jr. BD, Boyce D. Well fluid additive, well fluid made therefrom, method of treating a well fluid, method of circulating a well fluid. US Patent 6323158. Assigned to Bottom Line Industries Inc.; 2001.

Caenn R, Darley HCH, Gray GR. Composition and properties of drilling and completion fluids. 6th ed. Houston: Gulf Professional Publishing; 2011.

Candler J, Leuterman AJJ, Wong SYL. Sources of mercury and cadmium in offshore drilling discharges. SPE Drill Eng. 1992;7(4):279–83. doi:10.2118/20462-PA.

Cremeans KS. Lost circulation material and method of use. US Patent 6630429 B1; 2003.

Ghassemzadeh J. Lost circulation material for oilfield use. US Patent 7923413 B2. Assigned to Schlumberger Technology Corporation; 2011.

Gilmour A, Houwen O, Sanders M. US Patent No. 5,519,214. Washington, DC: U.S. Patent and Trademark; 1996.

Greaves LA, Eisen EA, Smith TJ, et al. Respiratory health of automobile workers exposed to metal working fluid aerosols. Am J Ind Med. 1997;32(5):450–9. doi:10.1002/(SICI)1097-0274(199711).

Green PC. Use of ground, sized cocoa bean shells as a lost circulation material in drilling fluid. US Patent 4474665. Assigned to WR Grace & Co.; 1984.

Growcock F, Harvey T. Drilling fluids: Chapter 2 drilling fluids processing handbook. Houston: Gulf Professional Publishing; 2005.

Kok MV, Alikaya T. Rheological evaluation of polymers as drilling fluids. Pet Sci Technol. 2003;21:113–23. doi:10.1081/LFT-120016930.

Lan Q, Li GR, Zhang JH, et al. Applications of bio-enzymatic completion fluid in the tight sand gas reservoirs of Ordos Daniudi gas field. In: International oil conference and exhibition in China, 8 June, Beijing; 2010. doi:10.2118/130974-MS.

Li FX, Jiang GC, Wang ZK, Cui MR. Drilling fluid from natural vegetable gum. Pet Sci Technol. 2014;32(6):738–44. doi:10.1080/10916466.2011.605092.

Lummus JL, Ryals JN. Preventing loss of drilling fluid to drilled formations. US Patent 3629102. Assigned to Pan American Petroleum Corporation; 1971.

Macquiod M, Skodack D. Method for using coconut coir as a lost circulation material for well drilling. US Patent 2004/0129460 A1; 2004.

Miller RW, Pesaran P. Effects of drilling fluids on soils and plants, complete fluid mixtures. J Environ Qual. 1980;9(4):552–6.

Morris JR. Drilling fluid. US Patent 3042607; 1962.

Murphy EC, Kehew AE. The effects of oil and gas drilling fluids on shallow groundwater in western North Dakota. Bismarck: Geological Survey; 1984.

Nestle AC. Drilling fluid. US Patent 2601050. Assigned to Texaco Development Corporation; 1952.

Sampey JA. Sugar cane additive for filtration control in well working compositions. US Patent 7094737B1; 2006.

Sharma VP, Mehto V. Studies on less expensive environmentally safe polymers for development of water based drilling fluids. In: SPE Asia Pacific oil and gas conference & exhibition; 11–13 September, Adelaide; 2004. doi:10.2118/100903-MS.

Wajheeuddin M, Hossain ME. An experimental study on particle sizing of natural substitutes for drilling fluid applications. J Nat Sci Sustain Technol. 2014;8(2):259–72.

Younkin WE, Johnson DL. The impact of waste drilling fluids on soils and vegetation in Alberta. In: Proceedings of symposium, research on environmental fate and effects of drilling fluids and cuttings, Lake Buena Vista, 1980; pp. 21–4.

Zhao S, Yan J, Wang J, et al. Water-based drilling fluid technology for extended reach wells in Liuhua Oilfield, South China Sea. Pet Sci Technol. 2009;27:1854–65. doi:10.1080/10916460802626372.

15

Development and evaluation of an electropositive wellbore stabilizer with flexible adaptability for drilling strongly hydratable shales

Wei-An Huang[1,2] · Zheng-Song Qiu[1] · Ming-Lei Cui[3] · Xin Zhao[1] · Jun-Yi Liu[1] · Wei-Ji Wang[1]

Abstract In order to overcome serious instability problems in hydratable shale formations, a novel electropositive wellbore stabilizer (EPWS) was prepared by a new approach. It has good colloidal stability, particle size distribution, compatibility, sealing property, and flexible adaptability. A variety of methods including measurements of particle size, Zeta potential, colloidal stability, contact angle, shale stability index, shale dispersion, shale swelling, and plugging experiments were adopted to characterize the EPWS and evaluate its anti-sloughing capacity and flexible adaptability. Results show that the EPWS has advantages over the conventional wellbore stabilizer (ZX-3) in particle size distribution, colloidal stability, inhibition, compatibility, and flexible adaptability. The EPWS with an average particle size of 507 nm and an average Zeta potential of 54 mV could be stable for 147 days and be compatible with salt tolerant or positive charged additives, and it also exhibited preferable anti-sloughing performance to hydratable shales at 77, 100, and 120 °C, and better compatibility with sodium bentonite than ZX-3 and KCl. The EPWS can plug micro-fractures and pores by forming a tight external mud cake and an internal sealing belt to retard pressure transmission and prevent filtrate invasion, enhancing hydrophobicity of shale surfaces by adsorption to inhibit hydration. The EPWS with flexible adaptability to temperature for inhibition and sealing capacity is available for long open-hole sections during drilling.

Keywords Shale · Wellbore stabilizer · Colloidal stability · Plugging · Hydrophobic modification

1 Introduction

The preservation of the wellbore stability has fundamental importance during oil and gas well drilling. About 75 % of the side walls of oil and gas wellbores consist of shale and mud rocks, which are responsible for 90 % of the wellbore instability problems (Corrêa and Nascimento 2005; Jiang et al. 2014; Wang et al. 2012a, b; Huang et al. 2007). Several types of shales with a significant amount of clay minerals are very reactive once they come into contact with water, and they can cause serious wellbore instability during drilling, like hole shrinkage and hole sloughing and caving (Hisham 2006; Wang et al. 2012a, b; Zhang et al. 2013a, b). Wellbore stability has been studied for a long time using mainly two quite different approaches. Some authors consider the problem exclusively from the point of view of rock mechanics and others from the point of view of chemical interactions between shales and fluids (Wu et al. 1993; Corrêa and Nascimento 2005). Based on the second approach, a variety of water-based drilling fluids and non-aqueous fluids have been applied; meanwhile, many types of wellbore stabilizers have been developed. These wellbore stabilizers have been classified into several groups, including inorganic salts, formate, polymers with special shale affinity, asphalts, sugars and sugar derivatives, glycerol, polyalcohol, and silicates (Van Oort 2003; Guo et al. 2006; Khodja et al. 2010; Jiang et al. 2011; Zhong et al. 2011). Among of them, asphalts as

✉ Wei-An Huang
masterhuang1997@163.com

[1] School of Petroleum Engineering, China University of Petroleum, Qingdao 266580, Shandong, China

[2] School of Mechanical and Chemical Engineering, The University of Western Australia, Crawley 6009, Australia

[3] Shandong Shengli Vocational College, Dongying 257097, Shandong, China

Edited by Yan-Hua Sun

one type of cheap and effective stabilizer can improve wellbore stabilization mainly through plugging or sealing, and they have been divided into four varieties according to modification methods: natural asphalt (no modification, mined from deposit), oxidized asphalt (oxidation by air blown to heighten softening point), sulfonated bitumen (changed into asphalt sulfonate partly to enhance water solubility), and emulsified asphalt (oil in water emulsion to improve dispersion in the drilling fluid and adsorption on the rock surface) (Sharma and Wunderlich 1987; Wang et al. 2005). All types of emulsified asphalts are mostly produced by shearing and thinning a mixture of asphalt, water, and emulsifier using an emulsion machine or a colloid mill. Ionic surfactants are often used as emulsifier, so the emulsified asphalt droplets are mostly charged and possess an electric double layer structure, according to the Stern model (Xie et al. 2005; Fan et al. 2014). Cationic emulsified asphalts have taken the place of anionic ones in drilling fluids to promote inhibition and adsorption (Shi et al. 2003; Wang and Xia 2006). In this paper, a novel cationic emulsified asphalt was investigated to overcome serious instability problems during drilling hydratable shale formations. It had good colloidal stability, particle size distribution, compatibility, sealing property, and flexible adaptability.

2 Experimental

2.1 Materials

Asphalts used in this study are listed in Table 1. 0# diesel was purchased from the Ruida Chemical Co. Ltd., Jinan, China. Liquid protecting agent (sorbitanoleate) and emulsifier (stearyltrimethyl ammonium chloride) were purchased from the First Chemical Co., Ltd., Dongying, China. Cationic polymer was supplied by the Nur Chemical Co. Ltd., Dongying, China. Hydrochloric acid, calcium chloride, potassium chloride, calcium oxide, and sodium hydroxide were all of analytical grade and commercially available. Sodium bentonite was purchased from the Boyou Bentonite Group Co., Ltd., China. Calcium bentonite was obtained from the Weifang Huawei Bentonite Group Co., Ltd., China, following the American Petroleum Institute (API) standard. Shale samples were taken from the Santai oilfield, Xinjiang, China. ZX-3 (cationic emulsified asphalt), DYFT-1

(sulfonated asphalt), SDT-108 (amphoteric polymer), CPAM (cationic polyacrylamide), SD-101 (sulfonated phenolic resin), CXB-3 (sulfonated lignite), PMHA-2 (amphoteric polymer), NPAN (polyacrylonitrile ammonium), SMP-1 (sulfonated phenolic resin), SPNH (sulfonated lignite), XY-27 (polymer viscosity reducer), and SD-506 (liquid lubricant) were of technical grade sourced from the Drilling Technology Research Institute of the Western Drilling Engineering Company, Xinjiang, China. Sand filter disks were purchased from the Geological Science Research Institute of Shengli Oilfield, Dongying, China.

2.2 Preparation of an electropositive wellbore stabilizer (EPWS)

Preparation of an electropositive wellbore stabilizer involved the following steps: after the addition of 0# diesel (80 g) in a 500-mL three-neck flask, the diesel was heated to 115 °C using a thermostat oil bath, adding plant asphalt A (20 g), petroleum asphalt B (20 g), asphalt C (40 g), and asphalt D (40 g) successively while stirring at 120 rpm. The stirring speed was then adjusted to 500 rpm, and the mixture was heated to 124 °C. The temperature was lowered to 95 °C after asphalts were dissolved completely (about 30 min), then the liquid protecting agent (12 g), emulsifier (6 g), and calcium oxide (7 g) were added together with deionized water (55 g), stirring for 30 min at 95 °C. The concentrated hydrochloric acid (12 g) was added, stirring for 40 min at 95 °C; 20 wt% calcium chloride solution (45 g), and 1 wt% cationic polymer solution (33 g) were finally added to the mixture, stirring for 20 min at 95 °C. Then stirring was stopped, and the temperature was lowered to no more than 55 °C. An electropositive wellbore stabilizer (EPWS) was finally obtained with pour-out of the oil in water emulsion from the three-neck flask. A schematic view of the asphalt emulsion is shown in Fig. 1.

2.3 Characterization and evaluation

2.3.1 Zeta potential and particle size distribution measurements

Zeta potentials and particle size distributions of test solutions were measured with a Zetasizer 3000 (Malvern

Table 1 Asphalts and their softening points	Sample	Softening point, °C	Source
	Plant asphalt A	51.7	Wanshun Chemical Factory
	Petroleum asphalt B	78.4	BEFAR GROUP
	Asphalt C	105.5	Wantong Petrochemical Co., Ltd.
	Asphalt D	131.0	Wantong Petrochemical Co., Ltd.

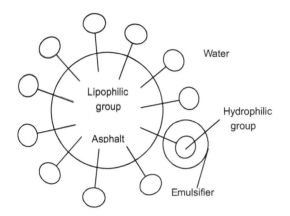

Fig. 1 Schematic view of the asphalt emulsion

Instrument, United Kingdom) (Fadaie et al. 2010). The equipment was turned on and preheated for 15 min. Then a diluted emulsion was added, and the particle size was measured. A total of 5 mL of the diluted emulsion were injected into the sample container, and the Zeta potential was determined.

2.3.2 Stability test

The stability of colloid, foam, and suspension was quantitatively evaluated with a Turbiscan stability analyzer (Formulation Company, France) (Wiśniewska 2010; Kang et al. 2011).

2.3.3 Fabric analysis

Some of shale fragments were first ground into powder. The diameter of some of them was less than 150 μm, and the others were no bigger than 58 μm.

X-ray diffraction (XRD) analysis was performed on 30 g of powdered shale sample (smaller than 58 μm) with an X-ray diffractometer (Rigaku D/max-IIIA, Japan) (Wang et al. 2009). A scanning electron microscope (Hitachi, S-4800) was used to observe the microstructure of the fresh appearance of several fragments (Klaver et al. 2012).

2.3.4 Shale dispersion

In this test, 50 g of shale cuttings (2–5 mm) was added to the test slurry in a conventional fluid cell. Then the fluid cell with cuttings was hot rolled in a conventional roller oven at different temperatures (77, 100, and 120 °C) for 16 h. After hot rolling, the cuttings were screened through a 1-mm size sieve and washed with tap water. After drying, the amount and percent recovered cuttings were determined (Zhong et al. 2012).

2.3.5 Shale swelling

A total mass of 10 g powders were added to a mold and compressed at 15 MPa for 5 min to prepare a core sample. The height of the core sample was measured and recorded as H. Then the mold was fixed onto a linear swell meter (NP-02A, Haitongda Company, China), the heights of the core sample were recorded 8 h after the test liquid had been added to the sample container.

2.3.6 Shale stability index

Shale stability index (SSI) values were determined to define the surface conditions of shale samples before and after exposure to test fluids. 180–250 μm shale particles were mounted in a holder and compressed. Then the hardness of shale samples was measured with a penetrometer before and after exposure to the test fluids.

2.3.7 Compatibility evaluation

The test slurry was a mixture (w/v) of sodium bentonite (2 %, 3 %, 4 %, 5 %, 6 %, or 7 %), 0.3 % NPAN, 0.3 % SDT-108, 3 % CXB-3, 3 % SD-101, 0.5 % $CaCl_2$, 1 % SD-506, 0.3 % XY-27, inhibitors (3 % KCl, 3 % ZX-3, 3 % EPWS separately), and deionized water. The mixture was stirred for 25 min at a high speed of 10,000 rpm and then aged for 24 h at room temperature. The base slurry was composed of 4 % (w/v) calcium bentonite, 3 % EPWS and fluid loss agents (1 % SDT-108, 1 % PMHA-2, 0.3 % CPAM, 0.3 % NPAN, 3 % SMP-1, 3 % SD-101, 3 % SPNH and 3 % CXB-3, respectively), and deionized water. This slurry was also stirred for 20 min at 10,000 rpm and then aged for 24 h at room temperature.

Aging experiments of muds were carried out in a XGRL-4 type rolling oven at 150 °C for 16 h. Mud properties were characterized before and after aging. According to API recommended practice of standard procedures (Recommended Practice 1988), the rheological parameters were calculated from 600 and 300 rpm readings, which were measured with a ZNN-D6 six type speed viscometer.

API filtration of each slurry was evaluated with a ZNS-2A filter press under a pressure of 689.5 kPa for 30 min.

2.4 Investigation of anti-sloughing mechanism

The particle size distribution of the test mud containing 4 % (w/v) sodium bentonite, 0.3 % NPAN, 0.3 % SDT-108, 3 % CXB-3, 3 % SD-101, 0.5 % $CaCl_2$, 1 % SD-506, 0.3 % XY-27, and inhibitors (3 % KCl, 3 % ZX-3, 3 % EPWS separately) was measured with a Bettersize 2000 laser particle size analyzer. Sealing performance of the test mud was evaluated using PPA-2A permeable plugging apparatus at

3.5 MPa for 30 min. The HTHP filtrate of each mud was tested using a GGS42-2 high pressure high temperature filter tester at a pressure difference of 3.5 MPa and different temperatures (70, 80, 90, 100, 110, 120, 130, 140, and 150 °C) for 30 min. The wettability of samples was determined from contact angle measurements by a JC2000D5M contact angle meter using plate-like shale cores.

3 Results and discussion

3.1 Characterization

Particle size distribution of ZX-3 and EPWS is presented in Fig. 2. The EPWS particles averaged 507 nm in diameter, ranging from 350 to 900 nm. Similar particle distribution was observed for cationic emulsified asphalt ZX-3, and the particles ranged from 700 to 1250 nm with an average of 844 nm. This indicates that EPWS was a more stable and better dispersing emulsion.

It can be seen from Fig. 3 that the Zeta potential of ZX-3 ranged from 25 to 70 mV with an average of 50 mV, while that of EPWS ranged from 30 to 75 mV and averaged 54 mV. The higher the positive Zeta potential of the emulsified asphalt, the stronger inhibition it would have by compressing diffuse double layers, adsorption onto the surfaces of clay and rock through electrostatic attraction, and worse compatibility with other additives and bentonite base slurry, because they are negatively charged, leading to poor stability and difficulty in controlling the viscosity and filtration. So the Zeta potential of EPWS was adjusted to about 55 mV to provide the stronger inhibition and the better compatibility at the same time.

The colloidal stability of the emulsified asphalt affects its storage and dispersion in drilling fluids. The colloidal stability of ZX-3 and EPWS is shown in Fig. 4. The transmitted and scattered light through the sample was recorded every 24 h. The test curves of ZX-3 and EPWS did not change with time extending for 90 days.

After the infrared scanning test, the samples were kept standing for a long time for further observation. No solid–liquid separation was observed for ZX-3 and EPWS after standing for 112 and 147 days, respectively.

Therefore, the EPWS exhibited excellent colloidal stability because the pre-dissolved cationic acrylamide with positive charges was added at the final stage after the oil in water emulsion had been formed (Li et al. 2012; Zhao et al. 2012).

3.2 Inhibition

3.2.1 Fabric and hydration properties of shale samples

Mineral composition (in wt%) of shale samples and the relative abundance of clay minerals (kaolinite, chlorite, illite, and smectite) are listed in Tables 2 and 3, separately. Rock samples from the Shuangfanggou Formation contained dominantly clay minerals (45.3 %) and quartz (28.7 %), followed by anorthoclase (15 %), potash feldspar (7 %), calcite (6 %), and hematite (4.3 %). Smectite was the main component of 4 samples (90 %–92 %), with an average of 92 %, followed by illite (3 %–5 %), kaolinite

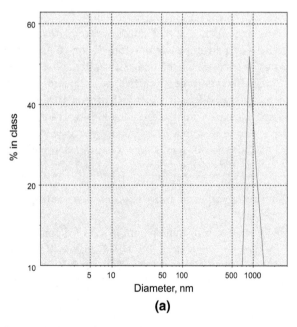

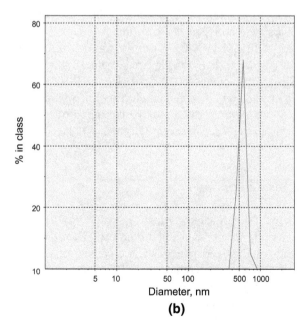

(a)   (b)

Fig. 2 Particle size distributions of **a** ZX-3 and **b** EPWS

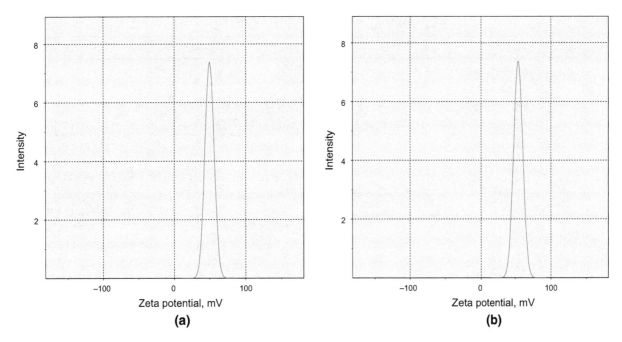

Fig. 3 Zeta potentials of **a** ZX-3 and **b** EPWS

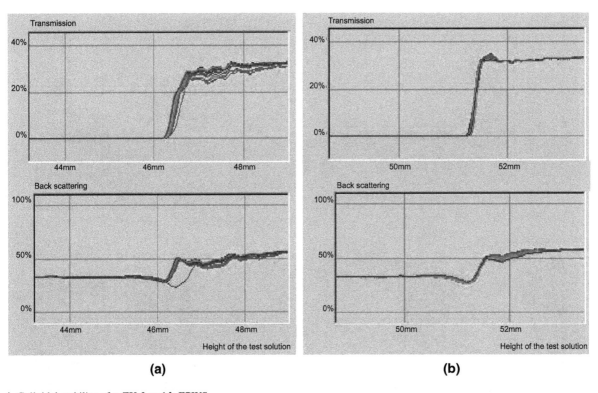

Fig. 4 Colloidal stability of **a** ZX-3 and **b** EPWS

(2 %–3 %) and chlorite (1 %–2 %). High proportions of smectite may lead to severe wellbore instability problems during drilling of this formation due to hydration, swelling, dispersion, and mudding of smectite. The shale sample from the Jiucaiyuan Formation contained only 30 % clay

minerals, but which is still mainly composed of smectite, making borehole wall collapse likely.

SEM images of the shale sample from well T_1s (2334–2337 m) are shown in Fig. 5. It can be seen from images that the microstructure of the shale was loose, and

Table 2 Mineral composition of shale samples determined by X-ray diffraction

Sample	Well name	Formation	Depth, m	Quartz, %	Potash feldspar, %	Anorthoclase, %	Calcite, %	Hematite, %	Clay, %
1	T_1s	Shuangfanggou	2053–2058	30	7	16		4	43
2	T_1s	Shuangfanggou	2248–2252	27	7	15	6	4	41
3	T_1s	Shuangfanggou	2334–2337	29		14		5	52
4	T_18	Jiucaiyuan	2396	36	7	21	6		30

Table 3 Relative abundance of clay minerals

Sample	Kaolinite, %	Chlorite, %	Illite, %	Smectite, %
1	2	2	4	92
2	2	1	3	94
3	3	2	5	90
4	4	2	4	90

micro-fractures and micro-pores were developed. The clay minerals were mixed with quartz crystal layers, filling intergranular pores. The micro-fractures and micro-pores supplied channels for invading filtrate and lost circulation of drilling fluids, leading to wellbore instability.

Recovery of shale samples from the Santai oilfield averaged 5.6 %, with a range from 4.3 % to 7.4 %. This indicates that the shale is easy to hydrate and disperse in water.

Linear swelling of shale samples was between 15.6 % and 18.1 % with an average of 17.3 %, indicating that the shale samples have high hydration capacity and swelling performance. The swelling was very high at the initial stage once the shale contacted water, and increased slowly after half an hour up to a final maximum at 8 h (Fig. 6). Therefore, the instability of shale formations in the Santai

(a)

(b)

(c)

(d)

Fig. 5 SEM images of the shale sample from well T_1s (2334–2337 m). **a** Outline; **b** Micro-fractures and pores; **c** Micro-fractures and pores; **d** Clay minerals mixed up with quartz

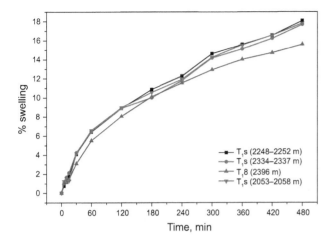

Fig. 6 Linear swelling of shale samples from the Santai oilfield

oilfield attributes to the micro-fractures and pores in shales, which provides enough channels and space for invasion of drilling fluids, and the strong hydration of shales. To prevent sloughing of this formation during operations, excellent inhibition and sealing characteristics of drill in fluids are required.

3.2.2 Comparative assessment of inhibition

As shown in Fig. 7, 10 % KCl solution, 3 % ZX-3 suspension, and 3 % EPWS suspension all exhibited better

capacity to inhibit swelling of shale powders and sodium bentonite compared to distilled water. The linear swelling of shale powders from well T_1s (2248–2252 m) within 8 h was reduced from 18.1 % to 15.4 % for 10 % KCl, 9.3 % for 3 % ZX-3 and 7.9 % for 3 % EPWS. The linear swelling of sodium bentonite was decreased from 38.4 % to 33.0 % by 10 % KCl, to 20.5 % by 3 % ZX-3 and to 17.9 % by 3 % EPWS. Moreover, the swelling of shale powders and sodium bentonite at the initial stage was clearly lowered by ZX-3 and EPWS. Among evaluated samples, the EPWS had the best capacity to inhibit hydration and swelling of shales.

Experimental results of the shale recovery and SSI measurements are presented in Fig. 8. The recovery and the SSI of samples in distilled water were very low, indicating easily hydration and dispersion in water and poor stability of the shale formation. The shale recovery and SSI reduced with an increase in temperature, suggesting that at high temperatures shale hydration would be promoted thus causing shale deterioration and wellbore instability. The recovery of shale particles was increased from 7.1 % to 29.9 % for 10 % KCl, to 35.7 % for 3 % ZX-3, and to 62.1 % for 3 % EPWS at 77 °C, while the SSI of shale powders rose from 24.1 to 40.9 mm for 10 % KCl, to 53.7 mm for 3 % ZX-3, and to 65.4 mm for 3 % EPWS. The recovery and SSI of shale samples were also increased significantly at 100 and 120 °C after adding KCl, ZX-3,

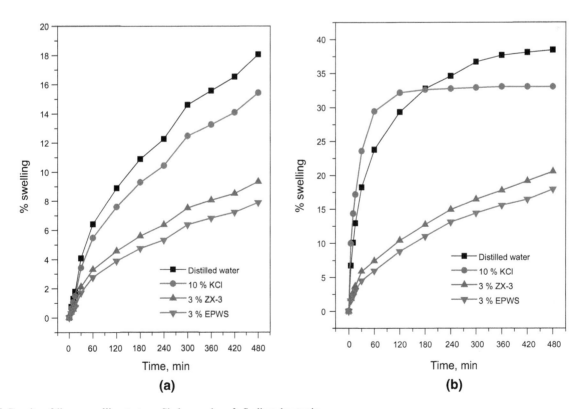

(a)   **(b)**

Fig. 7 Results of linear swelling tests. **a** Shale powders; **b** Sodium bentonite

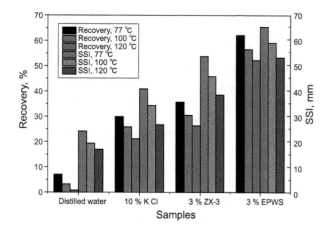

Fig. 8 Results of shale recovery and SSI measurements

and EPWS. The EPWS exhibited superlative performance in depressing hydration and improving wellbore stability at 77, 100, and 120 °C, corroborating its flexible adaptability to temperature.

3.3 Compatibility

Figure 9 reveals the effect of inhibitors on the plastic viscosity and filtration of drilling fluids containing different

amounts of sodium bentonite. Addition of 3 % KCl, 3 % ZX-3, or 3 % EPWS to the base slurry increased the plastic viscosity of the slurry in that they compressed the diffuse double layers of clay particles, leading to more face to edge connections in the system (Fig. 9a). The plastic viscosity of the slurry containing 3 % EPWS increased slowly as the content of sodium bentonite increased from 3 % to 7 %, showing its good compatibility with bentonite. The filtrate from the base slurry was smaller before adding 3 % KCl, 3 % ZX-3, or 3 % EPWS and reduced with an increase in the bentonite content (Fig. 9b). The effect of EPWS on filtration of the slurry containing different contents of sodium bentonite was lesser than ZX-3 and KCl, corroborating its preferable compatibility with bentonite and plugging property in drilling fluids.

It could be seen from Table 4 that SDT-108, CPAM, SD-101, and CXB-3 exhibited better ability to control fluid loss among the evaluated samples. SDT-108 is an amphoteric polymer with an intermediate molecular weight of about 1,000,000 and good salt tolerance. CPAM is one type of cationic polyacrylamide with a cationic degree of 5 %. SD-101 is a sulfonated phenolic resin of a high sulfonation degree, more than 25 %, and tolerant to saturated salt water. CXB-3 is a sulfonated lignite, and tolerant to salt. It can be concluded that the EPWS is compatible with

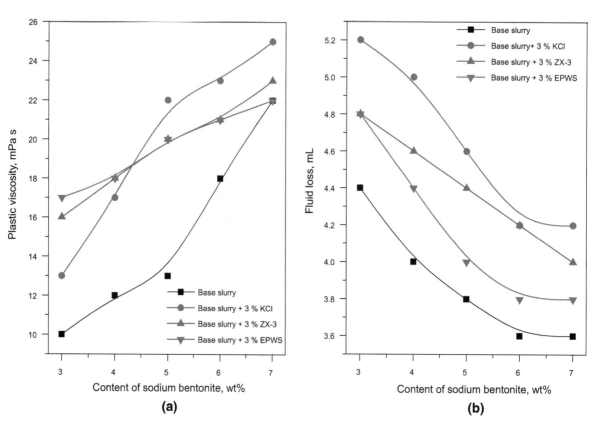

Fig. 9 Effect of inhibitors on intake capacity of drilling fluids. **a** Plastic viscosity; **b** Filtrate

Table 4 Compatibility evaluation of EPWS with fluid loss agents

Formula	Test condition	θ_6	θ_3	Apparent viscosity, mPa s	Plastic viscosity, mPa s	Yield point, Pa	Fluid loss, mL
Base slurry	Before aging	3	2.5	6	4	2	61
	After aging	5.5	5	6	3.5	2.5	96
+1 % SDT-108	Before aging	1	0.5	12.5	9	3.5	8.2
	After aging	1.5	1	19.2	17	2.2	9.2
+1 % PMHA-2	Before aging	12	11	41	18	23	7.6
	After aging	3	2	21.5	16.5	5	10.4
+0.3 % CPAM	Before aging	10	9	23.7	13	10.7	7.6
	After aging	5	4	20.7	13	7.7	12
+0.3 % NPAN	Before aging	1	0.5	3	2.5	0.5	17.2
	After aging	1	0.5	3.2	2.5	0.7	17.6
+3 % SMP-1	Before aging	1	0.5	5.2	5	0.2	8
	After aging	1	0.5	5.25	5	0.2	9.6
+3 % SD-101	Before aging	1	0.5	4.5	4	0.5	8.2
	After aging	1	0.5	5	5	0	9.2
+3 % SPNH	Before aging	1	0.5	3.7	3.5	0.2	8.4
	After aging	1	0.5	4.5	4	0.5	11.4
+3 % CXB-3	Before aging	1.5	1	5	3	2	7.8
	After aging	1	0.5	3.5	3	0.5	10.6

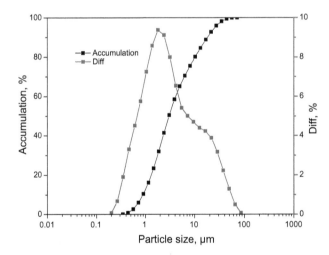

Fig. 10 Particle size distribution of the test slurry containing 3 % EPWS

those additives which are salt tolerant or positively charged.

3.4 Mechanism of stabilizing the borehole

3.4.1 Plugging

The particle size of the test slurry containing 3 % EPWS ranged from 0.34 to 71.5 μm, with an average diameter of 3.8 μm (Fig. 10). Other characteristic values included D_3

of 0.739 μm, D_6 of 0.914 μm, D_{10} of 1.135 μm, D_{16} of 1.468 μm, D_{25} of 1.975 μm, D_{75} of 10.03 μm, D_{84} of 16.12 μm, D_{90} of 22.80 μm. Therefore, the test slurry may penetrate and deposit in the pores and micro-fractures in the loose shale formations, whose micro-fracture and pore diameter ranges from 7.6 to 142 μm according the particle bridging rule. A sand filter disk with permeability of 457×10^{-3} μm^2, pore diameter of 6.4–125 μm was chosen to evaluate the sealing capacity of the EPWS (Fig. 11a). An external mud cake was formed at the sand filter disk surface during percolating and plugging, with a thickness of 100 μm (Fig. 11b). The thickness of the sealing belt near the external mud cake was about 130 μm, where the inner mud cake was formed. The compacted external mud cake and sealing belt may reduce and prevent the filtrate from invading into the deeper shale formations and retarding pressure transmission. Filtration of the test slurry containing 3 % DYFT-1, 3 % ZX-3, and 3 % EPWS separately was conducted at high temperatures and high pressures (HTHP), and the experimental results are presented in Fig. 12. The HTHP filtrate of the test slurry containing sulfonated asphalt DYFT-1 (softening point, 120 °C) decreased at the first half stage to a minimum value of 11 mL at 100 °C, then increased at the second half stage when the test temperature changed from 70 to 150 °C. When it comes to the slurries containing ZX-3, especially EPWS, the HTHP filtrate increased slowly with an increase in the test temperature.

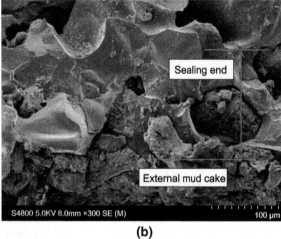

Fig. 11 SEM images of sand filter disk. **a** Before plugging; **b** After plugging

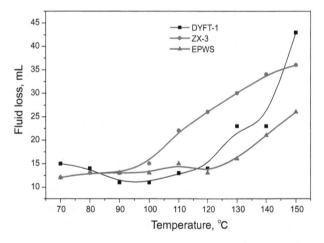

Fig. 12 Effect of temperature on the test slurries containing 3 % DYFT-1, ZX-3, and EPWS separately

The reason for this phenomenon lies in the sealing capacity of the sulfonated asphalt depends on its softening point and optimal plugging performance can be obtained at a temperature about 20 °C less than the softening point while that of the emulsified asphalt is affected by more factors such as the particle size of emulsion droplets, stability of the emulsion, and softening point of the asphalt used. The temperature generally increases with wellbore extension to deep formations, so the EPWS is more flexible than DYFT-1 and ZX-3 for plugging shale formations in long open-hole intervals.

3.4.2 Hydrophobic modification of shale surfaces

Contact angle results are illustrated in Fig. 13. It can be clearly seen that water spreads adequately on the original shale surface with a low contact angle of 7.6°. In contrast, the contact angles of the water droplet on the shale surface modified by suspensions containing DYFT-1, ZX-3, or EPWS were much higher, with the average contact angle of 26.3°, 38.6°, and 91.7°, respectively. Hydrophobic performance of the sulfonated asphalt mainly depends on its soluble parts, while that of the emulsified asphalt depends on more factors such as the emulsifier used, emulsion stability, asphalt properties, and the oil phase (diesel was used to prepare EPWS). The aqueous phase in the drilling fluid would be prevented from coming into contact with the shale surface after hydrophobic modification. Therefore, the EPWS could reduce filtrate invasion through modifying its surface wettability, thus inhibiting surface hydration of shales effectively.

4 Conclusions

1. Particle size distribution of cationic emulsified asphalt droplets was optimized by dissolving asphalt in diesel and emulsifying it with diesel and water together. Colloidal stability of cationic emulsified asphalt droplets was enhanced through increasing the viscosity of the emulsion using a cationic polymer solution. The Zeta potential of cationic emulsified asphalt droplets was adjusted to a suitable value by selecting the emulsifier and its dosage.
2. Newly synthesized EPWS with an average particle size of 507 nm and an average Zeta potential of 54 mV could be stable for 147 days, superior to the cationic emulsified asphalt ZX-3 prepared by the conventional method.

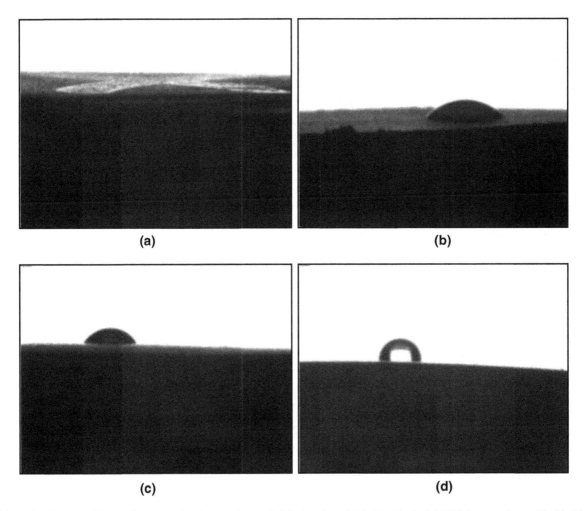

Fig. 13 Optical images of drops of water on the shale surface. **a** Original surface; **b** Modified by 3 % DYFT-1 suspension; **c** Modified by 3 % ZX-3 suspension; **d** Modified by 3 % EPWS suspension

3. The EPWS exhibited better flexible adaptability than ZX-3 and KCl to temperature at inhibition, sealing capacity, and preferable compatibility with sodium bentonite, salt tolerant, or positive charged additives.
4. The EPWS could plug micro-fractures and pores through forming a tight external mud cake and a sealing belt to retard pressure transmission and prevent filtrate invasion, enhancing hydrophobicity of the shale surface by adsorption to inhibit hydration.

Acknowledgments This work was financially supported by the National Science Foundation of China (No. 51374233), Shandong Province Science Foundation (No. ZR2013EEM032), the Fundamental Research Funds for the Central Universities (No. 13CX02044A) and the Project of China Scholarship Council (201306455021).

References

Corrêa CC, Nascimento RS. Study of shale-fluid interactions using thermogravimetry. J Therm Anal Calorim. 2005;79(2):295–8.

Fadaie S, Kashani MM, Maghsoudipour A, et al. Synthesis of yttria nanoparticles using NIPAM/AAc copolymer. Pigm Resin Technol. 2010;39(4):214–22.

Fan WY, Zhao PH, Kang JQ, et al. Application of molecular simulation technology to emulsified asphalt study. J China Univ Pet. 2014;38(6):179–85 (in Chinese).

Guo JK, Yan JN, Fan WW, et al. Applications of strongly inhibitive silicate-based drilling fluids in troublesome shale formations in Sudan. J Pet Sci Eng. 2006;50(3):195–203.

Hisham TE. Factors influencing determination of shale classification indices and their correlation to mechanical properties. Geotech Geol Eng. 2006;24(6):1695–713.

Huang WA, Qiu ZS, Xu JF, et al. Experimental study on sidewall instability mechanism of oil wells in the western Tuha Oilfield. Acta Pet Sin. 2007;28(3):116–20 (in Chinese).

Jiang GS, Ning FL, Zhang L, et al. Effect of agents on hydrate formation and low-temperature rheology of polyalcohol drilling fluid. J Earth Sci. 2011;22(5):652–7.

Jiang GC, Xuan Y, Li Y, et al. Inhibitive effect of potassium methylsiliconate on hydration swelling of montmorillonite. Colloid J. 2014;76(4):408–15.

Kang WL, Xu B, Wang YJ, et al. Stability mechanism of W/O crude oil emulsion stabilized by polymer and surfactant. Colloids Surf A. 2011;384(1):555–60.

Khodja M, Canselier JP, Bergaya F, et al. Shale problems and water-based drilling fluid optimization in the Hassi Messaoud Algerian oil field. Appl Clay Sci. 2010;49(4):383–93.

Klaver J, Desbois G, Urai J, et al. BIB-SEM study of the pore space morphology in early mature Posidonia Shale from the Hils area, Germany. International Journal of Coal Geology. 2012; 103:12–25.

Li HP, Zhao H, Liao KJ, et al. A study on the preparation and storage stability of modified emulsified asphalt. Pet Sci Technol. 2012;30(7):699–708.

Recommended Practice. Standard procedure for field testing drilling fluid. 12th ed. Recommended Practice, vol. 13. Washington, DC: API; 1988. pp 7–9.

Sharma MM, Wunderlich RW. The alteration of rock properties due to interactions with drilling-fluid components. J Pet Sci Eng. 1987;1(2):127–43.

Shi LS, Cao XX, Guo ZN, et al. Factors influencing the breaking time of cationic emulsifier bituminous mixture. J Shandong Univ (Eng Sci). 2003;33(1):97–100 (in Chinese).

Van Oort E. On the physical and chemical stability of shales. J Pet Sci Eng. 2003;38(3):213–35.

Wang DM, Xu YM, He DM, et al. Investigation of mineral composition of oil shale. Asia-Pac J Chem Eng. 2009;4(5): 691–7.

Wang Q, Zhou YC, Tang YL, et al. Analysis of effect factor in shale wellbore stability. Chin J Rock Mech Eng. 2012a;31(1):171–9 (in Chinese).

Wang Q, Zhou YC, Wang G, et al. A fluid-solid-chemistry coupling model for shale wellbore stability. Pet Explor Dev. 2012b; 39(4):508–13.

Wang SX, Xia SQ. The synthesize and performance research of a new kind of cationic bituminous emulsifier. Pet Asph. 2006;20(4): 30–4 (in Chinese).

Wang XY, Wei J, Li YQ. Study and application of new organic cationic gel drilling fluid. Drill Prod Technol. 2005;27(6):30–4 (in Chinese).

Wiśniewska M. Influences of polyacrylic acid adsorption and temperature on the alumina suspension stability. Powder Technol. 2010;198(2):258–66.

Wu TH, Randolph BW, Huang CS. Stability of shale embankments. J Geotech Eng. 1993;119(1):127–46.

Xie XW, Fu HQ, Hang H, et al. Synthesis of quaternary ammonium cationic emulsifier for asphalt. NianJie. 2005;26(2):39–41 (in Chinese).

Zhang SF, Qiu ZS, Huang WA, et al. A novel aluminum-based shale stabilizer. Pet Sci Technol. 2013a;31(12):1275–82.

Zhang SF, Qiu ZS, Huang WA, et al. Characterization of a novel aluminum-based shale stabilizer. J Pet Sci Eng. 2013b;103: 36–40.

Zhao H, Li HP, Liao KJ, et al. The anti-aging performance of emulsified asphalt. Pet Sci Technol. 2012;30(2):193–200.

Zhong HY, Qiu ZS, Huang WA, et al. Shale inhibitive properties of polyether diamine in water-based drilling fluid. J Pet Sci Eng. 2011;78(2):510–5.

Zhong HY, Qiu ZS, Huang WA, et al. Poly (oxypropylene)-amidoamine modified bentonite as potential shale inhibitor in water-based drilling fluids. Appl Clay Sci. 2012;67–68:36–43.

16

Forecasting volatility in oil prices with a class of nonlinear volatility models: smooth transition RBF and MLP neural networks augmented GARCH approach

Melike Bildirici[1] · Özgür Ersin[2]

Abstract In this study, the forecasting capabilities of a new class of nonlinear econometric models, namely, the LSTAR-LST-GARCH-RBF and MLP models are evaluated. The models are utilized to model and to forecast the daily returns of crude oil prices. Many financial time series are subjected to leptokurtic distribution, heavy tails, and nonlinear conditional volatility. This characteristic feature leads to deterioration in the forecast capabilities of traditional models such as the ARCH and GARCH models. According to the empirical findings, the oil prices and their daily returns could be classified as possessing nonlinearity in the conditional mean and conditional variance processes. Several model groups are evaluated: (i) the models proposed in the first group are the LSTAR-LST-GARCH models that are augmented with fractional integration and asymmetric power terms (FIGARCH, APGARCH, and FIAPGARCH); (ii) the models proposed in the second group are the LSTAR-LST-GARCH models further augmented with MLP and RBF type neural networks. The models are compared in terms of MSE, RMSE, and MAE criteria for in-sample and out-of-sample forecast capabilities. The results show that the LSTAR based and neural network augmented models provide important gains over the single-regime baseline GARCH models, followed by the LSTAR-LST-GARCH type models in terms of modeling and forecasting volatility in crude oil prices.

Keywords Volatility · Petrol prices · ARCH · STAR · Neural networks · LSTAR-LST-GARCH family

1 Introduction

The volatility of crude oil prices has received much attention recently because the crude oil is the most strategic and the most traded commodity in the world. Crude oil is traded internationally by many different players such as the oil producing nations, oil companies, individual refineries, oil importing nations, and speculators. Although crude oil price is basically determined by its supply and demand (Hagen 1994; Stevens 1995), it is also under the influence of many irregular events like stock levels, economic growth, political aspects, political instability, the decisions implemented by OPEC, and further psychological expectations of traders (Yu et al. 2008).

The volatility of oil prices is accepted to have important effects on economic activity. The fluctuations of the commodity market prices depend on the rise-and-fall of the oil price so that any sudden increase or decrease in oil prices cause economic slowdown and price fluctuations in other commodity prices. As a result, crude oil price forecasting is a very important field of research, and modeling/forecasting oil prices is hindered by its intrinsic difficulties such as the high volatility (Wang et al. 2005). As the crude oil spot price series are usually considered as a nonlinear and nonstationary time series, which is interactively affected by many factors, predicting crude oil price accurately is rather challenging (Yu et al. 2008).

Oil prices may not always adjust instantaneously to new information, on the other hand, low liquidity and infrequent trading in imperfect markets could lead to a delay in response to new information (McMillan and Speight 2006;

✉ Melike Bildirici
  bildiri@yildiz.edu.tr

[1] Department of Economics, Yıldız Technical University, Istanbul, Turkey

[2] Department of Economics, Beykent University, Istanbul, Turkey

Edited by Xiu-Qin Zhu

Monoyios and Sarno 2002; Lee et al. 2008). In this perspective, there is a significant literature focusing on improving the capabilities of econometric models to model oil prices. A fraction of the studies investigate the path followed by the oil prices by utilizing various GARCH models. In addition, many studies focus on the economic factors that have had strong impacts on the increasing volatility especially for the periods with regime changes. One point to be taken into consideration is the fact that regime changes caused by many economic factors decrease the forecast capabilities of the single-regime models drastically. Furthermore, economic factors have important effects on the business cycles by disturbing the processes followed by economic time series. As an example, the 1st and 2nd oil shocks in 1974 and 1979 had significant impacts on the performances of the econometric models.

Consequently the traditional volatility models which do not take into consideration the regime switching characteristics of the factors such as oil shocks became obsolete in modeling volatility in petrol prices.

In this paper, the volatility in oil prices is aimed to be investigated. In accordance with this purpose, the oil prices will be modeled by improving both the conditional mean and the conditional variance with nonlinear time series and neural network models to achieve possible gains in forecasting and modeling capabilities. The business cycles in the economies reveal different dynamics under different regimes so that a traditional GARCH model becomes insufficient once volatility is encountered. The motive behind the usage of LSTAR structure (one of the nonlinear time series modeling) is to improving forecasting and modeling power of the GARCH models for policy purposes. In addition to the expectation of augmenting the GARCH models with LSTAR models, augmenting these models with MLP and RBF type neural networks would likely to bring increase in the forecasting capabilities. With this respect, our study aims to incorporate regime switching and neural networks with GARCH models. With this purpose, GARCH structure will be incorporated with LSTAR and multi-layer perceptron (MLP) and radial basis function (RBF) models. These three approaches consider the characteristics of oil prices which exhibit strong regime changes in addition to nonlinear volatility. Accordingly, at the first step, ST-GARCH models will be extended to LSTAR-LST-GARCH. Afterwards, by incorporating fractional integration and asymmetric power properties, four models will be evaluated: the LSTAR-LST-GARCH, LSTAR-LST-FIGARCH, LSTAR-LST-APGARCH, and LSTAR-LST-FIAPGARCH. At the second step, models will be augmented with RBF and MLP type neural networks. Hence, the LSTAR type, nonlinearity is introduced in the conditional mean and conditional variance processes of neural networks models to obtain the LSTAR-LST-GARCH-MLP and LSTAR-LST-GARCH-RBF models. Similarly, models are augmented with asymmetric

power terms and fractional integration. As a result, at the second step, MLP and RBF neural networks are introduced to obtain LSTAR-LST-GARCH-MLP and LSTAR-LST-GARCH-RBF models. Following the asymmetric power and fractional integration augmentations, models are denoted as LSTAR-LST-GARCH-MLP and LSTAR-LST-GARCH-RBF, LSTAR-LST-APGARCH-MLP and LSTAR-LST-APGARCH-RBF, LSTAR-LST-FIGARCH-MLP and LSTAR-LST-FIGARCH-RBF, and LSTAR-LST-FIAPGARCH-MLP and LSTAR-LST-FIAPGARCH-RBF models.

A literature review is given in Part 2. Econometric methodology is given in Part 3 where both the newly proposed LSTAR-LST-GARCH family and neural network-based LSTAR-LST-GARCH-NN family of models are evaluated. Empirical application to oil prices is given in Part 4 and concludes in Part 5.

2 Literature review

Hamilton (1983, 1985) are among the early studies that drew attention on the relationship between energy prices and showed that oil prices have important effects on the economy. Barone et al. (1998) suggested a semi-parametric modeling technique for oil price forecasting. Further, Alvarez et al. (2008) showed in their research that the random walk-type behavior in energy futures prices, thus the autocorrelation in oil prices diminishes over time. Adrangi et al. (2001) tested the presence of low-dimensional chaotic structure in crude oil, heating oil, and unleaded gasoline futures prices with their sample starting by the early 1980s. In their study, they pointed at chaotic structure and high persistence in the series investigated and concluded that chaotic structure and high persistence in the data would create strong implications for regulators and short-term trading strategies. Ewing and Maliq (2013) employs univariate and bivariate GARCH models to examine the volatility of gold and oil futures and showed that incorporating structural breaks is important for empirical analysis focusing on oil prices.

Ye et al. (2002, 2005, 2006) defined an econometric model for evaluating WTI prices by using variables such as the OECD petroleum inventory levels, relative inventories, and high and low-inventory variables. Lanza et al. (2005) analyzed crude oil and oil products' prices by using error correction models.

Further, many papers demonstrated that the prediction performance might be very poor if the traditional statistical and econometric models such as linear regressions are employed (Weigend and Gershenfeld 1994). The main reason behind is the phenomenon that the traditional statistical and econometric models are built on linear assumptions, which, as a result, fail to capture the nonlinear patterns hidden in the crude oil price series (Yu et al. 2008). It is the fact that the oil prices may not always adjust instantaneously to the newly available information. Low liquidity and infrequent trading that occur under imperfect markets could cause to delays in response, following the availability of new information (McMillan and Speight 2006; Monoyios and Sarno 2002; Lee et al. 2008).

To overcome the difficulty in terms of forecast accuracy encountered in forecasting crude oil prices, Abramson and Finizza (1991) study is among the early studies that followed the neural network approach to model the crude oil market. Elsharkawy (1998) showed that radial basis function type neural network model had better forecast accuracy than the conventional methods in terms of predicting the oil formation volume factor, solution gas–oil ratio, oil viscosity, saturated oil density, under-saturated oil compressibility, and evolved gas gravity. Kaboudan (2001) showed that though both neural networks and genetic programming proved better results compared to the random walk to model oil prices, genetic programing provided superior results than the neural networks. Similarly, Mirmirani and Li (2004) showed that genetic algorithms and ANN models provided better results in forecasting oil prices. Tang and Hammoudeh (2002) derived a conclusion that shows the importance of both nonlinearity and heteroscedasticity in oil prices. They showed that by taking the GARCH effects into consideration, a nonlinear regression model with GARCH-type errors provide significant gains in modeling OPEC oil prices. Malliaris and Malliaris (2005) showed that the nonlinear models derived by the neural network provided superior forecasting in the majority of different oil types; namely for crude oil, heating oil, gasoline, and natural gas; with propane, the neural network gave the least accurate prediction. Yu et al. (2007) followed neural network modeling methods to forecast crude oil prices and showed that NN models provide significant gains in terms of different error criteria. Yu et al. (2007, 2008) proposed an efficient EMD-based neural network ensemble learning algorithm that uses feed-forward neural networks for modeling and forecasting world crude oil spot prices. Qunli et al. (2009) used radial basis functions (RBF) and showed that a RBF type neural network that benefits from wavelet transformations provided better results than the linear approaches in modeling monthly crude oil prices. Alizadeh and Mafinezhad (2010) used neural network models that utilized a predefined crisis variable to model and forecast Brent petrol prices and showed that the model is capable in forecasting both in normal and critical conditions. Bildirici and Ersin (2013) modeled the oil prices with newly introduced LST-LST-GARCH-MLP models.

3 Econometric methodology

Econometric modeling of volatility in the autoregressive conditional heteroscedasticity (ARCH) specification of conditional volatility gained significance especially following the important paper of Engle (1982). Further, the model is extended to generalized ARCH (GARCH) model by Bollerslev (1986), a model which has found many significant applications to capture the distributional aspects such as volatility clustering, heavy tails or excess kurtosis, and non-normal distribution. Additionally, the asymmetric power GARCH (APGARCH) model developed by Ding et al. (1983) further augments the model with power transformations without simple squared shocks and conditional variances as in the traditional GARCH models. Further, Baillie et al. (1996) and Bollerslev and Mikkelsen (1996) proposed the fractionally integrated GARCH (FIGARCH) model that encounters for the short-run dynamics of the conditional mean process modeled following ARMA process in the standard GARCH model. An important finding shown by Baillie et al. (1996) is that financial macroeconomic time series are subject to long memory characteristics such that volatility shows strong persistency. Chung (1999) and Conrad and Haag (2006) showed that the long-run persistence decays with hyperbolic rates. Tse and Tsui (1997) followed by Tse (1998) propose a combination of the FIGARCH and APGARCH model and the obtained FIAPGARCH model incorporates fractional integration with asymmetric power terms to capture the above-mentioned distributional aspects in financial time series.

Further, one of the expectations is that by integrating fractional integration with GARCH models, certain improvements in terms of forecasting accuracy and volatility modeling of oil prices could be achieved. Additionally, augmenting GARCH models with asymmetric power terms to obtain APGARCH models and merging fractional integration with APGARCH models to obtain FIAPGARCH models provide improvements in terms of volatility modeling. Considering the aspects such as intervention in oil prices and sudden changes in prices, FIGARCH and FIAPGARCH models provide important tools to analyze the finite persistence in oil prices though the models maintain assuming single-regime architectures without taking regime switches or regime transitions.

3.1 LSTAR type nonlinearity in the conditional mean and variance

To model nonlinearity in GARCH processes, Franses and van Dijk (2000) evaluated the smooth transition GARCH (ST-GARCH), where regime changes are governed with transition functions similar to the modeling and evaluation techniques of STAR models for the conditional mean processes developed by Terasvirta (1994). Lundberg and Terasvirta (1998) developed the STAR-ST-GARCH model that allows nonlinearity in both the conditional mean and the conditional variance processes of a time series. Chan and McAleer (2002, 2003) evaluated the statistical properties in context of estimation of STAR-GARCH models. In the study, we will allow models to follow STAR type nonlinearity both in the conditional mean and the conditional variance which are evaluated under LSTAR-LST-GARCH architecture. LSTAR-LST-GARCH models are LSTAR-LST-GARCH, LSTAR-LST-FIGARCH, LSTAR-LST-APGARCH, and LSTAR-LST-FIAPGARCH models and possess both ST-GARCH (Lundberg and Terasvirta 1998) and STAR-GARCH characteristics since both the conditional mean and the conditional variance are allowed to follow STAR type nonlinearity. ST-GARCH model shares similarities but have differences with the GJR-GARCH (Glosten et al. 1993) and TGARCH (Zakoian 1994) models in terms of the transition function since ST-GARCH models allow smooth transition functions instead of threshold function in defining regime changes.

Further, the artificial neural network ARCH process (ANN-GARCH) developed by Donaldson and Kamstra (1997) augments the GJR model with multi-layer perceptron-based neural network architecture with logistic squashing functions to capture nonlinearity by utilizing the universal approximation property (Cybenko 1989) of ANN models. In pursuit of these concepts, many papers developed neural network models. Lai and Wong (2001) contributed to the nonlinear time series modeling methodology by making use of single-layer neural networks; further, modeling of NN models for estimation and prediction for time series has important contributions governed by Weigend et al. (1991), Weigend and Gershenfeld (1994), White (1992), Hutchinson et al. (1994), Gencay and Liu (1997), Gencay and Stengos (1997, 1998) and Refenes et al. (1997) which contributed to financial analysis and stock market returns estimation, to pattern recognition and optimization. Dutta and Shekhar (1998) provided applications of neural networks for bond ratings. NN modeling methodology is applied successfully by Wang et al. (2005) for forecasting the value of a stock index. Bildirici and Ersin (2009) modeled NN-GARCH family models to forecast daily stock returns for short- and long-run horizons and they showed that GARCH models under NN architecture provide significant forecasting performance.

3.1.1 LSTAR-LST-GARCH model

Chan and McAleer (2002) discussed the STAR-GARCH model that has STAR type nonlinearity in the conditional

mean process. Franses and van Dijk (2000) discuss the STAR-STGARCH model that allows STAR type nonlinearity in both the conditional mean and the conditional variance and is developed based on the Terasvirta (1994) type LSTAR model with logistic transition function governing the dynamics of the transition between different regimes. In the paper, the LSTAR-LST-GARCH model will be extended to RBF type neural networks in addition to MLP type variants proposed in Bildirici and Ersin (2013).

At the first stage, the conditional mean process, y_t, is assumed to follow a LSTAR(p) process with two regimes as,

$$
y_t = \left(\theta_{1,0} + \sum_{i=1}^{r} \theta_{1,i} y_{t-i} \right) \times (1 - H(\varepsilon_{t-1}; \gamma, n))
$$
$$
+ \left(\theta_{2,0} + \sum_{i=1}^{r} \theta_{2i} y_{t-i} \right) \times H(\varepsilon_{t-1}; \gamma, n) + \varepsilon_t. \quad (1)
$$

The conditional variance follows a smooth transition LST-GARCH process,

$$
\sigma_t^2 = \left(w_{1,0} + \sum_{p=1}^{p} \alpha_{1,p} \varepsilon_{t-p}^2 + \sum_{q=1}^{q} \beta_{1,q} \sigma_{t-q}^2 \right)
$$
$$
\times (1 - H(\varepsilon_{t-1}; \gamma, n))
$$
$$
+ \left(w_{2,0} + \sum_{p=1}^{p} \alpha_{2,p} \varepsilon_{t-p}^2 + \sum_{q=1}^{q} \beta_{2,q} \sigma_{t-q}^2 \right)
$$
$$
\times H(\varepsilon_{t-1}; \gamma, n), \quad (2)
$$

where, the transition between regimes is defined with a logistic function,

$$
H(\varepsilon_{t-1}; \gamma, n) = \frac{1}{1 + e^{-\gamma(\varepsilon_{t-1} - n)}}. \quad (3)
$$

In the LSTAR-LST-GARCH model given in Eq.'s (1) and (2) with the transition function Eq. (3), the speed of transition function is determined by the estimate of the γ parameter and the n is the threshold parameter. The logistic function $H(\varepsilon_{t-1}; \gamma, n)$ is a twice differentiable continuous function bounded between [0, 1] lower and upper bounds for different values of the transition variable ε_{t-1} and its distance to the threshold n (see Bildirici and Ersin 2013). The transition is observed to be relatively slow for low values of γ, though the transition between regimes speeds up as γ takes larger values. The ARCH and GARCH parameter estimates $\alpha_{1,p}$, $\beta_{1,q}$ approach to $\alpha_{2,p}$, $\beta_{2,q}$ depending on the transition between regimes defined with $H(\varepsilon_{t-1}; \gamma, n)$. The stability condition, $(\alpha + \beta) < 1$, could vary for different values taken by the $H(\varepsilon_{t-1}; \gamma, n)$: as $H(\varepsilon_{t-1}; \gamma, n) \rightarrow 1$ for $\varepsilon_{t-1} > n$ innovations larger than the threshold, based on the regime dynamics the stability structure of the model approaches $(\alpha_{1,p} + \beta_{1,q}) \rightarrow (\alpha_{2,p} + \beta_{2,q})$. Further, for positive $\varepsilon_{t-1} > n$, as $\varepsilon_{t-1} \rightarrow +$ values, $H(.) \rightarrow 1$; while, for negative values of the transition variable and as long as $\varepsilon_{t-1} < n$, $\varepsilon_{t-1} \rightarrow -$ negative large values, transition function approaches $H(.) \rightarrow 0$ zero. The speed of transition is determined by the parameter γ and the above-mentioned characteristics of transition between volatility dynamics of the two regimes are highly influenced by the values taken by ε_{t-1}, n and γ. The inflection point for the transition function occurs at $H(\varepsilon_{t-1}; \gamma, n) = 1/2$ if the transition function is equal to the threshold parameter, $\varepsilon_{t-1} = n$ where the stability condition holds if $\left\{ ((\alpha_{1,p} + \alpha_{2,p}) + (\beta_{1,q} + \beta_{2,q}))/2 \right\} < 1$. Additionally, for very large values of the transition variable $\gamma \rightarrow \infty$; and $H(.)$ transition function behaves like the identity function that gives sudden shifts between two regimes, i.e., for $\gamma = \infty$ and $\varepsilon_{t-1} < n$, $H(.) = 0$ and, for $\varepsilon_{t-1} > n$, $H(.) = 1$. As a result, the ST-GARCH process reduces to the TAR-TGARCH model for very large values of γ. Further, if $\gamma = 0$, transition function becomes $H(.) = 0.5$, and hence the process reduces to a single-regime AR-GARCH(p, q) process.

3.1.2 LSTAR-LST-FIGARCH model

Fractionally Integrated GARCH (FIGARCH(1, d, 1)) model is developed under these findings by Bollersev and Mikkelsen (1996) and Baillie et al. (1996) as an extension of the GARCH model to account for long memory. In this section, we will first evaluate fractional integration in a GARCH setting to evaluate long memory in conditional variance. Afterwards, smooth transition type nonlinearity setting will be introduced to the evaluated FIGARCH and FIAPGARCH models. The LSTAR-LST-FIGARCH model generalizes the LSTAR-LST-GARCH type nonlinearity to account for long memory in the conditional variance process,

$$(1 - \beta_i L)\sigma_t^2 = \left(\omega_1 + \left((1 - \beta_1 L) - (1 - \phi_1 L)(1 - L)^{d,1}\right) \right. $$
$$\times \left(|\varepsilon_{t-1}| - \theta_1 \varepsilon_{t-1}\right)^2 \right)(1 - H(\varepsilon_{t-1}; \gamma, n))$$
$$+ \left(\omega_2 + \left((1 - \beta_2 L) - (1 - \phi_2 L)(1 - L)^{d,2}\right) \right.$$
$$\times \left(|\varepsilon_{t-1}| - \theta_2 \varepsilon_{t-1}\right)^2 \right)(H(\varepsilon_{t-1}; \gamma, n))$$

$$(4)$$

with the transition function defined in Eq. (3). The range of the cluster of the volatility depends on the transition function and changes between $H(\varepsilon_{t-1}; \gamma, n) = 0$ and $H(\varepsilon_{t-1}; \gamma, n) = 1$. Further, γ is the speed of transition parameter and $\gamma > 0$ ensures that the transition between the regimes follows a nonlinear sigmoid type transition in modeling the dynamics of the conditional volatility. The constant term takes on values between $\varphi = \omega/(1 - \alpha)$ and $\varphi = \omega/(1 - \beta)$ based upon whether the conditional volatility is the regime dictated by $H(.) = 0$ and $H(.) = 1$. Similar to the ST-GARCH model, the constant term ranges

between the extreme regimes, the level of conditional volatility will change in different regimes (For ST-FIG-ARCH models, readers are referred to Kılıç (2011) and Bildirici and Ersin (2013).

3.1.3 LSTAR-LST-APGARCH model

The Asymmetric Power GARCH (APGARCH) model is developed by Ding et al. (1983). The model is based on different power transformations without simple squared shocks and conditional variances as in the traditional GARCH models. The STAR-ST-APGARCH model possesses nonlinear dynamics with smooth transition functions to allow different asymmetric power terms in two regimes with the following conditional variance process,

$$\sigma_t^{\delta,i} = \left(\omega_{0,1} + \sum_{p=1}^{p} \alpha_{p,1}\left(|\varepsilon_{t-p,1}| - \gamma_{p,1}\varepsilon_{t-p,1}\right)^{\delta,1} \right.$$
$$+ \sum_{q=1}^{q} \beta_{q,1}\sigma_{t-q,1}^{\delta,1} \right) \times (1 - H(\varepsilon_{t-1}; \gamma, n))$$
$$+ \left(\omega_{0,2} + \sum_{p=1}^{p} \alpha_{p,2}\left(|\varepsilon_{t-p,2}| - \gamma_{p,2}\varepsilon_{t-p,2}\right)^{\delta,2} \right.$$
$$+ \sum_{q=1}^{q} \beta_{q,2}\sigma_{t-q,2}^{\delta,2})H(\varepsilon_{t-1}; \gamma, n). \quad (5)$$

Similarly, the transition function is defined by Eq. (3) to obtain the LSTAR-LST-APGARCH model. Accordingly, the obtained model possesses such dynamics that both the conditional mean and the conditional variance follows nonlinear dynamics restricted to have two regimes between which the transition is defined by a smooth and continuously differentiable logistic function. The obtained model is defined as the *logistic smooth transition asymmetric power GARCH* model in which δ represents the asymmetric power parameter to be estimated by nonlinear least squares or maximum likelihood methods as in the APGARCH methodology of Conrad et al. (2010). The estimation of the threshold parameter n and the γ parameter that defines the speed of transition is conducted through a grid search following the Terasvirta (1994).

3.1.4 LSTAR-LST-FIAPGARCH model

Following the LST-FIGARCH model structure, smooth transition fractionally integrated asymmetric power GARCH model denoted as LST-FIAPGARCH which is obtained by allowing the smooth transition type nonlinearity between two FIAPGARCH processes in two regimes as,

$$
\begin{aligned}
(1 - \beta_i L)\sigma_t^{\delta,i} = & \Big(\omega_{0,1} + \big((1 - \beta_1 L) - (1 - \phi_1 L)(1 - L)^{d,1}\big) \\
& \times \big(|\varepsilon_{t-1}| - \theta_1 \varepsilon_{t-1}\big)^{\delta,1}\Big) \times \big(1 - (H(\varepsilon_{t-1}; \gamma, n))\big) \\
& + \big(\omega_{0,2} + ((1 - \beta_2 L) - (1 - \phi_2 L) \\
& \times left(1 - L)^{d,2}\big)\big(|\varepsilon_{t-1}| - \theta_2 \varepsilon_{t-1}\big)^{\delta,2}\big) \\
& \times (H(\varepsilon_{t-1}; \gamma, n)).
\end{aligned}
\tag{6}
$$

As previously, the transition function $H(.)$ is defined as a logistic function bounded between 0 and 1. The obtained model is defined as the *logistic smooth transition fractionally integrated asymmetric power GARCH*(LST-FIAPGARCH) model in which δ represents the asymmetric power parameter to be estimated (Conrad et al. 2010; Bildirici and Ersin 2013).

3.2 Neural network models, nonlinear GARCH models, and ANN augmented GARCH models

3.2.1 Neural networks: an overview

Artificial neural networks models (ANN) are functional models that provide well-known approximation properties applied in many fields such as finance, medicine, and engineering. In economics and business literature, the early studies could be given as Tam and Kiang (1992), Do and Grudnitski (1992) which used neural networks for banking failure detection and residential property appraisals. Freisleben (1992) and Refenes et al. (1997) utilized ANN models for stock prediction. Hutchinson et al. (1994) showed that the learning networks could be used efficiently for pricing and hedging in securities markets. Studies such as Gencay and Stengos (1997, 1998), Gencay and Liu (1997), Kanas (2003), Kanas and Yannopoulos (2001), Shively (2003) and Bildirici and Ersin (2009) applied ANN models to stock market return forecasting and financial analysis.

The MLP model is evaluated as an important class of neural network models. MLP consists of a set of sensory units based on three layers, while a common application of such ANN model possesses mostly a single hidden layer. Hence, a MLP consists of the input layer, one or more hidden layers and an output layer.

Estimation in the LSTAR-LST-GARCH-MLP and LSTAR-LST-GARCH-RBF models is conducted with conjugant gradient-based backpropagation algorithm. The learning and model selection processes are gathered to improve forecast accuracy. During the learning process, *weight decay* is conducted to further improve the model eliminating the insignificant coefficients (Weigend et al. 1991; Bartlett 1997; Krogh and Hertz 1995). For details regarding weight decay in learning process, an

investigation is given by Gupta and Lam (1998). The algorithm cooperation and early stopping for NN-GARCH processes are given in Bildirici and Ersin (2009). The algorithm used in the study could be taken as estimating neural network models with LSTAR type nonlinear structures with different number of neurons in the hidden layer, this means estimating models with different architecture variations. Once the optimum architecture is selected, the model is re-estimated with early stopping and weight decay k times. To save CPU time, k is preferred as 20 in the study. Further, the number of neurons is allowed to vary ranging from 2 to 20 considering the sample size. Neurons are constrained as being logistic activation functions, which have similar transition properties as the LSTAR models in the hidden layer. The output layer is restricted to have linear functions. The models estimated are utilized for out-of-sample forecasting. Each model architecture is estimated $k = 20$ times, and because there are eight different neural network-based model architecture to be estimated in the study, the total number of estimated models is 160; whereas, the selected 8 models (based on the lowest MSE error criteria) will be reported in the study. The methodology is as follows. Model estimation is gathered through utilizing backpropagation algorithm and the parameters are updated with respect to a quadratic loss function; whereas, the weights are iteratively calculated with *weight decay* method to achieve the lowest error. Alternative methods include Genetic Algorithms (Goldberg 1989) and second-order derivative-based optimization algorithms such as Conjugate Gradient Descent, Quasi-Newton, Quick Propagation, Delta-Bar-Delta, and Levenberg–Marquardt, which are fast and effective algorithms but may be subject to over-fitting (see Patterson 1996; Haykin 1994; Fausett 1994). In the study, we followed a two-step methodology. At the first step, all models were trained over a given training sample vis-à-vis checking for generalization accuracy in the light of MSE criteria in test sample. The approach is repeated for estimating each model for 100 times with different number of sigmoid activation functions in the hidden layer. To obtain parsimony, best model is further selected with respect to the AIC information criterion (see Faraway and Chatfield 1998). For estimating NN-GARCH models with early stopping combined with algorithm corporation, readers can refer to Bildirici and Ersin (2009).

The models below represent the architectures to be evaluated in the analysis for estimating the daily petrol prices in the application section. Each model possesses LSTAR type nonlinear structure in the conditional mean and conditional variance processes. The models are optimized to forecast the conditional variance of the petrol prices. Therefore, the LSTAR-LST-GARCH-MLP and

LSTAR-LST-GARCH-RBF structures of these models will be introduced as follows.

3.2.2 LSTAR-LST-GARCH-MLP model

Following Bildirici and Ersin (2013), a MLP neural network model hints a similar architecture with a logistic smooth transition GARCH (LST-GARCH) process. A *logistic smooth transition GARCH neural network model* is a LSTAR-LST-GARCH model with MLP type neural networks in each regime,

$$
\sigma_t^2 = \left(\omega_{0,1} + \sum_{p=1}^{p} \alpha_{p,1} \varepsilon_{t-p}^2 + \sum_{q=1}^{q} \beta_{q,1} \sigma_{t-q}^2 \right.
$$
$$
+ \sum_{h=1}^{h} \xi_{1,h} \psi_1 \left(z_{1,h} \lambda_{1,h} \right) \Bigg) \left(1 - H(\varepsilon_{t-1}; \gamma, n) \right)
$$
$$
+ \left(\omega_{0,2} + \sum_{p=1}^{p} \alpha_{p,2} \varepsilon_{t-p}^2 + \sum_{q=1}^{q} \beta_{q,2} \sigma_{t-q}^2 \right.
$$
$$
\left. + \sum_{h=1}^{h} \xi_{2,h} \psi_2 \left(z_{2,h} \lambda_{2,h} \right) \right) H(\varepsilon_{t-1}; \gamma, n), \tag{7}
$$

where, n is the threshold, γ is the parameter that defines the speed of transition in the logistic transition function,

$$
H(\varepsilon_{t-1}; \gamma, n) = \frac{1}{1 + e^{-\gamma(\varepsilon_{t-1} - n)}}. \tag{8}
$$

MLP neural networks that possesses h hidden neurons in each $i = 1, 2$ regimes are,

$$
\psi_i \left(z_{i,h} \lambda_{i,h} \right) = \frac{1}{1 + \exp\left(-\left(\lambda_{i,h} + \sum_{d=1}^{d} \left[\sum_{h=1}^{h} \lambda_{h,d,i} z_{t-d,i,h} \right] \right) \right)}, \tag{9}
$$

where $\psi_i(z_{i,h}, \lambda_{i,h})$ is a log-sigmoid activation function. The inputs are are normalized and are defined as follows:

$$
z_{t-d} = \left[\varepsilon_{t-d} - E(\varepsilon) \right] / \sqrt{E(\varepsilon^2)}, \tag{10}
$$

$$
\frac{1}{2} \lambda_{h,d,i} \sim \text{uniform} \, [-1, +1]. \tag{11}
$$

The LSTAR-LST-GARCH-MLP model given in Eqs. (7)–(11) is a neural network augmented version of the LSTAR-LST-GARCH model given in Eq. (2).

The model proposed above will be augmented with asymmetric power term in the conditional variance to obtain LSTAR-LST-APGARCH-MLP model.

3.2.3 LSTAR-LST-APGARCH-MLP model

By augmenting the LSTAR-LST-GARCH-MLP model with asymmetric power terms, the LSTAR-LST-

APGARCH-MLP model is obtained. This model is a two-regime nonlinear model where both the conditional mean and the conditional variance follow a nonlinear process in the fashion of Terasvirta (1994) LSTAR model. The model also benefits from the well-known generalization properties of neural networks in the fashion of Cybenko (1989). The LSTAR-LST-APGARCH-MLP model allows smooth transition between two regimes of the conditional variance defined as a LST-APGARCH process with neural network augmentations as follows:

$$
\sigma_t^{\delta,i} = \left(\omega_{0,1} + \sum_{p=1}^{p} \alpha_{p,1} \left(\left| \varepsilon_{t-p,1} \right| - \theta_{p,1} \varepsilon_{t-p} \right)^{\delta,1} \right.
$$
$$
\left. + \sum_{q=1}^{q} \beta_{q,1} \sigma_{t-q,1}^{\delta,1} + \sum_{h=1}^{h} \xi_{h,1} \psi_1 \left(z_{1,h} \lambda_{1,h} \right) \right) \left(1 - H(\varepsilon_{t-1}; \gamma, n) \right)
$$
$$
+ \left(\omega_{0,2} + \sum_{p=1}^{p} \alpha_{p,2} \left(\left| \varepsilon_{t-p,2} \right| - \theta_{p,2} \varepsilon_{t-p} \right)^{\delta,2} \right.
$$
$$
\left. + \sum_{q=1}^{q} \beta_{q,2} \sigma_{t-q,2}^{\delta,2} + \sum_{h=1}^{h} \xi_{h,2} \psi_2 \left(z_{2,h} \lambda_{2,h} \right) \right) H(\varepsilon_{t-1}; \gamma, n). \tag{12}
$$

Accordingly, Eq. (12) is a hybrid model consisting of two regime LSTAR process in the conditional mean with residuals following a nonlinear neural network model for the conditional variance with multi-layer perceptrons in each regime of the conditional variance process. For the estimation and the statistical properties of the model, readers are referred to Bildirici and Ersin (2009, 2013) and to the neural network section of this paper.

3.2.4 LSTAR-LST-FIGARCH-MLP model

LSTAR-LST-FIGARCH-MLP model is a fractionally integrated volatility model augmented with two regime MLP neural networks in the conditional variance,

$$
(1 - \beta_i L) \sigma_t^2 = \left(\omega_{0,1} + \left((1 - \beta_1 L) - (1 - \phi_1 L)(1 - L)^{d,1} \right) \right.
$$
$$
\times \left(\left| \varepsilon_{t-1} \right| - \theta_1 \varepsilon_{t-1} \right)^2 + \sum_{h=1}^{h} \xi_{h,1} \psi_1 \left(z_{1,h} \lambda_{1,h} \right) \Bigg)
$$
$$
\times \left(1 - H(\varepsilon_{t-1}; \gamma, n) \right) + \left(\omega_{0,2} + \left((1 - \beta_2 L) \right. \right.
$$
$$
\left. - (1 - \phi_2 L)(1 - L)^{d,2} \right) \left(\left| \varepsilon_{t-1} \right| - \theta_2 \varepsilon_{t-1} \right)^2
$$
$$
\left. + \sum_{h=1}^{h} \xi_{h,2} \psi_2 \left(z_{2,h} \lambda_{2,h} \right) \right) \left(H(\varepsilon_{t-1}; \gamma, n) \right). \tag{13}
$$

It should be noted that the LSTAR-LST-FIGARCH-MLP model reduces to LSTAR-LST-GARCH-MLP following the application of the restriction on the fractional integration parameter $d = 0$.

3.2.5 LSTAR-LST-FIAPGARCH-MLP model

The model is defined as follows:

$$
\begin{aligned}
(1 - \beta L)\sigma_t^{\delta,i} = & \left(\omega_{0,1} + \left((1 - \beta_1 L) - (1 - \phi_1 L)(1 - L)^{d,1} \right) \right. \\
& \times \left(|\varepsilon_{t-1}| - \theta_1 \varepsilon_{t-1} \right)^{\delta,1} + \sum_{h=1}^{h} \xi_{h,1} \psi_1 \left(z_{1,h} \lambda_{1,h} \right) \right) \\
& \times (1 - H(\varepsilon_{t-1}; \gamma, n)) + \left(\omega_{0,2} + \left((1 - \beta_2 L) \right. \right. \\
& - (1 - \phi_2 L)(1 - L)^{d,2} \right) \left(|\varepsilon_{t-1}| - \theta_2 \varepsilon_{t-1} \right)^{\delta,2} \\
& \left. + \sum_{h=1}^{h} \xi_{h,2} \psi_2 \left(z_{2,h} \lambda_{2,h} \right) \right) H(\varepsilon_{t-1}; \gamma, n).
\end{aligned}
\tag{14}
$$

The model assumes regime-dependent asymmetry based on $\delta(i)$, where $i = 1, 2$ for the two regime structure. By applying $d = 0$ restriction, Eq. (14) reduces to LSTAR-LST-APGARCH-MLP and additional restriction on the asymmetry parameters $\delta(i) = 2$, leads the model further reduce to LSTAR-LST-GARCH-MLP.

In fact, asymmetry is also introduced in the model through the ARCH terms. As a typical, the ARCH term is written as, $\left(|\varepsilon_{t-1}| - \theta_i \varepsilon_{t-1} \right)^{\delta,i}$ in each i = 1, 2 regimes. To reduce the APGARCH process to baseline GARCH, one need also specify the ARCH term in a way to eliminate the deviations from the absolute innovations $|\varepsilon_{t-1}| = 0$ to obtain $(-\theta_i \varepsilon_{t-1})^{\delta,i}$. Further by restricting the model as $\theta_i = (-1)$ so that $\theta_i (\varepsilon_{t-1})^{\delta,i}$ followed by $\delta = 2$, the APGARCH model reduces to a GARCH process. One possibility is that, for modeling time series, it is possible to obtain different types of GARCH processes in each regime, i.e., a time series could follow a GARCH process in one regime, while following an APGARCH or FIAPGARCH process in the second regime. The study restricts the models to follow the same type of GARCH processes for simplicity. On the other hand, the fractional integration parameters are regime specific and allow different dynamics to be modeled simultaneously. The parameter d could be estimated as less than 0.5 in regime 1 and more than 0.5 in regime 2, suggesting different long memory dynamics and stationarity processes in each regime occurring below and above the threshold, n. As a result, assuming same GARCH structure in two regimes produce interesting findings and an approach that allows regime-wise comparative analysis (See Bildirici and Ersin 2013).

3.3 RBF neural network augmentations of the nonlinear GARCH models

The RBF neural network is represented as a composition of three layers of nodes; first, the input layer that feeds the input data to each of the nodes in the second or hidden layer; the second layer that differs from other neural networks in that each node represents a data cluster which is centered at a particular point and has a given radius and in the third layer, consisting of one node (Bishop 1995).

Wright (2003) discusses the radial basis function interpolation and shows the developments in RBF networks. Liu and Zhang (2010) combined RBF neural network models with the Markov switching model to merge Markov switching Neural Network model based on RBF models. RBF neural network in their models are trained to generate both time series forecasts and certainty factors. Santos et al. (2010) developed a RBF-GARCH model that possesses a modeling structure that assumes a RBF type neural network in the conditional mean, where the residuals follow a GARCH process. Further, Coelho and Santos (2011) extended their RBF-GARCH approach and provided an application to Spanish energy pool prices and showed that RBF-GARCH approach provided significant improvement in future forecasts. It should be noted that, their approach is similar in one way to the STAR-GARCH approach of Chan and McAleer (2003) that assumes STAR type nonlinearity in the conditional mean process only. However, one important fact is that RBF-GARCH approaches of Santos et al. (2010) and Coelho and Santos (2011) benefit from different NN learning algorithms.

Our approach is differentiated than the above-mentioned studies in three ways. First, similar to the MLP-based models given in this paper, the proposed LSTAR-LST-GARCH-RBF models utilize neural network architectures in the conditional variance processes. It should be noted that heteroscedasticity is a strong factor that diminishes the forecast capabilities of the model. Second, models are estimated with neural network learning algorithms and the estimation of the models benefits from algorithm cooperation weight decay and early stopping. Third, our models follow STAR type division of the regression space both in the conditional mean and in the conditional variance with an expectation that this approach provides improvement in the modeling and forecasting capabilities as will be evaluated in Sect. 4.

3.3.1 LSTAR-LST-GARCH-RBF model

LSTAR-LST-GARCH-RBF model is defined as follows:

$$
\begin{aligned}
\sigma_t^2 = & \left(\omega_{0,1} + \sum_{p=1}^{p} \alpha_{p,1} \varepsilon_{t-p}^2 + \sum_{q=1}^{q} \beta_{q,1} \sigma_{t-q}^2 \right. \\
& \left. + \sum_{h=1}^{h} \xi_{1,h} \phi_1 \left(\| z_t - \mu_1 \| \right) \right) (1 - H(\varepsilon_{t-1}; \gamma, n)) \\
& + \left(\omega_{0,2} + \sum_{p=1}^{p} \alpha_{p,2} \varepsilon_{t-p}^2 + \sum_{q=1}^{q} \beta_{q,2} \sigma_{t-q}^2 \right. \\
& \left. + \sum_{h=1}^{h} \xi_{2,h} \phi_1 \left(\| z_t - \mu_2 \| \right) \right) H(\varepsilon_{t-1}; \gamma, n).
\end{aligned}
\tag{15}
$$

A Gaussian basis function for the hidden units given as $\phi(x)$ for $x = 1, 2,..., X$, where the activation function is defined as Gaussian function,

$$\phi(x) = \exp\left(\frac{-\|x - \mu_j\|^2}{2\rho^2}\right) \tag{16}$$

with p defining the width of each function. z_t is a vector of lagged explanatory variables, $\alpha + \beta < 1$ is essential to ensure stationarity. Networks of this type can generate any real-valued output, but in their applications where they have a priori knowledge of the range of the desired outputs, it is computationally more efficient to apply some nonlinear transfer function to the outputs to reflect that knowledge. The conditional variance is subject to smooth transition based on the logistic function, $H(\varepsilon_{t-1}; \gamma, n) = 1/\left(1 + e^{-\gamma(\varepsilon_{t-1} - n)}\right)$, where the speed of transition is given by γ. For the two regime model, $i = 1, 2$, the inputs are subject to,

$$z_{t-d} = [\varepsilon_{t-d} - E(\varepsilon)]\Big/ \sqrt{E(\varepsilon^2)} \tag{17}$$

$$\frac{1}{2}\lambda_{h,d,i} \sim \text{Uniform } [-1, +1]. \tag{18}$$

3.3.2 LSTAR-LST-APGARCH-RBF model

Radial basis functions are three layer neural network models with linear output functions and nonlinear activation functions defined as Gaussian functions in hidden layer utilized to the inputs in light of modeling a radial function of the distance between the inputs and calculated value in the hidden unit. The output unit produces a linear combination of the basis functions to provide a mapping between the input and output vectors.

$$\sigma_t^{\delta,i} = \left(\omega_{0,1} + \sum_{p=1}^{p}\alpha_{p,1}\left(|\varepsilon_{t-p,1}| - \theta_{p,1}\varepsilon_{t-p}\right)^{\delta,1} + \sum_{q=1}^{q}\beta_{q,1}\sigma_{t-q,1}^{\delta,1}\right.$$

$$+ \sum_{h=1}^{h}\xi_{h,1}\phi_1(\|z_t - \mu_1\|)\right)(1 - H(\varepsilon_{t-1}; \gamma, n))$$

$$+ \left(\omega_{0,2} + \sum_{p=1}^{p}\alpha_{p,2}\left(|\varepsilon_{t-p,2}| - \theta_{p,2}\varepsilon_{t-p}\right)^{\delta,2}\right.$$

$$+ \sum_{q=1}^{q}\beta_{q,2}\sigma_{t-q,2}^{\delta,2} + \sum_{h=1}^{h}\xi_{h,2}\phi_2(\|z_t - \mu_1\|)\right)H(\varepsilon_{t-1}; \gamma, n), \tag{19}$$

where, $i = 1, 2$ is the number of regimes. Similar to the LSTAR-LST-APGARCH-MLP model, the LSTAR-LST-APGARCH-RBF model nests several models. Equation (19) reduces to the LSTAR-LST-GARCH-RBF model if the power term $\delta = 2$ and $\theta_{p,i} = 0$, to the LSTAR-GARCH-RBF model for $\theta_{p,i} = 0$, and to the LSTAR-GJR-

RBF model if $\delta = 2$ and $0 \leq \theta_{p,i} \leq 1$ restrictions are allowed. The model may be shown as LSTAR-TGARCH-RBF model if $\delta = 1$ and $0 \leq \theta_{p,i} \leq 1$.

3.3.3 LSTAR-LST-FIAPGARCH-RBF model

LSTAR-LST-FIAPGARCH-RBF model is defined as follows:

$$(1 - \beta L)\sigma_t^{\delta,i} = \left(\omega_{0,1} + \left((1 - \beta_1 L) - (1 - \phi_1 L)(1 - L)^{d,1}\right)\right.$$

$$\times \left(|\varepsilon_{t-1}| - \theta_1\varepsilon_{t-1}\right)^{\delta,1} + \sum_{h=1}^{h}\xi_{h,1}\phi_1(\|z_t - \mu_1\|)\right)$$

$$\times \left(1 - H(\varepsilon_{t-1}; \gamma, n)\right) + \left(\omega_{0,2} + \left((1 - \beta_2 L)\right.\right.$$

$$- (1 - \phi_2 L)(1 - L)^{d,2})\left(|\varepsilon_{t-1}| - \theta_2\varepsilon_{t-1}\right)^{\delta,2}$$

$$+ \sum_{h=1}^{h}\xi_{h,2}\phi_2(\|z_t - \mu_2\|)\right)H(\varepsilon_{t-1}; \gamma, n), \tag{20}$$

where, h is neurons defined with Gaussian function as in Eq. (16). The LSTAR-LST-FIAPGARCH-RBF model is a variant of the LSTAR-LST-APGARCH-RBF model with fractional integration. To obtain the model with short memory characteristics, $d = 0$ restriction on the fractional integration parameter should be imposed. As a result, the model reduces to LSTAR-APGARCH-RBF model. Additionally, by applying $d = 0$ with the restrictions discussed above, models with no fractional integration discussed above could be easily achieved. In addition to $d = 0$ restriction, Eq. (20) reduces to LSTAR-LST-GARCH-RBF with the restriction on the asymmetry parameters $\delta(i) = 2$ after eliminating the deviations from the absolute innovations with $|\varepsilon_{t-1}| = 0$ and $\theta_i = (-1)$.

4 Econometric results

4.1 Data

In order to test forecasting performance of the above-mentioned models, Brent crude oil spot prices were used for oil price volatility. We take the daily data from January 20, 1986 to January 30, 2013, excluding public holidays, data are converted into daily returns by taking first differenced logarithms as $y = \ln(P_t/P_{t-1})$. In the process of model estimation, the sample is divided between training, test, and out-of-sample (forecasting) samples with the percentages of 80 %, 10 %, and 10 %, respectively. The descriptive statistics are reported in Table 1 below. Accordingly, the daily return series are subject to

Table 1 Descriptives of Brent crude oil daily returns, January 20th, 1986 to January 30th, 2013

Mean	Median	Max	Min	SD	Skewness	Kurtosis	JB	SW	ARCH
8.13e−05	0.000320	0.08317	−0.176495	0.011105	−0.759616	17.742	63300.56 [0.0000]	0.91201 [0.000]	77.86521 [0.0000]

JB and the SW are the Jarque–Berra and Shapiro–Wilk normality tests

ARCH test is the ARCH-type heteroscedasticity test in the residuals of the AR(1) model selected by SIC information criterium. The probability values for the reported tests are given in brackets

Table 2 GARCH family results

	1. GARCH	2. APGARCH	3. FIGARCH	4. FIAPGARCH
Cst(M)	0.000120 (0.00010935)	0.0002159** (0.000061)	0.0002579*** (0.00008093)	0.0002549*** (0.00008215)
Cst(V)	0.012561*** (0.0039790)	0.619456* (0.33203)	1.097080* (0.56766)	5.319243* (3.0849)
d-Figarch	–	–	0.474443*** (0.053529)	0.438994*** (0.042718)
ARCH	0.096773*** (0.018932)	0.077441*** (0.0087413)	0.219596*** (0.067838)	0.238516*** (0.071217)
GARCH	0.898413*** (0.017768)	0.929974*** (0.0082689)	0.590660*** (0.095018)	0.577437*** (0.089258)
APARCH (Gamma1)	–	0.166567*** (0.060379)	–	0.089858* (0.058840)
APARCH (Delta)	–	1.125234*** (0.10841)	–	1.764091*** (0.081130)
Log$L$	22,372.768	22,625.828	22,583.532	22,589.793
AIC	−6.466831	−6.539123	−6.527185	−6.528416
SIC	−6.462875	−6.532200	−6.521251	−6.620504
JB	3203.4	5381.8	2987.5	3821.6
Kurtosis	6.2693	7.2522	6.1596	6.5782

Standard errors are given in parentheses

*, **, and *** denote 10 %, 5 %, and 1 % significance levels

Log$L$ is the Log-likelihood statistic. AIC and SIC denote the Akaike and Schwarz information criteria

leptokurtic distribution with the kurtosis statistic being 17.74 and skewness statistic calculated as −0.76. Jarque–Berra and Shapiro–Wilk tests suggest that the null hypothesis of normal distribution for daily returns can be rejected at the 5 % significance level. Further, the ARCH-type heteroscedasticity cannot be rejected for the daily returns series.

4.2 Econometric results: model evaluation

At the first stage, the GARCH family models were taken as baseline models and are estimated for evaluation purposes. Results are given in Table 2. The models given in the table have different characteristics to be evaluated: namely, fractional integration, asymmetric power, and fractionally integrated asymmetric power models, namely, GARCH, APGARCH, FIGARCH, and FIAPGARCH models. By hybridization of two groups of nonlinear models, we obtained STAR-ST-GARCH models that allow for STAR type nonlinearity in both the conditional mean and variance.

The LSTAR-LST-GARCH models are reported in Table 3. The results show significant improvements of LSTAR-LST-GARCH models over their single-regime

variants reported in Table 2. The log-likelihood statistics are also high as AIC and SIC criteria report similar conclusions for the in-sample results.[1] Models have similar performances in the in-sample modeling. Further, models will be evaluated for out-of-sample forecasting capabilities with MSE, MAE, and RMSE statistics.

After allowing the GARCH processes to follow LST type nonlinearity, the dynamics are strikingly different in the light of the estimated parameters. In the LSTAR-LST-FIGARCH model, d parameters are estimated as 0.437 and 0.822 for regime 1 and 2, respectively, suggesting strong persistence in the second regime. For the LSTAR-LST-FIAPGARCH, after the inclusion of the asymmetric power terms, the d parameters are estimated as 0.44 and 0.45. The results also suggest that different conclusions could be derived due to the parametric specification of the analyzed GARCH models in addition to possible neglected nonlinearity.

The RBF and MLP type neural network augmented versions of the models will be analyzed. The model architectures of the proposed ANN models and their

---

[1] Models have similar performances in the in-sample modeling. For model performances, the models will be evaluated in terms of out-of-sample forecasting statistics.

Table 3 The LSTAR-LST-GARCH family models: estimates

	LSTAR-LST-GARCH		LSTAR-LST-APGARCH		LSTAR-LST-FIGARCH		LSTAR-LST-FIAPGARCH	
	Regime 1	Regime 2	Regime 1	Regime 2	Regime 1	Regime 2	Regime 1	Regime 2
Cst(M)	−0.000038*** (0.2684e−017)	0.00056*** (0.5350e−005)	−0.0000276 *** (0.4935e−005)	0.00009*** (0.3389e−006)	0.0000757*** (0.3974e−006)	0.0000962*** (.4932e−005)	0.000050*** (0.0011653)	0.000099*** (7.6142e−006)
Cst(V)	−0.00002*** (0.0000012)	0.00011*** (.1627e−005)	0.0036114*** (0.1555e−005)	0.001824*** (0.8884e−005)	0.031261*** (0.00014335)	0.031463*** (0.010639)	0.030331 (0.032575)	0.040000*** (0.0035798)
d-Figarch	—	—	—	—	0.437453*** (0.0067429)	0.821958*** (0.14722)	0.443713*** (0.2050)	0.450001*** (0.097121)
ARCH	0.070244*** (0.004856)	0.05804*** (0.02902)	0.157681 *** (0.019837)	0.154319* (0.09616)	0.098575*** (0.0048703)	0.160380*** (0.04430)	0.099359*** (0.009024)	0.100000*** (0.043737)
GARCH	0.955639*** (0.001756)	0.934448*** (0.20991)	0.914455*** (0.000928)	0.818426 *** (0.0061486)	0.402214*** (0.00286)	0.326375*** (0.1646)	0.401029*** (0.0662)	0.400000*** (0.20677)
APARCH (Gamma1)	—	—	0.576147*** (0.02203)	0.107065*** (0.041587)	—	—	0.006459*** (0.0003546)	0.010004*** (0.002871)
APARCH (Delta)	—	—	2.036870*** (0.022905)	2.006556*** (0.023098)	—	—	2.006473*** (0.29007)	2.000000*** (0.040708)
ARCH in mean	—	—	−0.001073 (0.00131)	−0.000818 (0.0389)	−0.000769*** (0.0001)	−0.00019** (0.0001)	−0.00008*** (0.000012)	−0.00006*** (0.00002)
Diagnostic tests								
LogL	23,310.192		16,613.846		13,393.367		13,189.971	
AIC	−12.653782		−15.734425		−16.621534		−16.934091	
SIC	−12.556380		−15.716555		−16.60163		−16.88337	
Kurtosis	8.9001		8.8815		7.864		6.2124	
ARCH (1–2)	1158.7 [0.00]		367.67 [0.00]		5.6985 [0.00]		3.3956 [0.00]	
ARCH (1–5)	470.78 [0.00]		164.13 [0.00]		2.6327 [0.00]		1.6217 [0.00]	

Standard errors are given in parentheses

*, **, and *** denote 10 %, 5 %, and 1 % significance levels

LogL is the Log-likelihood statistic. AIC and SIC denote the Akaike and Schwarz information criteria. For the ARCH type, remaining heteroscedasticity tests, calculated p values are reported in brackets

Table 4 Neural networks augmented LSTAR-LST-GARCH-MLP and RBF models: architecture and training results

MLP-based ANN models and architectures

Training statistics	1. LSTAR-LST-GARCH-MLP(5:2:2:6:1)	2. LSTAR-LST-APGARCH-MLP(5:2:2:11:1)	3. LSTAR-LST-FIGARCH-MLP(5:2:2:8:1)	4. LSTAR-LST-FIAPGARCH-MLP(5:2:2:4:1)
Training ρ	0.928664	0.951848	0.908285	0.905530
Test ρ	0.946714	0.952677	0.917540	0.912975
Training MSE	0.001284	0.0007911	0.001035	0.001139
TEST MSE	0.000989	0.0007660	0.000946	0.001122
Training algorithm (convergence)	BFGS (10)	BFGS (12)	BFGS (22)	BFGS (24)

RBF-based ANN models and architectures

Learning results	1. LSTAR-LST-GARCH-RBF (5:2:2:28:1)	2. LSTAR-LST-APGARCH-RBF (5:2:2:26:1)	3. LSTAR-LST-FIGARCH-RBF (5:2:2:30:1)	4. LSTAR-LST-FIAPGARCH-RBF (5:2:2:27:1)
Training ρ	0.877056	0.930455	0.862266	0.879122
Test ρ	0.902086	0.939193	0.870629	0.887971
Training MSE	0.002152	0.001129	0.001516	0.001437
TEST MSE	0.001789	0.000974	0.001446	0.001423
Training algorithm (convergence)	RBFT	RBFT	RBFT	RBFT

ρ denotes Pearson's correlation statistic calculated for the targets and forecasts. MSE represents training and test sample mean squared errors. BFGS is the Broyden–Fletcher–Goldfarb–Shanno nonlinear optimization algorithm. The epoch shows the step number the algorithm converged

All models are restricted to have logistic activation functions in the hidden layer and identity activation functions in the output layers. Model architectures are given in parenthesis. As a typical, LSTAR-GARCH-MLP(5:2:2:6:1) model is a nonlinear model with 5 input variables in the input layer modeled as a 2 regime LSTAR process in the conditional mean with 2 regimes following LST-GARCH conditional variance processes passing through 6 neurons to the output layer connected to the output layer to produce 1 output

training results are reported in Table 4 in terms of MSE and ρ correlation statistics for the training and test samples.

Among the LSTAR-GARCH-NN models, the training and test MSE errors are calculated comparatively lower for the LSTAR-LST-GARCH-MLP models. Training MSE statistics for the LSTAR-LST-GARCH-RBF, LSTAR-LST-APGARCH-RBF, LSTAR-LST-FIGARCH-RBF, and LSTAR-LST-FIAPGARCH-RBF models are 0.001789, 0.000974, 0.001446, and 0.001423, respectively. On the other hand, the MSE statistics calculated for their MLP variants are 0.000989, 0.0007660, 0.000946, and 0.001122, respectively.

Radial Basis Function augmented versions of the LSTAR-LST-GARCH family models provided small deviation from the results obtained for their MLP variants in terms of training performances. As a typical, the highest training ρ is obtained as 0.93 for the LSTAR-LST-APGARCH-RBF and is higher than 3 out of 4 MLP-based models. Among the MLP-based models, ρ statistic is calculated as 0.95 for the LSTAR-LST-APGARCH-MLP. Overall, the MLP- and RBF-based models provide improvement over the GARCH and LSTAR-LST-GARCH

two regime variants. The results at this stage showed a general improvement of the RBF- and MLP-based models over the LSTAR-LST-GARCH models. To obtain conclusions, the out-of-sample forecasting capabilities of the models should be evaluated. One-step-ahead forecast results are given in Table 5.

The one-step-ahead forecast RMSE is the lowest for the LSTAR-LST-APGARCH-MLP (RMSE = 0.00000091651 514) followed by the LSTAR-LST-FIGARCH-MLP (RMSE = 0.00000118743421), LSTAR-LST-FIAPGARCH-MLP (RMSE = 0.00000399624824), and LSTAR-LST-GARCH-MLP (RMSE = 0.00000474763099) models. Compared to the LSTAR-LST-GARCH models, the models provided significant improvement. Overall result is that, LSTAR-LST-GARCH-MLP models provided the highest one-step-ahead forecast accuracy followed by the LSTAR-LST-GARCH-RBF models.

Models are evaluated for their generalization capabilities in the larger out-of-sample horizons in terms of the MSE, RMSE, and MAE criteria. Results are given in Table 6 for a total of 16 models. The forecast horizon is selected as 2, 10, and 40 days ahead to evaluate the models' performances in longer horizons.

Table 5 One-step-ahead forecast results

	1. LSTAR-LST-GARCH-MLP(5:2:2:6:1)	2. LSTAR-LST-APGARCH-MLP(5:2:2:11:1)	3. LSTAR-LST-FIGARCH-MLP(5:2:2:8:1)	4. LSTAR- LST-FIAPGARCH-MLP(5:2:2:4:1)
MSE	0.00000000002254	0.00000000000084	0.00000000000141	0.00000000001597
MAE	0.00000332503469	0.00000069276605	0.00000093842087	0.00000301391256
MRSE	0.01202169456515	0.00035311744060	0.00074167725712	0.00942662726642
MRAE	0.07334809470262	0.01433928143676	0.02207744574166	0.07055172460999
ρ	0.93236330386609	0.99869776882616	0.99667226091915	0.90707791817108
RMSE	0.00000474763099	0.00000091651514	0.00000118743421	0.00000399624824
	1. LSTAR-LST-GARCH-RBF(5:2:2:28:1)	2. LSTAR-LST-APGARCH-RBF (5:2:2:26:1)	3. LSTAR-LST-FIGARCH-RBF (5:2:2:30:1)	4. LSTAR-LST-FIAPGARCH-RBF (5:2:2:27:1)
MSE	0.00000000003826	0.00000000002264	0.00000000002363	0.00000000002017
MAE	0.00000467340330	0.00000364627551	0.00000367533719	0.00000346294812
MRSE	0.01807061475042	0.01117553124815	0.01527277725031	0.01142473379382
MRAE	0.09949911206105	0.07941845168250	0.08899030858628	0.07972759121250
ρ	0.88209622896126	0.93220341293641	0.86380964347867	0.88099209829884
RMSE	0.00000618546684	0.00000475815090	0.00000486106984	0.00000449110231

MSE mean squared error, *MAE*, mean absolute error, *MRSE* mean relative absolute error, *MRAE* mean relative absolute error, *RMSE* root mean squared error. ρ shows the Pearson's correlation coefficient

In Table 6, models with the lowest RMSE, is denoted in bold within each model group for the above-mentioned out-of-sample forecast horizons. As a typical, for two days ahead, within the single regime GARCH type models, the lowest RMSE error is achieved with the FIGARCH model. At the second part, where the LSTAR-LST augmented two regime variants are evaluated, the lowest RMSE is achieved with the LSTAR-LST-GARCH model. Further, among the MLP neural networks augmented LSTAR-LST-GARCH-MLP models, the lowest RMSE is obtained with the LSTAR-LST-GARCH-MLP model. Additionally, among the RBF neural networks augmented variants, the lowest RMSE is achieved by the LSTAR-LST-GARCH-RBF model. Therefore, the models denoted with a RMSE value in bold represent the lowest RMSE achieved "within" the relevant model group only that consists of 4 different models only. Furthermore, a total of 16 different models, the baseline GARCH, their LSTAR-LST augmented two regime variants, the neural networks arhitecture and learning algorithm augmented models (i.e. RBF and MLP based 8 models) are ranked starting from the 1st towards the 4th model in terms of RMSE again. The models that take the 1st, 2nd, 3rd and 4th places are denoted accordingly. Following this procedure, "the best 4" are reported seperately for 3 different forecast horizons i.e. for 2, 10, and 40 days to evaluate the estimated models for their forecast capabilities and to determine if the improvements exist not only in short horizons such as the 2 days ahead forecasts, but also in longer horizons.

The models in the first column have the GARC1H architecture in common followed by its nonlinear LSTAR and MLP, RBF augmentations. A significant decrease in RMSE, MAE, and MSE criteria is achieved as we move from single-regime GARCH model to LSTAR-LST-GARCH, LSTAR-LST-GARCH-MLP, and LSTAR-LST-GARCH-RBF. The RMSE reported for GARCH model is 0.000027 and 0.0000047 for the LSTAR-LST-GARCH, showing a 82.6 % decrease in the RMSE compared to the baseline GARCH for 2 days ahead forecasts. For the MLP augmented LSTAR-LST-GARCH-MLP, the RMSE is calculated as 0.00000134 which shows a 95 % decrease compared to the single-regime GARCH model. Hence, the LSTAR-GARCH model without neural networks provides improvement over the baseline GARCH for 2 days ahead. The LSTAR-GARCH-RBF model has a RMSE = 0.000005003, and performs almost equal to the LSTAR-LST-GARCH model.

For the baseline GARCH, RMSE = 0.0000519 and 0.000051 for 10 and 40 days ahead forecasts, while for the LSTAR-LST-GARCH, the RMSE statistics are calculated as 0.0000504 and 0.0000497. Accordingly, LSTAR-LST-GARCH performs better compared to the baseline GARCH, however, the improvement is limited. The results show that the predictive gains from the LSTAR-LST-GARCH suffer for longer horizons in the out-of-sample forecasts.

The LSTAR-LST-GARCH-RBF and MLP models aim at augmenting the forecasting capabilities of the LSTAR-LST-GARCH in long forecast horizons. For 10 days ahead, the RMSE is calculated as 0.0000016 for the LSTAR-LST-GARCH-MLP and is 0.0000047 for the LSTAR-LST-GARCH-RBF models. For 40 days ahead, the LSTAR-

Table 6 Out-of-sample forecast statistics

	GARCH	APGARCH	FIGARCH	FIAPGARCH
2 days				
RMSE	0.00002665333000	0.00002516346558	**0.00002442130218**	0.00002661578479
MAE	0.00002603000000	0.00002454000000	0.00002377000000	0.00002602000000
MSE	0.00000000071040	0.00000000063320	0.00000000059640	0.00000000070840

	LSTAR-LST-GARCH	LSTAR-LST-APGARCH	LSTAR-LST-FIGARCH	LSTAR-LST-FIAPGARCH
2 days				
RMSE	**0.00000470106371 (3rd)**	0.00000752329715	0.00002251888097	0.00002191506331
MAE	0.00001306620000	0.00001233097000	0.00004970591220	0.00006855640160
MSE	0.00000000002210	0.00000000005660	0.00000000050710	0.00000000048027

	LSTAR-LST-GARCH-MLP	LSTAR-LST-APGARCH-MLP	LSTAR-LST-FIGARCH-MLP	LSTAR-LST-FIAPGARCH-MLP
2 days				
RMSE	**0.00000134164079 (1st)**	0.00000237907545 **(2nd)**	0.00000622334315	0.00000671043963
MAE	0.00000130964370	0.00000226002346	0.00000557109994	0.00000618080882
MSE	0.00000000000180	0.00000000000566	0.00000000003873	0.00000000004503

	LSTAR-LST-GARCH-RBF	LSTAR-LST-APGARCH-RBF	LSTAR-LST-FIGARCH-RBF	LSTAR-LST-FIAPGARCH-RBF
2 days				
RMSE	**0.00000500299910 (4th)**	0.00000966488489	0.00000600999168	0.00000666933280
MAE	0.00000480348034	0.00000951497399	0.00000600977507	0.00000666288849
MSE	0.00000000002503	0.00000000009341	0.00000000003612	0.00000000004448

	GARCH	APGARCH	FIGARCH	FIAPGARCH
10 days				
RMSE	**0.00005195190083**	0.00005236410985	0.00005220153254	0.00005235456045
MAE	0.00004212000000	0.00004099000000	0.00004127000000	0.00004116000000
MSE	0.00000000269900	0.00000000274200	0.00000000272500	0.00000000274100

	LSTAR-LST-GARCH	LSTAR-LST-APGARCH	LSTAR-LST-FIGARCH	LSTAR-LST-FIAPGARCH
10 days				
RMSE	0.00005047771786	0.00011618950039	**0.00002413917977**	0.00005954829972
MAE	0.00003265000000	0.00011570000000	0.00002169000000	0.00004441000000
MSE	0.00000000254800	0.00000001350000	0.00000000058270	0.00000000354600

	LSTAR-LST-GARCH-MLP	LSTAR-LST-APGARCH-MLP	LSTAR-LST-FIGARCH-MLP	LSTAR-LST-FIAPGARCH-MLP
10 days				
RMSE	**0.00000162788206 (1st)**	0.00000544242593 **(4th)**	0.00000561515806	0.00000611800621
MAE	0.00000157805413	0.00000442900285	0.00000462606822	0.00000508919208
MSE	0.00000000000265	0.00000000002962	0.00000000003153	0.00000000003743

	LSTAR-LST-GARCH-RBF	LSTAR-LST-APGARCH-RBF	LSTAR-LST-FIGARCH-RBF	LSTAR-LST-FIAPGARCH-RBF
10 days				
RMSE	**0.00000470850295 (2nd)**	0.00000558121850	0.00000477388730 **(3rd)**	0.00000682129020
MAE	0.00000335796420	0.00000448305125	0.00000391396811	0.00000610994458
MSE	0.00000000002217	0.00000000003115	0.00000000002279	0.00000000004653

Table 6 continued

	GARCH	APGARCH	FIGARCH	FIAPGARCH
40 days				
RMSE	0.00005116639522	0.00004876474136	0.00005105878964	**0.00004737087713**
MAE	0.00004242000000	0.00003854000000	0.00004182000000	0.00003692000000
MSE	0.00000000261800	0.00000000237800	0.00000000260700	0.00000000224400
	LSTAR-LST-GARCH	LSTAR-LST-APGARCH	LSTAR-LST-FIGARCH	LSTAR-LST-FIAPGARCH
40 days				
RMSE	0.00004972926704	0.00014869431731	**0.00013345411196**	0.00013667479651
MAE	0.00003266120910	0.00014800000000	0.00008539000000	0.00010590000000
MSE	0.00000000247300	0.00000002211000	0.00000001781000	0.00000001868000
	LSTAR-LST-GARCH-MLP	LSTAR-LST-APGARCH-MLP	LSTAR-LST-FIGARCH-MLP	LSTAR-LST-FIAPGARCH-MLP
40 days				
RMSE	**0.00000243926218 (1st)**	0.00000425440948 **(2nd)**	0.00000442605920 **(3rd)**	0.00000475078941
MAE	0.00000190278855	0.00000340258977	0.00000370453230	0.00000387005690
MSE	0.00000000000595	0.00000000001810	0.00000000001959	0.00000000002257
	LSTAR-LST-GARCH-RBF	LSTAR-LST-APGARCH-RBF	LSTAR-LST-FIGARCH-RBF	LSTAR-LST-FIAPGARCH-RBF
40 days				
RMSE	0.00000511370707	**0.00000449332839 (4th)**	0.00000565420198	0.00000606217783
MAE	0.00000394412289	0.00000375403891	0.00000460309756	0.00000507094872
MSE	0.00000000002615	0.00000000002019	0.00000000003197	0.00000000003675

Statistics are defined as follows. *RMSE* root mean squared error, *MAE* mean absolute error. Models are ordered from the lowest error criteria (for both RMSE and MAE) to the highest

The rank of each model is given in () brackets. Models are evaluated in terms of their capability in forecasting the conditional mean and variance separately

LST-GARCH-RBF has a RMSE of 0.0000051, while the RMSE for the LSTAR-LST-GARCH-MLP is 0.0000024 and is almost halve of that obtained for LSTAR-LST-GARCH and its RBF variant. The results show that RBF-based model failed to provide significant improvements over the LSTAR-LST-GARCH, though MLP-based variant had the lowest RMSE, MSE, and MAE statistics. However, at this stage, the conclusions only show that by keeping the GARCH architecture constant, the MLP model showed significant forecast accuracy gains over the models in the first column. Note that, the models with FIGARCH, APGARCH, and FIAPGARCH architectures provided different results.

If an overlook is to be presented, as the forecast horizon is enlarged to 10 and 40 days ahead, the results provide a drastic improvement in longer horizons for the MLP-based models followed by the RBF-based variants. For comparative purposes, the models in each row model group are evaluated among themselves and the model with the lowest RMSE and MSE statistics is denoted in bold for 2, 10, and 40 days ahead forecasts. Additionally, the models are

ranked according to the RMSE statistics from lowest to highest to simplify the evaluation. For 2 days ahead, the FIGARCH model has the best forecast accuracy among the single-regime GARCH models (RMSE = 0.0000244). Among the two regime models, the LSTAR-LST-GARCH model has the best forecast capability (RMSE = 0.0000047). Among the MLP-based models, the LSTAR-LST-GARCH-MLP model has the highest forecast accuracy (RMSE = 0.00000134). Among the RBF-based variants, the LSTAR-LST-GARCH-RBF model has the best forecast accuracy (RMSE = 0.0000050). For 10 days ahead, the GARCH, the LSTAR-LST-FIGARCH, the LSTAR-LST-GARCH-MLP, and the LSTAR-LST-GARCH-RBF are the models with the lowest RMSE (and lowest MSE) among their own model group. For 40 days ahead, the FIAPGARCH, the LSTAR-LST-FIGARCH, the LSTAR-LST-GARCH-MLP, and the LSTAR-LST-APGARCH-RBF have the lowest RMSE statistics among their own model group. For different horizons, as the horizon moves from 2 to 40 days ahead MLP-based models showed significant improvement followed by the RBF-based GARCH models. For simplicity, the first 4 models are to be reported. For 2 days ahead, the LSTAR-LST-GARCH-MLP is the 1st, while the LSTAR-LST-APGARCH-MLP is the 2nd, the LSTAR-LST-GARCH is the 3rd, and the LSTAR-LST-GARCH-RBF is the 4th. For 10 days ahead forecasts, the LSTAR-LST-GARCH-MLP is the 1st, the LSTAR-LST-GARCH-RBF is the 2nd, the LSTAR-LST-GARCH-RBF is the 3rd and the LSTAR-LST-APGARCH-MLP model is the 4th. For 40 days ahead, among 16 models estimated, the LSTAR-LST-GARCH-MLP takes the 1st place with the lowest RMSE (0.0000024), followed by the LSTAR-LST-APGARCH-MLP taking the 2nd place (RMSE = 0.00000425). LSTAR-LST-FIGARCH-MLP takes the 3rd place (RMSE = 0.000004426) and the LSTAR-LST-APGARCH-RBF model is the 4th model (RMSE = 0.00000449).

Results supported the following conclusions for modeling and forecasting volatility in crude oil prices: (i) The nonlinear volatility models with STAR type nonlinearity namely, LSTAR-LST-GARCH family provided significant gains in terms of in-sample (one-step-ahead) forecasting accuracy and these models provided significant improvement over their single-regime GARCH variants. Further, for short horizons, the LSTAR-LST-GARCH family provided significant forecasting gains over their single-regime variants. (ii) RBF- and MLP-based neural networks augmentations of the LSTAR-LST-GARCH family models provided improved modeling capabilities for the crude oil prices that are subject to nonlinearity, asymmetry, and leptokurtic distribution. (iii) LSTAR-LST-GARCH-MLP and LSTAR-LST-GARCH-RBF showed gains in forecast capabilities which concentrate especially on the out-of-sample forecasting. Among the RBF and MLP augmented LSTAR-LST-GARCH family, the MLP-based models augmented the forecast accuracy of the LSTAR-LST-GARCH models followed by the RBF models. Additionally, the fractional integration and asymmetric power terms increased the forecast accuracy separately, though still the GARCH- and APGARCH-based MLP and RBF models provide satisfactory results, while the FIAPGARCH specification provided comparatively low gains in terms of forecast capabilities. It should be noted that the results are gathered for the daily Brent oil data set and cannot be generalized to all financial time series. Since certain financial time series such as the stock index returns possess strong asymmetric power effects and fractional integration, the LSTAR-LST-FIAPGARCH and its MLP/RBF variants may provide improved forecasting capabilities and therefore to obtain generalized results, the models should be evaluated for different financial time series in the developed and developing markets. The overall result of the empirical analysis suggests that nonlinear augmentations of

GARCH models for forecasting crude oil prices with the neural network architectures and nonlinear econometric techniques provide gains for the researchers and policy makers that aim at evaluating the paths followed by oil price time series.

4.3 Policy implications

The petrol price is an important variable for explaining business cycles and economic growth. As a result, petrol prices exhibits a large volatility not only through the channels of supply and demand, but also through political factors in addition to OPEC decisions. Volatility in petrol prices has strong impacts on economic variables such as economic growth, industrial production, and employment decisions in labor markets, not to mention its effect on the current account deficits and financial markets through various channels, since crude oil is also a financial commodity traded in spot and future markets. The results obtained in the study through the LSTAR-LST-GARCH-RBF and MLP models showed that, the adjustment process in oil prices do not occur instantaneously to new information. As shown by McMillan and Speight (2006) and Monoyios and Sarno (2002), low liquidity and infrequent trading in imperfect markets cause delays in the adjustment process after new information in financial markets. The results coincided with the fact that, increases in volatility are generally short lived; however, due to the persistence in oil prices, these effects may lead to long-lived effects in terms of persistence. The positive and large fractional coefficient estimates, in addition to large estimates of asymmetric power terms in both regimes justify the fact that shocks have relatively persistence effects; hence, within a political perspective, the governments should evaluate the oil prices and global factors very cautiously and simultaneously, policy makers should keep the interventions at the modest levels to avoid large fluctuations in petrol prices. Further, the long memory characteristics accelerate the expected temporary effects of these shocks, thus the persistency might increase the impacts of the oil shocks. Therefore, policy interventions should be kept at very modest levels to avoid large fluctuations.

If the results are to be summarized, the oil price possesses important characteristics such as nonlinearity, asymmetry, and transition effects, in addition to its fractionally integrated persistence effects. The policy maker and the researcher should evaluate the policies to be applied with great care. However, the nonlinear volatility models that incorporate fractional integration and power terms capture the data generating process more effectively; therefore, might be utilized important tools for policies. On the other hand, within a political perspective, the policies focusing on stabilization of volatility of this crucial commodity may have destabilizing effects on the production and on the financial markets. Since crude oil prices are interlinked to various financial assets, this result translates itself to different derivatives and the economy, and this destabilization effect is largely under the influence of persistence in oil prices and also in the external shocks that oil prices are subject to. As a result, policies possible destabilizing effects without taking persistence into account result in additional effects in various markets. Secondly, the estimation sample in the study corresponded to a period with large oil shocks and economic crises periods. On the other hand, following the general methodology, the out-of-sample results are obtained for a period corresponding to year 2014, a relatively stable period. The forecasting practice in the paper showed significant gains in terms of forecast accuracy. On the other hand, through incorporating nonlinearity in the GARCH processes, the performance of these models would improve under unexpected changes in oil prices. Further, the utilization of the nonlinear models helps the policy maker by evaluating threshold characteristics of these models. As a result, the nonlinear models that incorporate neural networks' forecast capabilities with nonlinear econometric techniques are to be considered as tools for the investors and policy makers. However, the evaluation of the estimates provided by nonlinear models should always be evaluated with caution not to mention many external factors that lead to fluctuations and sudden/sharp changes in oil prices.

5 Conclusion

The study aimed at evaluating a new group of nonlinear models that combine the forecasting capabilities of MLP and RBF type neural networks with GARCH type volatility models and that augment LSTAR type nonlinear econometric time series models proposed by Luukkonen et al. (1988) and Terasvirta (1994). The proposed LSTAR-LST-GARCH-MLP and LSTAR-LST-GARCH-RBF family models aim at modeling not only the conditional mean processes but also the conditional variance simultaneously with STAR type nonlinearity, allowing the transition between the regimes to be captured with logistic transition functions. Accordingly, at the first stage, crude oil prices were modeled with baseline GARCH models with fractional integration and asymmetric power terms. At the second stage, LSTAR type nonlinear architecture was introduced to the baseline models to obtain the LSTAR-LST-GARCH models. At the third stage, LSTAR-LST-GARCH models were augmented with RBF and MLP neural networks to improve the modeling and forecasting capabilities of the researcher aiming at forecasting crude

oil prices. Accordingly, the models were compared in terms of MSE, RMSE, and MAE error criteria for in-sample and out-of-sample forecasts. The results showed that the LSTAR based and neural network augmented models provided significant gains in terms of modeling the daily returns of oil prices when compared with the results of the baseline GARCH family models. The results also showed that the LSTAR-LST-GARCH-RBF and LSTAR-LST-GARCH-MLP models provided significant gains in modeling petrol prices and in forecasting out-of-sample oil price returns. Following the findings of the paper, the future studies should aim at modeling different financial series that are subject to nonlinearity and volatility to test the forecasting capabilities of neural network algorithms and architectures.

The crude oil daily returns are evaluated as a result of their characteristics which could be classified as possessing strong nonlinearity, volatility defined with excess kurtosis. The nonlinearity inherited in crude oil prices has strong implications for regulators and short-term trading strategies. The oil prices possess important characteristics such as nonlinearity, asymmetry, transition effects, fractionally integrated, and persistence effects that should lead the policy maker and the researcher to evaluate the policies to be applied with great care; hence, the nonlinear volatility models that incorporate fractional integration and power terms capture the data generating process more effectively, therefore, provide important tools for policy makers.

Acknowledgments We thank anonymous referees for their useful comments.

References

Abramson B, Finizza A. Using belief networks to forecast oil prices. Int J Forecast. 1991;7(3):299–315.

Adrangi B, Chatrath A, Dhanda KK, et al. Chaos in oil prices? Evidence from futures markets. Energy Econ. 2001;23(4): 405–25.

Alizadeh A, Mafinezhad K. Monthly Brent oil price forecasting using artificial neural networks and a crisis index. In: Electronics and information engineering (ICEIE), international conference. 2010;2:465–8.

Alvarez-Ramirez J, Alvarez J, Rodriguez E. Short-term predictability of crude oil markets: a detrended fluctuation analysis approach. Energy Econ. 2008;30:2645–56.

Baillie R, Bollersev T, Mikkelson HO. Fractionally integrated generalized autoregressive conditional heteroskedasticity. J Econom. 1996;74:3–30.

Barone-Adesi G, Bourgoin F, Giannopoulos K. Don't look back. Risk. 1998;11:100–3.

Bartlett PL. For valid generalization, the size of the weights is more important than the size of the network. In: Mozer MC, Jordan MI, Petsche T, editors. Advances in neural information processing systems. Cambridge: MIT Press; 1997. p. 134–40.

Bildirici M, Ersin Ö. Forecasting oil prices: smooth transition and neural network augmented GARCH family models. J Pet Sci Eng. 2013;109:230–40.

Bildirici M, Ersin Ö. Improving forecasts of GARCH family models with the artificial neural networks: an application to the daily returns in Istanbul Stock Exchange. Expert Syst Appl. 2009;36:7355–62.

Bishop CM. Neural networks for pattern recognition. Oxford: Oxford University Press; 1995.

Bollerslev T. Generalized autoregressive conditional heteroscedasticity. J Econom. 1986;31:307–27.

Bollerslev T, Mikkelsen HO. Modeling and pricing long memory in stock market volatility. J Econom. 1996;73(1):151–84.

Chan F, McAleer M. Estimating smooth transition autoregressive models with GARCH errors in the presence of extreme observations and outliers. Appl Fin Econ. 2003;13(8):581–92.

Chan F, McAleer M. Maximum likelihood estimation of STAR and STAR-GARCH models: theory and Monte Carlo evidence. J Appl Econom. 2002;17(5):509–34.

Chung C. Estimating the fractionally integrated GARCH model, Working paper. Taiwan: National Taiwan University; 1999.

Coelho LS, Santos A. RBF neural network model with GARCH errors: application to energy price forecasting. Elect Power Syst Res. 2011;81(1):74–83.

Conrad C, Haag BR. Inequality constraints in the fractionally integrated GARCH model. J Fin Econom. 2006;4(3):413–49.

Conrad C, Rittler D, Rotfuss W. Modeling and explaining the dynamics of European Union allowance prices at high-frequency, Working Paper 0497. Heidelberg: University of Heidelberg, Department of Economics; 2010.

Cybenko G. Approximations by superpositions of sigmoidal functions. Math Control Signals Syst. 1989;2:303–14.

Ding Z, Granger CWJ, Engle RF. A long memory property of stock market returns and a new model. J Emp Fin. 1983;1:83–106.

Do AQ, Grudnitski GA. Neural network approach to residential property appraisal. Real Estate Apprais. 1992;Dec:38–45.

Donaldson RG, Kamstra M. An artificial neural network-GARCH model for international stock return volatility. J Emp Fin. 1997;4:17–46.

Dutta S, Shekhar S. Bond rating: a non-conservative application of neural networks. IEEE international conference on neural networks. 1998;443–58.

Elsharkawy A. Modeling the properties of crude oil and gas systems using RBF network. In: SPE Asia Pacific oil and gas conference and exhibition, 12–14 Oct 1998, Perth, Australia. 1998. doi:10.2118/49961-MS:1-12.

Engle RF. Autoregressive conditional heteroscedasticity with estimates of the variance of United Kingdom inflation. Econometrica. 1982;50:987–1007.

Ewing BT, Maliq F. Volatility transmission between gold and oil futures under structural breaks. Int Rev Econ Financ. 2013;25:113–21.

Faraway J, Chatfield C. Time series forecasting with neural networks: a comparative study using airline data. Appl Stat. 1998;47(2):231–50.

Fausett L. Fundamentals of neural networks. Englewood Cliffs: Prentice Hall; 1994.

Franses PH, van Dijk D. Non-linear time series models in empirical finance. Cambridge: Cambridge University Press; 2000.

Freisleben B. Stock market prediction with backpropagation networks. In: Proceeding of the 5th international conference on industrial and engineering. Applications of applied intelligent and expert system, 9–12 June 1992. Berlin: Springer; 1992. p. 451–60.

Gencay R, Liu T. Nonlinear modelling and prediction with feedforward and recurrent networks. Physica D. 1997;108:119–34.

Gencay R, Stengos T. Moving average rules, volume and the predictability of security returns with feed-forward networks. J Forecast. 1998;17:401–14.

Gencay R, Stengos T. Technical trading rules and the size of the risk premium in security returns. Stud Nonlin Dyn Econom. 1997;2:23–34.

Glosten LR, Jagannathan R, Runkle D. On the relation between the expected value and the volatility of the nominal excess return on stocks. J Fin. 1993;48:1779–801.

Goldberg DE. Genetic algorithms and walsh functions: part II, deception and its analysis. Complex Syst. 1989;3:153–71.

Gupta A, Lam M. The weight decay backpropagation for generalizations with missing values. Ann Oper Res. 1998;78:165–87.

Hagen R. How is the international price of a particular crude determined? OPEC Rev. 1994;18(1):145–58.

Hamilton JD. Historical causes of postwar oil shocks and recessions. Energy J. 1985;6:97–116.

Hamilton JD. Oil and the macroeconomy since world war II. J Polit Econ. 1983;91:228–48.

Haykin S. Neural networks. Macmillan: A Comprehensive Foundation; 1994.

Hutchinson JM, Lo AJ, Poggio T. A nonparametric approach to pricing and hedging derivative securities via learning networks. J Fin. 1994;49:851–89.

Kaboudan MA. Compumetric forecasting of crude oil prices. IEEE Congr on Evol Comp. 2001;1:283–7.

Kanas A. Non-linear forecasts of stock returns. J Forecast. 2003;22(4):299–315.

Kanas A, Yannopoulos A. Comparing linear and nonlinear forecasts for stock returns. Int Rev Econ Fin. 2001;10(4):383–98.

Kılıç R. Long memory and nonlinearity in conditional variances: a smooth transition FIGARCH model. J Emp Fin. 2011;18(2):368–78.

Krogh A, Hertz J. A simple weight decay can improve generalization. In: Moody J, Hanson S, Lippmann R, editors. Advances in neural information processing systems 4. San Mateo: Morgan Kauffmann Publishers; 1995.

Lai TL, Wong SP. Stochastic neural networks with applications to nonlinear time series. J Am Stat Assoc. 2001;96:968–81.

Lanza A, Manera M, Giovannini M. Modeling and forecasting cointegrated relationships among heavy oil and product prices. Energy Econ. 2005;27(6):831–48.

Lee YH, Liu HC, Chiu CL. Nonlinear basis dynamics for the brent crude oil markets and behavioral interpretation: a STAR-GARCH approach. Int Res J Fin Econom. 2008;14:51–60.

Liu D, Zhang L. China stock market regimes prediction with artificial neural network and markov regime switching. World Congr Eng (WCE). 2010;1:378–83.

Lundbergh S, Terasvirta T. Modeling economic high-frequency time series with STAR-STGARCH models. Stockholm School of Economics: Dep Econ; 1998.

Luukkonen R, Saikkonnen P, Terasvirta T. Testing linearity against smooth transition autoregressive models. Biometrika. 1988;75:491–9.

Malliaris M, Malliaris S. Forecasting energy product prices. In: Proceedings of international joint conference on neural networks. Montreal, Canada. 2005;5:3284–9.

McMillan DG, Speight AEH. Nonlinear dynamics and competing behavioral interpretations: evidence from intra-day FTSE-100 index and futures data. J Futur Mark. 2006;26(4):343–68.

Mirmirani S, Li HC. A comparison of VAR and neural networks with genetic algorithm in forecasting price of oil. Adv Econom. 2004;19:203–23.

Monoyios M, Sarno L. Mean reversion in stock index futures markets: a nonlinear analysis. J Futur Mark. 2002;22:285–314.

Patterson D. Artifical neural networks. Singapore: Prentice Hall; 1996.

Qunli W, Ge H, Xiaodong C. Crude oil price forecasting with an improved model based on wavelet transform and RBF neural network. Proc Int Forum Inf Technol Appl. 2009;1:231–4.

Refenes AP, Burgess AN, Bentz Y. Neural networks in financial engineering: a study in methodology. IEEE Trans Neural Netw. 1997;8:1222–67.

Santos AAP, Coelho LS, Klein CE. Forecasting electricity prices using a RBF neural network with GARCH errors. In: 2010 international joint conference on neural networks (IJCNN), 18–23 July 2010. p. 1–8.

Shively PA. The nonlinear dynamics of stock prices. Q Rev Econ Fin. 2003;43(3):505–17.

Stevens P. The determination of oil prices 1945–1995. Energy Policy. 1995;23(10):861–70.

Tam KY, Kiang MY. Managerial applications of neural networks: the case of bank failure predictions. J Manag Sci. 1992;38(7):926–47.

Tang L, Hammoudeh S. An empirical exploration of the world oil price under the target zone model. Energy Econ. 2002;24:577–96.

Terasvirta T. Specification, estimation, and evaluation of smooth transition autoregressive models. J Am Stat Assoc. 1994;89:208–18.

Tse Y. The conditional heteroscedasticity of the yen-dollar exchange rate. J Appl Econom. 1998;13:49–55.

Tse YK, Tsui AKC. Conditional volatility in foreign exchange rates: evidence fro the Malaysian ringgit and Singapore dollar. Pacific-Basin Fin J. 1997;5:345–56.

Wang SY, Yu L, Lai KK. Crude oil price forecasting with TEI@I methodology. J Syst Sci Complex. 2005;18(2):145–66.

Weigend AS, Rumelhart DE, Huberman BA. Generalization by weight-elimination with application to forecasting. In: Lippmann RP, Moody J, Touretzky DS, editors. Advances in neural information processing systems 3. San Mateo: Morgan Kaufmann; 1991.

Weigend AS, Gershenfeld NA. Time series prediction: forecasting the future and understanding the past. Reading: Addison-Wesley; 1994.

White H. Artificial neural networks: approximation and learning theory. Oxford: Blackwell; 1992.

Wright GB. Radial basis function interpolation: numerical and analytical developments, Thesis. University of Colorado. 2003.

Ye M, Zyren J, Shore J. Forecasting short-run crude oil price using high and low-inventory variables. Energy Policy. 2006;34:2736–43.

Ye M, Zyren J, Shore J. A monthly crude oil spot price forecasting model using relative inventories. Int J Forecast. 2005;21:491–501.

Ye M, Zyren J, Shore J. Forecasting crude oil spot price using OECD petroleum inventory levels. Int Adv Econ Res. 2002;8:324–34.

Yu L, Lai KK, Wang SY, et al. Oil price forecasting with an EMD-based multiscale neural network learning paradigm, vol. 4489., Lecture Notes in Computer Science Berlin: Springer; 2007. p. 925–32.

Yu L, Wang S, Lai KK. Forecasting crude oil price with an EMD-based neural network ensemble learning paradigm. Energy Econ. 2008;30:2623–35.

Zakoian JM. Threshold heteroscedastic models. J Econ Dyn Control. 1994;18:931–55.

17

Numerical analysis of hydroabrasion in a hydrocyclone

Mehdi Azimian[1] · Hans-Jörg Bart[1]

Abstract The velocity profiles and separation efficiency curves of a hydrocyclone were predicted by an Euler–Euler approach using a computational fluid dynamics tool ANSYS-CFX 14.5. The Euler–Euler approach is capable of considering the particle–particle interactions and is appropriate for highly laden liquid–solid mixtures. Predicted results were compared and validated with experimental results and showed a considerably good agreement. An increase in the particle cut size with increasing solid concentration of the inlet mixture flow was observed and discussed. In addition to this, the erosion on hydrocyclone walls constructed from stainless steel 410, eroded by sand particles (mainly SiO_2), was predicted with the Euler–Lagrange approach. In this approach, the abrasive solid particles were traced in a Lagrangian reference frame as discrete particles. The increases in the input flow velocity, solid concentration, and the particle size have increased the erosion at the upper part of the cylindrical body of the hydrocyclone, where the tangential inlet flow enters the hydrocyclone. The erosion density in the area between the cylindrical to conical body area, in comparison to other parts of the hydrocyclone, also increased considerably. Moreover, it was observed that an increase in the particle shape factor from 0.1 to 1.0 leads to a decrease of almost 70 % in the average erosion density of the hydrocyclone wall surfaces.

Keywords Hydrocyclone · Computational fluid dynamics (CFD) · Separation efficiency · Erosion rate · Erosion impact parameters

1 Introduction

Hydrocyclones have been applied for many engineering processes for more than a century. Currently they are broadly used in industry for removal, classification or separation of particles in many mechanical and chemical processes such as reactors, dryers, removal of catalyst from liquids, wet flue gas desulfurization processes, treatment of waste water streams and advanced coal utilization such as fluidized bed combustion (Milin et al. 1992).

Desander hydrocyclones are used to provide reliable and efficient separation of sand and solid particles from water, condensate flows, and gas streams. They have proven to be a valuable part of many oil and gas production facilities in the petroleum industry. The desander cyclones are pressure-driven separators that require a pressure drop across the unit to cause separation of the solids from the bulk phase (water, oil, gas, etc.). It is typical to collect the solids in a closed underflow container or vessel, and periodically dump these solids. Desanded water/oil/gas in the central core section reverses direction and is forced out through the central vortex finder towards the overflow at the top of the cyclone (Process Group Pty Ltd. 2012).

Particles are exposed to a strong centrifugal force field sometimes up to $3000\times$ gravitational acceleration (g). The flow enters the hydrocyclone with a linear motion; a high intensity force is then generated by movement of the slurry in a curving path through the cylindrical body. The hydraulic residence time in hydrocyclones is in a range of 1–2 s. This is an advantage when compared to the

✉ Hans-Jörg Bart
bart@mv.uni-kl.de

[1] Department of Mechanical and Process Engineering, University of Kaiserslautern, 67663 Kaiserslautern, Germany

Edited by Yan-Hua Sun

traditional gravity separators, which have residence times in the range of some couple of minutes. The solid particles are exposed to centrifugal acceleration and therefore, the settlement rate of particles will be varied in accordance to their various sizes, densities, and shapes. Due to an open overflow and high centrifugal forces, an air core may form in the hydrocyclone, which will increase the turbulent fluctuation and decrease separation efficiency. Although the cyclone geometries are quite simple, the flow behavior inside the cyclones is relatively complex. Occurrence of a strong air core in the center of the hydrocyclone causes the measurement of tangential and axial velocities to be a difficult task (Huang et al. 2009). However, with virtual experiments based on CFD techniques one should overcome these problems.

Pericleous and Rhodes (1986) are one of the first who predicted successfully the flow in a hydrocyclone. The improved Prandtl mixing length model was applied by them in order to simulate a separator with a diameter of 200 mm. They compared the velocity distribution with experimental results gained by laser Doppler velocimetry (LDV). Solutions to the equation of turbulent flow motion and a comparison of the solutions with LDV measured flow pattern of a 75-mm hydrocyclone was successfully performed by Hsieh and Rajamani (1991). A series of works on the simulation of hydrocyclones using the incompressible Navier–Stokes equations, complemented by an adequate turbulence model, have proven to be suitable for modeling the flow in hydrocyclones (Cullivan et al. 2004). The application of developed CFD techniques will alleviate some of the difficulties and problems when using models with empirical correlations (Vieira et al. 2011). Investigation of numerical methods for simulating a hydrocyclone was performed by Xu et al. (2009). They concluded that the re-normalization group (RNG) k–ε turbulence model was not appropriate for modeling flow in hydrocyclones. On the other hand, they found out that the Reynolds stress model (RSM) and the large eddy simulation (LES) model were able to provide better results in comparison with the experimental data. Ghadirian et al. (2013) modeled the air core using a transient two-phase simulation, with the results of the single phase runs as the initial input data. Afterwards, they performed final simulations involving particles and found out that the use of the LES turbulence model provides the best results matched to their experimental data. Monredon et al. (1992) developed mathematical models based on fluid mechanics involving some simplifying assumptions. Furthermore, Ipate and Căsăndroiu (2007) investigated the difference between the behavior of a series of various geometric configurations of hydrocyclones on the fluid dynamics. Chu and Chen (1993) applied the particle dynamics analyzer (PDA) to measure the size distribution of solid particles, concentration, and

velocity profiles in a hydrocyclone. Wang et al. (2007) simulated the air core with the volume of fluid multiphase model (VOF) that tracks the interface between air and water. The VOF method is known for its ability to conserve the mass of the traced fluid, also when the fluid interface changes its topology. This change is traced easily, so the interfaces can for example join or break apart. The focus of the current paper is on erosion aspects of the hydrocyclone wall surfaces and therefore, only water was modeled as the primary phase and solid particles were modeled as the secondary phase. Erosion in hydrocyclones is an important phenomenon, which is not studied to our best knowledge in the literature in detail. Only Utikar et al. (2010) investigated the erosion caused by solid particles entrained in the gas flow in a cyclone separator. In the present study, beside the fluid dynamics study of a hydrocyclone and determination of separation efficiency curves using an Euler–Euler (Eu–Eu) approach, the particle tracking and erosion intensity inside the hydrocyclone are investigated in detail, applying the Euler–Lagrange (Eu–La) approach. The motivation for using Eu–Eu approach is that it can consider the particle–particle interactions by introducing granular quantity parameters such as temperature, pressure or viscosity, which can be derived from the kinetic theory. Nevertheless, prediction of modeling uncertainties is challenging and the computational costs for solving the additional transport equations are quite high, particularly if multiple particle classes are present in the flow. On the other hand, application of Lagrangian tracking involves the integration of particle paths through the discretized domain. Individual particles can be tracked from the injection area at the inlet until they escape from the domain. The Eu–La model has difficulties in considering the particle–particle interactions, especially if too many particles are present.

In the literature, there is a lack of experimental results for erosion of internal surfaces of a hydrocyclone. Therefore, in the current study, the experimental results concerning flow behavior and separation efficiency are used initially to validate the Eu–Eu simulation results. This has validated the application of a proper turbulence model and correct input conditions for further simulations with the Eu–La technique.

2 Hydrocyclone geometry

A typical hydrocyclone as depicted in Fig. 1 consists of a cylindrical body, a conical body, open discharge at the bottom called the underflow outlet, a tangential inlet to the cylindrical part of the cyclone body, and an overflow discharge part. The cylindrical body is closed on its top with a plate through which passes an axially located overflow

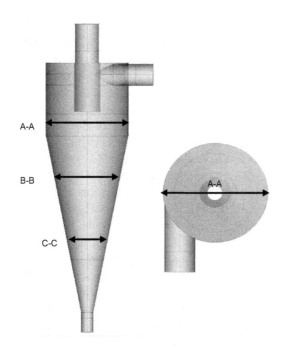

Fig. 1 Hydrocyclone with a cylindrical inlet

pipe, which is called the vortex finder (Vieira et al. 2010). In Fig. 1, A–A, B–B, and C–C are axial positions at 60, 120, and 180 mm distances from the top of the hydrocyclone, respectively. The experimental database of hydrocyclone flow of Hsieh (1988) (geometrical data see Table 1), providing velocity profiles within the hydrocyclone measured by the LDV method and also results concerning the particle classification were used to validate our CFD simulation results. Hsieh (1988) used a pumped recirculation system to measure the particle classification efficiency of hydrocyclones using a microtrac particle analyzer (Milin et al. 1992). The Euler number of the cyclone is the well-known pressure loss factor defined by the following equation:

$$Eu = \frac{2\Delta p}{\rho v_i^2} \qquad (1)$$

where Δp is the pressure drop across the cyclone; v_i is the superficial velocity in the cyclone body as the characteristic

Table 1 Hydrocyclone geometry

Part	Value
Hydrocyclone diameter (A–A), mm	75
Inlet diameter, mm	22
Vortex finder diameter, mm	22
Vortex finder length, mm	50
Cylindrical section length, mm	75
Spigot diameter, mm	11
Included cone angle, degree	20

velocity; and ρ is the density of the liquid phase. Based on the various input conditions and the resulting pressure loss, the Euler number for the hydrocyclone is simply calculated using Eq. (1). Moreover, the separation efficiency of the hydrocyclone for various input solid concentrations is presented later in this work.

The hydrocyclone geometry and subsequently the mesh were generated with ANSYS-ICEM (14.5). The structured mesh (hexahedral elements) was created using the O-grid technique as depicted in Fig. 2 with a considerably high quality. Grid study was performed to find an optimal mesh size and quality in order to have mesh-independent CFD results. The grid study was first performed by simulating the hydrocyclone in ANSYS-CFX without coupling the erosion model. It was observed that taking the outlet velocity components at overflow and underflow as the grid study criterions, a fine mesh with around 150,000 elements would be sufficient to obtain mesh independent results with a low deviation of about 0.3 %, which is presented in detail in Table 2. The deviation values in percentage as presented in Tables 2 and 3 were calculated using the following equation:

$$d_n = \left| \frac{Z_n - Z_{n-1}}{Z_{n-1}} \right| \times 100 \% \qquad (2)$$

where d_n is the deviation; Z_n and Z_{n-1} define the value of a simulation output with the mesh steps n and $n-1$, respectively. Additionally, a grid study was performed with coupling an erosion model with CFD simulations. The average erosion rate on hydrocyclone walls was considered here as the grid study parameter. It was observed that a fine mesh with around 700,000 elements, specifically with mesh refinement next to the wall surfaces, would be sufficient to obtain mesh independent results with a low deviation of 0.56 %, which is presented in detail in Table 3. Therefore,

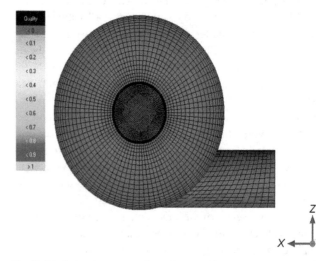

Fig. 2 Mesh density near walls and at overflow

Table 2 Grid study without consideration of an erosion model

Mesh elements	Water velocity, m/s		Deviation, %	
	Overflow	Underflow	Overflow	Underflow
10,000	2.79951	2.73324	–	–
20,000	2.92017	2.73604	4.31	0.10
50,000	2.93838	2.84202	0.62	3.87
100,000	2.93987	2.94375	0.05	3.58
150,000	2.92767	2.99231	0.41	1.65
200,000	2.93462	3.00313	0.24	0.36

Table 3 Grid study with consideration of an erosion model

Mesh elements	Average erosion, kg/(m^2 s)	Deviation, %
150,000	3.16e−07	–
300,000	2.96e−07	6.23
450,000	2.91e−07	1.74
700,000	2.88e−07	0.93
900,000	2.87e−07	0.56

the model with 150,000 mesh elements was used in Eu–Eu simulations for only the flow behavior studies and the model with 700,000 mesh elements was used in Eu–La simulations for the particle tracking and erosion studies.

3 CFD simulations

3.1 Euler–Euler approach

In the present study, an Eu–Eu approach was applied to model the highly laden liquid–solid mixture in the hydrocyclone, to predict the particle–particle interactions, the flow behavior in the hydrocyclone, and the separation efficiency. Here, the focus was on the full Eulerian approach employing momentum transport equations for each phase since it is more accurate. However, the computational effort is increased compared to the mixture model. A separate set of mass, momentum, and energy conservation equations is solved for each phase. The continuity equations for the both phases considering isothermal condition are given as Eqs. (3) and (4) for the liquid (fluid) phase and the particle phase, respectively.

$$\frac{\partial(\rho_f \varepsilon_f)}{\partial t} + \nabla \cdot (\rho_f \varepsilon_f u_f) = 0 \tag{3}$$

$$\frac{\partial(\rho_p \varepsilon_p)}{\partial t} + \nabla \cdot (\rho_p \varepsilon_p u_p) = 0 \tag{4}$$

The incompressible momentum equations for the liquid and particle phases are derived as follows:

$$\rho_f \varepsilon_f \left[\frac{\partial u_f}{\partial t} + (u_f \cdot \nabla u_f) \right] = -\varepsilon_f \nabla P + \varepsilon_f \nabla \cdot \tau_f - \varepsilon_f M \tag{5}$$

$$\rho_p \varepsilon_p \left[\frac{\partial u_p}{\partial t} + (u_p \cdot \nabla u_p) \right] = -\varepsilon_p \nabla P + \varepsilon_p \nabla \cdot \tau_p - \nabla P_p \\ + \varepsilon_p M \tag{6}$$

where ρ_f and ρ_p are the density of the fluid and the particle, respectively; u_f and u_p are the velocity of the fluid and the particle, respectively; ε_f and ε_p are the volume fraction of the fluid and particle, respectively; M is the particle–fluid interaction source term; ∇P is the pressure gradient term; ∇P_p is the particle–particle interaction term; and τ indicates the viscous shear stress tensor. Forces that impacted the particles were namely the drag force which was modeled by the Gidaspow drag model, the lift force which was modeled by the Saffman–Mei lift model, the pressure force, and the added mass force. For modeling the turbulence, the turbulence models of k–ε, k–ω, Menter's shear stress transport (SST), and baseline Reynolds stress model (BSL-RSM) were applied for the liquid (continuous) phase. For the solid phase, the zero equation model was applied, which computes the turbulence fluctuations via the data from the fluid phase.

3.2 Euler–Lagrange approach

In this approach water is treated as the continuum phase and the abrasive particles are traced in a Lagrangian reference frame. This approach is known as a discrete particle or Eu–La model. Since each particle is tracked from its injection location to the final destination point, the tracking procedure is applicable to the steady state flow analysis. In the second part of the current study, the Eu–La approach was applied to track the abrasive solid particles and to predict the location and intensity of erosion on hydrocyclone wall surfaces. For erosion computations, the particles should be traced inside the domain by considering the particles as the discrete phase. Moreover, the application of the Eu–La approach is essential when multiple particle classes are present. A simplified set of incompressible Navier–Stokes equations is solved in this approach for the liquid phase consisting of mass conservation:

$$\nabla \cdot u_f = 0 \tag{7}$$

and the momentum conservation:

$$\frac{\partial}{\partial t}(\rho_f u_f) + \nabla \cdot (\rho_f u_f^2) = -\nabla P + \nabla \cdot \tau_f + F_{D,s} \tag{8}$$

The momentum source due to the drag force of particles, $F_{D,s}$, is monitored and accumulated in every cell as the particles are passing. The particle trajectories are predicted based on the liquid phase velocity field by evaluation of a local momentum balance as follows:

$$\frac{Du_{\mathrm{p}}}{Dt} = F_{\mathrm{D}}(u_{\mathrm{f}} - u_{\mathrm{p}}) + g\left(\frac{\rho_{\mathrm{p}} - \rho_{\mathrm{f}}}{\rho_{\mathrm{p}}}\right) \qquad (9)$$

In Eq. (9), the particle acceleration is due to the drag force and the gravitational acceleration. The drag coefficient, C_{D}, can be derived from the following equation:

$$C_{\mathrm{D}} = a_1 + \frac{a_2}{\mathrm{Re}_{\mathrm{p}}} + \frac{a_3}{\mathrm{Re}_{\mathrm{p}}^2} \qquad (10)$$

where the coefficients a_{1-3} are given for various ranges of particle Reynolds number (Re_{p}) from 0.1 to 50,000 by Morsi and Alexander (1972). Twelve transport equations have to be solved when using the Eu–La model. An additional particle class can be added without the need of a further transport equation. Therefore, for consideration of different particle classes, the Eu–La approach is much cheaper than the Eu–Eu approach.

4 Boundary conditions

In the CFD simulations, the Reynolds averaged Navier–Stokes equations (RANS) for both Eu–Eu and Eu–La methods were supplemented by applying the k–ε, k–ω, SST, and BSL-RSM turbulence models for comparison. The numerical solution of the equations was based on the finite volume method (FVM). The correct selection of turbulence models is an important and challenging task in predicting the turbulent features of the flow inside the hydrocyclone.

4.1 Inlet boundary conditions

At the inlet boundary of the hydrocyclone, the fluid inlet velocity was set according to the experimental data of Hsieh (1988). The injection position of solid particles was the inlet surface. The velocity of particles was assumed to be equal to the fluid inlet velocity (Wan et al. 2008). A number of 100,000 particles were fed into the hydrocyclone. The particles used were fine fire-dried sand, whose density was 2300 kg/m³, needed as the value for the modeling. The pressure level was set in the hydrocyclone feed according to experimental operating conditions. In Eu–Eu simulations for separation efficiency computations, diameters of the particles were given as single-input values in the range of 3–180 μm and the sand loading C_{s} was varied from 14.4 wt% to 55.8 wt% according to the experimental conditions of Hsieh (1988).

On the other hand, for erosion modeling, the diameters of sand particles used in our hydroabrasion studies have been measured experimentally with a HORIBA particle size analyzer (Retsch Technology, LA950 model). In Eu–

La simulations for the erosion studies, instead of a single particle size, a particle size distribution as depicted in Fig. 3 was given as an input. The four main characteristic parameters for defining a particle size distribution curve are the minimum diameter, maximum diameter, mean diameter, and the standard deviation value. Based on the distribution of measured size in Fig. 3, the minimum diameter was set to 67.5 μm, the maximum diameter to 678.5 μm, the mean diameter to 263.9, and finally the standard deviation value was set to 105.4 μm. The sand concentration was applied from 5 wt% to 15 wt%.

4.2 Outlet boundary conditions

At the hydrocyclone underflow and overflow parts, the pressure outlet boundary condition was set at a given standard atmospheric pressure condition.

4.3 Wall boundary conditions

On hydrocyclone walls, a no-slip condition was assumed. The grid nodes in the vicinity of walls were approximated and treated using the wall function. The wall roughness was given as 1.5 μm in the simulations for the stainless steel 410 as the wall material.

4.4 Coefficient of restitution

The coefficient of restitution was also applied on the wall surfaces of the hydrocyclone. The restitution coefficient for an orthogonal contact of a particle against a rigid surface is defined as the ratio of the rebound velocity to the impingement velocity. For a non-orthogonal contact as shown in Fig. 4, the behavior of the center of the particle after the rebound can be defined through the normal (R_{n}) and tangential restitution coefficients (R_{t}) with Eqs. (11) and (12), respectively.

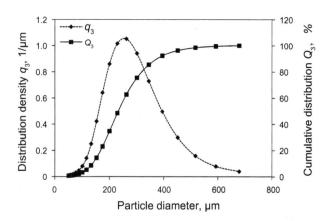

Fig. 3 Measured size distribution of sand particles

$$R_n = \frac{V_{nr}}{V_{ni}} \qquad (11)$$

$$R_t = \frac{V_{tr}}{V_{ti}} \qquad (12)$$

where V_{ni} and V_{nr} are the normal velocity components of the particle before and after the impact with the surface, respectively. V_{ti} and V_{tr} are the tangential velocity components of the particle before and after the impact with the surface, respectively. The restitution coefficient depends on a considerable number of parameters, mainly on particle material properties, particle size, particle shape, impact velocity, impact angle, and target surface material properties. The particles rebound dynamics can only be described statistically since the particles in practical applications are irregular in shape and vary in size. When the hardness of particles is higher than the target material, their continuous impact causes the surface to become pitted with craters after some incubation period. When the impact duration becomes longer, it causes ripple patterns to be created on the eroded surface. Consequently, the local impact angle may vary for each particle as it impacts the eroded surface (Tabakoff et al. 1996).

Wan et al. (2008) defined the coefficient of restitution for the cyclone walls using a trial and error method. They adopted different coefficients of restitution at different wall positions. When the calculated separation efficiency of cyclone showed a good agreement with the experimental result using a certain coefficient of restitution, it was adopted. By assuming that the particle size and shape remain unchanged through the passage from the inlet to the spigot, the impact velocity and impact angle of particles with respect to the cyclone wall are changing from top to bottom of the hydrocyclone. For the current study, according to Wan et al. (2008), the particle coefficient of restitution was set as 1.0–0.9 at the annular space. From top to bottom of the separation space (cylindrical body part), the particle coefficient of restitution was set as 0.9–0.6, and at the dust hopper (conical part), the particle coefficient of restitution was set as 0.5–0.05 as a linear function of height.

4.5 Solver control

A small physical time scale is required when having large regions of separated flow or multiphase flow. The option of a local timescale factor allows different time scales to be applied in different regions of the simulation domain. Smaller time scales are used in regions of the flow where the local time scale is quite short such as for fast flows. The bigger time scales are applied for regions where the time scales are locally large, similar to the slow flows. Local timescale factor is a different approach for controlling how fast the equations in the non-linear loop will be solved. The simulations were run in a steady state mode. Convergence was defined by RMS residuals less than 10^{-5} and imbalances (conservation) less than 10^{-2} for all variables. Under these simulation conditions, the residues were enough to guarantee the convergence.

5 Flow behavior and separation efficiency

Efficiency of separation, η, is a common measure of the efficiency of screening or cleaning of the cyclone. A theoretical relation exists between the fluid variable and physical characteristics of the system. The classification performance of a hydrocyclone is influenced by the design variables such as the hydrocyclone dimension and the operating variables such as the feed pressure and physical properties of solid particles and also the resulting mixture flow at the inlet. The separation efficiency curve expresses the relationship between the weight fraction or percentage of each particle size of the feed flow and the underflow discharge. The separation efficiency η is formulated as

$$\eta = \frac{w_{p,inlet} - w_{p,overflow}}{w_{p,inlet}} \times 100 \,\% \qquad (13)$$

where $w_{p,inlet}$ is the percent by weight of particles at the inlet; and $w_{p,overflow}$ is the weight percentage of particles at overflow.

The particle size is commonly used to represent the performance of the hydrocyclone. The tangential motion of the liquid flow generates centrifugal forces; since the particles have a different density from the liquid density, a radial velocity with reference to the liquid will be generated on the particles (Rietema 1961). The direction of the radial velocity is outwards and when the centrifugal forces are relatively strong, the particle can reach the cyclone walls through its motion towards the downward discharge and is then separated. A particle may not reach the wall under the three following conditions:

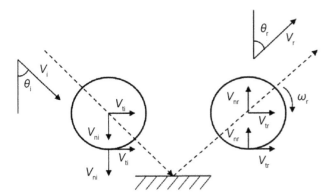

Fig. 4 Particle impact and rebound velocity components

(a) If the radial velocity of the liquid which is directed inwards is large and thus causes it to be entrained towards the center of the hydrocyclone.

(b) If it enters into the hydrocyclone at a great distance from the walls.

(c) If the particle residence time is too short in comparison with the average residence time in hydrocyclones (Dwari et al. 2004).

The addition of solid particles increases the viscosity of the slurry flow significantly compared to the viscosity of the pure fluid. For diluted mixtures, the fluid can still be considered as a Newtonian fluid; however, the effect of solid particles in the flow through the modification of the viscosity must be considered while applying single-phase simulations (Romero et al. 2004). The slurry viscosity can be described as relative to the viscosity of the liquid phase. Depending on the size and concentration of the solid particles, several models exist that describe the relative viscosity as a function of volume fraction C_v of solid particles. In this study, Eq. (14) expressed by Thomas (1965) was applied to calculating the viscosity term of the slurry flow. Afterwards, this was used to find out an assumed value for the viscosity term for solid particles as an input parameter for material definition in Eu–Eu simulations.

$$\frac{\mu_m}{\mu_f} = 1.0 + 2.5C_v + 10.05C_v^2 + 0.00273\exp(16.6C_v) \tag{14}$$

where μ_f is the fluid viscosity; μ_m is the viscosity of the mixture; and C_v is the volume fraction of the solid particle given as

$$C_v = \frac{C_p \cdot \rho_m}{\rho_p} \tag{15}$$

where C_p is the concentration of the solid particles in weight fraction; ρ_m, ρ_p, and ρ_f are the densities of the mixture, the solid particle phase, and the fluid phase, respectively. The density of the mixture ρ_m can be described by the following equation:

$$\frac{1}{\rho_m} = \frac{C_p}{\rho_p} + \frac{1 - C_p}{\rho_f} \tag{16}$$

By defining the mixture viscosity μ_m with Eq. (14), the assumed viscosity of the sand particles, μ_p, was finally calculated from Eq. (17), and used for definition of physical properties of the sand particles in the liquid flow.

$$\mu_p = \frac{\mu_m - (1 - C_p)\mu_f}{C_p} \tag{17}$$

The water flow streamlines inside the hydrocyclone obtained by the Eu–Eu method are shown in Fig. 5. Under the combined effects of centripetal buoyancy and fluid drag

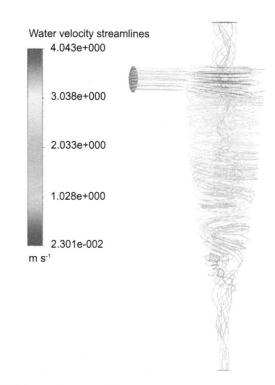

Water velocity streamlines
4.043e+000
3.038e+000
2.033e+000
1.028e+000
2.301e-002
m s⁻¹

Fig. 5 Water flow streamlines

force, a high amount of water, which is the light phase, will discharge as depicted in Fig. 5 from the overflow outlet and the suspension is withdrawn at the underflow with a relatively high solid concentration. In Fig. 6, three single sand particles with different diameter sizes of 20, 100, and 600 μm were tracked inside the hydrocyclone from the inlet feed to the outlets by consideration of the fully coupled interactions of the two phases. The selection of these three diameters was chosen in order to show the difference among the streamlines of various particle sizes inside the hydrocyclone. It was observed that due to the centrifugal force field, the 20-μm particles were carried out through the overflow but the bigger particles discharged through the spigot at the underflow outlet.

The k–ε turbulence model, SST model, two equation k–ω model, and the BSL Reynolds stress model were used to simulate the flow inside the hydrocyclone. The modeling results were used to calculate the separation efficiency curves for comparison with experimental results of Hsieh (1988). For the k–ε, SST, and k–ω turbulence models, two transport equations were used; however for the BSL-RSM, five additional transport equations had to be applied. The ANSYS-CFX tool was used to solve the governing set of partial differential equations for the mixture flow and turbulent components. The CFD results using the k–ε model deviate a lot from the experimental results and therefore, they are not presented here. The cut size d_{50} is the particle size at which the efficiency of hydrocyclone is 50 %. The cut

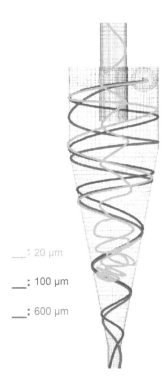

Fig. 6 Tracking of sand particles with various sizes

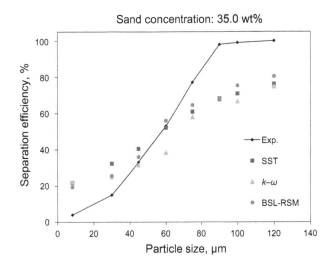

Fig. 8 CFD/Exp. comparison of separation efficiency with 35 wt% sand concentration

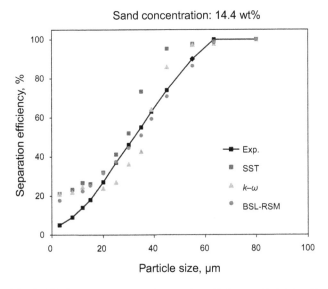

Fig. 7 CFD/Exp. comparison of separation efficiency with 14.4 wt% sand concentration

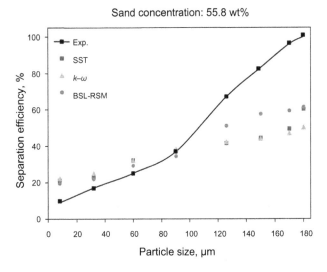

Fig. 9 CFD/Exp. comparison of separation efficiency with 55.8 wt% sand concentration

1.41 and 1.97 kg/s, respectively, when the solid concentrations are 35 % and 55.8 %.

Satisfactory results are obtained by applying the k–ω and SST turbulence models but only for relatively low solid particle loadings in the feed flow. Results obtained with three different turbulence models of SST, k–ω, and BSL Reynolds stress model are shown in Fig. 7. All the CFD results are in fair agreement with experimental results. However, the simulation results obtained with the BSL RSM fit very well to the experimental data in Fig. 7, especially for particles larger than the cut size. Deviation may be due to the absence of a curvature term. This term is important for the strong swirling flows inside the hydrocyclone (Stephens and Mohanarangam 2009). These deviations increase with an increase in the solid concentration in the feed flow, especially for larger particle sizes (bigger

size is predicted from the CFD simulations and is approximately 33 μm as depicted in Fig. 7, when the solid concentration is 14.4 wt% and the total input mass flow is 1.19 kg/s. Each simulation point in Figs. 7, 8, and 9 represents a simulation run with a specified constant mean diameter for the particle stream ranging from 3 μm to 180 μm. The cut size, d_{50}, increases with an increase in the solid concentration. For example, the cut size with a solid concentration of 55.8 wt% is approximately 106 μm (see Fig. 9). The total mass flow is

Sand velocity vector

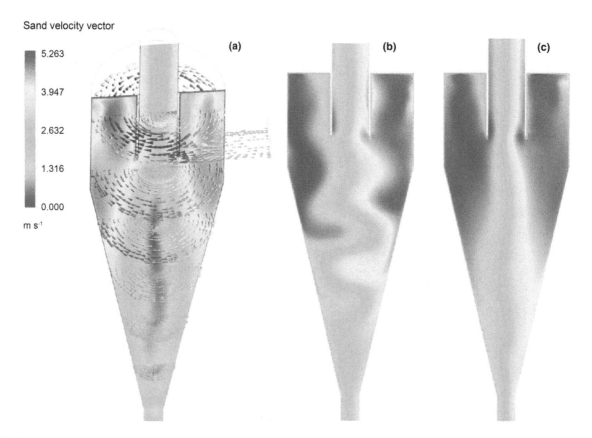

Fig. 10 Sand velocity vectors at different heights from the top (**a**), and tangential velocity profiles inside the cyclone with timescale factor of 1 (**b**), and 0.4 (**c**)

than 100 μm) as presented in Figs. 8 and 9. As it was mentioned, the presence of solid particles increases the viscosity of the slurry flow significantly compared to the viscosity of the pure fluid. Therefore, for high solid concentrations, as it is the case for the simulation with 35 wt% and 55.8 wt% sand concentration, the consideration of the suspension as a Newtonian fluid is not any longer a valid assumption. Moreover, as the solid concentration increases, the importance of modified and precise modeling of particle–particle interactions will arise. The results of all the turbulence models at higher solid loading, especially when the particle size is relatively large, are not in very good agreement with the experimental data. Overall, it was observed that the CFD results obtained by the BSL-RSM, which is a seven differential equation turbulence model, are in better agreement with experimental data compared with the CFD results obtained by the two equation turbulence models of SST and k–ω. However, the computational time for the BSL-RSM is relatively higher than the other two turbulence models. The improvement obtained by the BSL-Reynolds stress model is still far lower than the error between the experimental data and simulation results for high solid loadings as shown in Figs. 8 and 9. It should be noted that at higher concentrations, interaction among the particles increases drastically. Therefore, it is essential to perform four-way coupled

simulations instead of two-way coupled simulations in order to capture the precise flow behavior of particulate flows with a high solid loading. Four-way coupled simulation takes the particle–particle interactions also into account in comparison with two-way coupled simulation, which takes only the fluid on particles and particles on fluid effects. On the other hand, for higher concentrations, the viscosity of the slurry flow increases significantly compared to the viscosity of the pure fluid. Therefore, the consideration of the suspension as a Newtonian fluid leads to large deviations for higher concentrations as shown in Fig. 9.

In Fig. 10a, the velocity vectors of sand particles are shown at various heights from the top of the hydrocyclone. When a particle initially flowed down due to the external downward flow, at the same time the radial velocity directing to the center of hydrocyclone forces the particle to be moved inward (Wang et al. 2007). When the particle is small, the resulting inward drag force is higher than the centrifugal force, which causes the particle to be caught by the upward inner flow and discharged through the vortex finder. The bigger or heavier particles flow towards the downward discharge, as the centrifugal force acting on the particle is higher than the inward drag force.

The timescale factor is an essential parameter for steady state simulation of flow inside the cyclones. Applying a

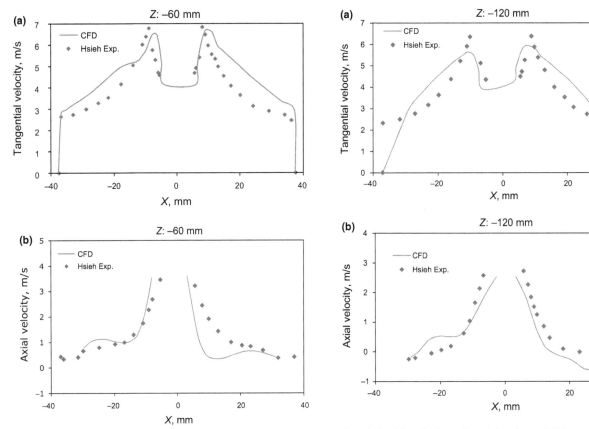

Fig. 11 Particle velocity at the axial position of 60 mm from the top of the cyclone. **a** Tangential velocity; **b** Axial velocity

Fig. 12 Particle velocity at the axial position of 120 mm from the top of the cyclone. **a** Tangential velocity; **b** Axial velocity

timescale factor of for example 1.0 as shown in Fig. 10b would not be sufficient to simulate the central core in a steady way as simulated by a smaller timescale factor of 0.4 depicted in Fig. 10c. The relatively low fluctuation of the central core in Fig. 10c matches with the findings in the literature such as that by Cullivan et al. (2004). Figures 11, 12, 13 show comparisons of the simulated tangential and axial velocity profiles at three different axial positions of 60, 120, and 180 mm from the top of the hydrocyclone, defined as the A–A, B–B, and C–C planes in Fig. 1, respectively. As shown in Figs. 11, 12, 13, the results obtained by CFD simulations are in good agreement with experimental results of Hsieh (1988).

According to Panton (2013), the tangential velocity profile could be defined as a Rankine vortex having a quasi-forced vortex in the inner part and a quasi-free vortex in the outer section. It is possible to say that the tangential velocity profile has a trend like an inverse W and the axial velocity profile has a trend like an inverse V at the distinct axial locations inside the hydrocyclone. The axial velocity distributions at all locations of 60, 120, and 180 derived from the CFD runs are shown in Fig. 14. As it is depicted, the axial velocities of particles decrease as the distance

from the top increases. The velocity magnitudes even reach negative values near the walls or in the outer part of the flow, while the slurry flow travels towards the bottom of the hydrocyclone.

6 Erosion modeling

Impact erosion is characterized by individual particles contacting the surface at a certain angle and velocity. Removal of material over time occurs through small-scale deformation, cutting, fatigue cracking or a combination of these, depending on the properties of the erodent material and the target surface. The required information on particle dynamics and rebound characteristics has been developed such that the particle trajectories inside the hydrocyclone can be predicted. The material removal process can then be calculated using this trajectory and impact data. The equations developed for predicting the erosion rates are homogeneous in nature, this means they can be applied to calculate the material loss resulting from a given particle entering the system at a given location and rebounding in a given manner (Grant and Tabakoff 1975).

(a)

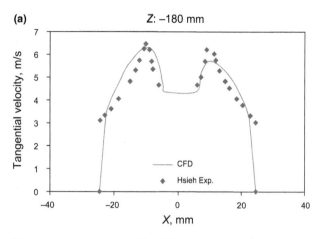

(b)

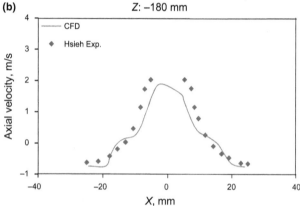

Fig. 13 Particle velocity at the axial position of 180 mm from the top of the cyclone. **a** Tangential velocity; **b** Axial velocity

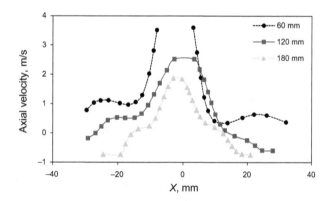

Fig. 14 A comparison of axial velocities at different locations driven from CFD runs

Grant and Tabakoff (1973) have experimentally proved that the wear of a material is mostly dependent on the impingement angle and the impact velocity of abrasive particles. The uniqueness of this model compared to other erosion models of its kind is that it contains the particle impact restitution coefficient as an influencing parameter and in particular the tangential restitution coefficient R_t as the decisive parameter affecting the erosion rate.

Table 4 Coefficients for stainless steel 410 surface eroded by sand particles

Parameter	Stainless steel 410-sand particles
k_{12}	0.293
k_1	$6.53 \times 10^{-8} \; \frac{\text{g/g}}{(\text{m/s})^2}$
k_3	$8.94 \times 10^{-13} \; \frac{\text{g/g}}{(\text{m/s})^4}$
Angle of max. erosion (β_0)	$30°$

Fig. 15 Sand particle tracking

Equation (18) was developed by Grant and Tabakoff (1973) to predict the erosion of ductile metals and alloys and is defined as the ratio of the eroded mass of target material to the mass of impinging solid particles. They assumed that the erosion process is dependent on two mechanisms; one at low angles of impingement; one at high angles of impingement; and a combination of the two at intermediate impact angles. The relationship for erosion was derived as Eq. (18). The first term of this expression predominates at low angles of attack, whereas the second term predominates at normal impact where the tangential velocity approaches zero.

$$E = k_1 f(\beta_i)(V_{ti}^2 - V_{tr}^2) + f(V_{ni}) \qquad (18)$$

where k_1 is the material constant; $f(\beta_i)$ is the empirical function of particle impact angle; V_{ti} and V_{tr} are the tangential components of incoming particle velocity and of rebounding particle velocity, respectively; $f(V_{ni})$ is the component of erosion due to the normal component of velocity. By inserting $V_{ti} = V_i \cos \beta_i$, the erosion equation can be written as

$$E = k_1 f(\beta_i) V_i^2 \cos^2 \beta_i (1 - R_t^2) + f(V_{ni}) \qquad (19)$$

The particle impact angle influences the erosion rate in a very special way, as it was found that the maximum erosion rate occurs at an impact angle of 30°, whereas only a residual amount of erosion results from the normal impact. The effect of the particle impact angle is lumped into $f(\beta_i)$, and a strict empirical approach is used to predict its behavior (Grant and Tabakoff 1973). The result of this analysis yields the following expression:

$$f(\beta_i) = \left[1 + k_2 k_{12} \sin \left(\frac{\pi \beta_i}{2\beta_0} \right) \right]^2 \qquad (20)$$

where β_0 is the angle of attack where the maximum erosion occurs; k_{12} is a material constant and

$$k_2 : \begin{cases} 1 & \beta_i \leq 2\beta_0 \\ 0 & \beta_i > 2\beta_0 \end{cases} \qquad (21)$$

The component of erosion resulting from normally impacting particles is expressed as

$$f(V_{ni}) = k_3 (V_i \sin \beta_i)^4 \qquad (22)$$

The investigated hydrocyclone in this study was constructed from stainless steel 410 faced to the two-phase water–sand flow. Stainless steel 410 is a basic martensitic grade, which contains 11.5 % chromium, offering both exceptional wear and corrosion resistance. The constants and the angle of maximum erosion required for the Grant–Tabakoff erosion model for the stainless steel 410, eroded by sand particles (SiO_2) are given in Table 4 based on the experimental work of Clevenger and Tabakoff (Clevenger and Tabakoff 1975). Total erosion rate of the wall due to the solid particles interactions with the wall is finally computed from the following equation:

$$\dot{E} = \dot{N} \cdot m_p \cdot E \qquad (23)$$

where $\dot{N}$ is the number rate of solid particles; and m_p is the particle mean mass.

As it has already been mentioned, the Eu–La approach was applied to predicting the erosion rate or the material loss of the hydrocyclone wall surfaces. The erosion leads to physical damage and functional failure, since even small deformations on the hydrocyclone wall can interrupt the pressure profile and cause a decrease in the separation efficiency. As depicted in Fig. 15, the particles followed a distinct path rather than a tangled movement on hydrocyclone walls while flowing down. According to the particle tracking inside the hydrocyclone, the erosion location and its density rate on the wall surfaces were predicted with the Grant-Tabakoff erosion model and presented in Fig. 16 as an

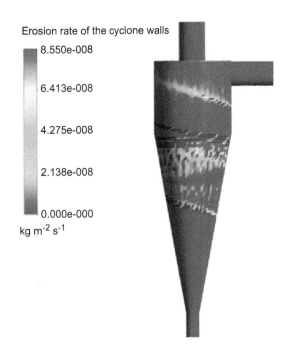

Fig. 16 Erosion density and locations on the wall surfaces

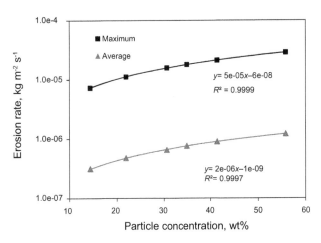

Fig. 17 Effect of the particle concentration on the material loss of hydrocyclone walls

illustrating example. The input conditions for the results in Figs. 15 and 16 were the solid concentration of 15 wt%, input flow velocity of 2.7 m/s, particle size distribution as given in Fig. 3, assumption of spherical sand particles, and the isothermal assumption of the flow in the hydrocyclone. The most eroded part was at the intersection between the cyclone cylindrical part and the conical part, which was in good agreement with the findings of Utikar et al. (2010). At the region close to the underflow discharge, the erosion is quite low although due to the area restriction and rise of the solid concentration, a different outcome could be expected. The reason is that in this region, almost all the particles are sliding on the walls and therefore the impact angle is close to 0°. The resulting erosion is therefore less than the erosion in central parts where the particles impact the walls with various angles, causing the chipping and cutting mechanisms to occur. The particle size distribution used in the simulations is that of the standard sand being used in our hydroabrasion experiments presented in Fig. 3.

6.1 Particle concentration effect

The particle concentration is very often interpreted as the percentage content by weight or by volume of the particles in a gaseous or liquid medium. With an increase in the particle concentration, more particles strike the target surface and therefore enhance the erosion. In Fig. 17, the effect of solid particle concentration on the average and maximum erosion rate in the hydrocyclone is depicted. As expected, the erosion rate or the material loss of hydrocyclone walls increases with an increase in the sand concentration in the range of 15–55 wt%. The total mass flow rate for 15 wt% solid concentration was 1.19 kg/s. The constant input conditions here were the input flow velocity of 2.7 m/s, the particle size distribution as given in Fig. 3, the assumption of spherical sand particles, and the isothermal assumption of the flow in the hydrocyclone. The erosion rate as a function of the solid concentration is expressed by a linear curve in the logarithmic scale of y axis. When the sand concentration is quite small, the interaction among particles can be ignored and the erosion increases linearly with the sand loading. However, when the sand concentration is large enough, the interaction among particles has to be taken into account. It should be mentioned that the erosion rates shown in Figs. 17, 18, 19, 20, 21 and 22 give a global measure of the erosion within the whole internal surfaces of the hydrocyclone.

In Fig. 18, the erosion rate and locations of material losses on hydrocyclone walls for two sand particle concentrations of 5 wt% and 15 wt% are shown and compared with each other. The constant input conditions here were the input flow velocity of 2.7 m/s, the particle size distribution as given in Fig. 3, the assumption of spherical sand particles, and the assumption of isothermal flow in the hydrocyclone. As depicted, an erosion field at the upper part of the cylindrical body of the hydrocyclone appeared and developed, when the solid concentration was increased. The erosion density at the intersection of cylindrical to conical body was also relatively increased.

6.2 Flow velocity effect

In Fig. 19, the effect of the inlet feed velocity on the erosion is presented. The constant input conditions here were the sand concentration of 15 wt%, the particle size distribution as given in Fig. 3, the assumption of spherical sand particles, and the assumption of isothermal flow in the hydrocyclone. The trend of erosion shows an increase as a potential function in a logarithmic scale as the feed velocity increases. The erosion rate increases with the impact velocity to a power m, varying between 1.5 and 3.5 in experiments carried out in different laboratories. The exponent m averages around 2 for wear on metal plates by a water-sand jet, between 1 and 4 in slurry pipes, and from 2.2 to 3 in centrifugal pumps. The dispersion of values for m can be explained by the diversity of mechanisms and the mean flow parameters, as well as differences in experimental methods (Shook and Roco 1991). However, this effect will be up to a certain velocity and afterwards the erosion rate would increase only slightly. The power coefficient m changes with particle breakage, which depends on the erodent material. Brittle particles break intensely upon impact with the target material, which causes a sharp decline in the value of exponent m.

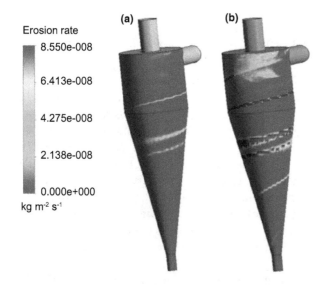

Fig. 18 Variation in the erosion rate and locations for different solid concentrations. **a** 5 %; **b** 15 % concentration

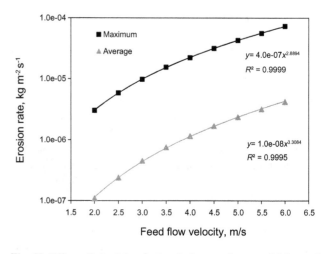

Fig. 19 Effect of the inlet feed velocity on the material loss of hydrocyclone walls

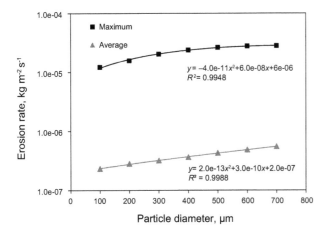

Fig. 20 Effect of the sand particle size on the material loss of hydrocyclone walls

6.3 Particle size effect

An increase in the particle size at a constant solid concentration decreases the number of particles suspended in the mixture and at the same time increases the kinetic energy per particle impact. Thus the increase in wear with an increase in particle size is generally attributed to the increase in energy available for erosion (Desale et al. 2009). Grant and Tabakoff (1975) found that the increase in the particle size did not influence the erosion after a certain threshold value was reached.

The effect of the sand particle size is shown in Fig. 20. The constant input conditions here were the sand concentration of 15 wt%, the input flow velocity of 2.7 m/s, the assumption of spherical sand particles, and the assumption of isothermal flow in the hydrocyclone. The increasing trend of erosion or material loss in this case can be described by a polynomial function in a logarithmic scale, when the particle size increases in the range of 100–700 μm. As the particle size increases, there are fewer particles to impact the surface and this results in lower impact rate. However, each particle will be heavier and thus will have greater kinetic energy. Therefore, it is difficult to quantify the sand size effect as the competing effects between kinetic energy of the particle and impact rates must to be taken into account (Rajahram et al. 2011).

In Fig. 21, the erosion rate and location of material loss of hydrocyclone walls for two various particle sizes of 20 μm and 100 μm are shown and compared with each other. As it is expected, the erosion rate caused by solid particles with smaller size is lower than the erosion caused by bigger particles. For the 20 μm particles, there is no signs of erosion at the lower body parts of the hydrocyclone while most of the particles are discharged from the upper outlet. However, the erosion at the vortex finder through the upper discharge appears and is developed in the way which the particles travel

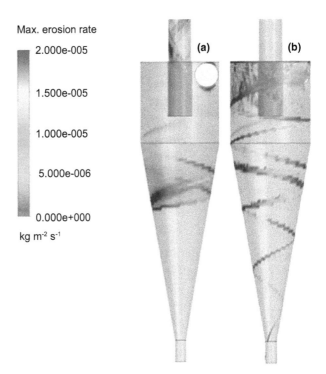

Fig. 21 Erosion rate and location for different particle sizes. a 20 μm; b 100 μm

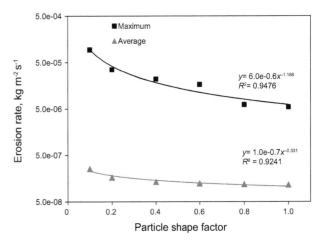

Fig. 22 Effect of the particle shape factor on the material loss of hydrocyclone walls

towards the upper outlet (overflow). As the larger particles roll down the conical part, the centrifugal force on the particle increases. The radius of the hydrocyclone conical part decreases as the flow passes towards underflow. However, the tangential velocity of the particle remains almost the same.

6.4 Particle shape effect

The effect of the particle shape factor on erosion was also investigated and the simulated results are given in Fig. 22. The constant input conditions here were sand concentration of 15 wt%, input flow velocity of 2 m/s, the particle size distribution as given in Fig. 3, and the assumption of isothermal flow in the hydrocyclone. The shape factor is determined using the following equation:

$$f_s = \frac{4\pi A}{p_e^2} \qquad (24)$$

where A is the projected area of the particle; and p_e is the perimeter of the particle (Desale et al. 2008). The shape factor of 0 is assumed for a disk-shaped particle and 1.0 for an ideal spherical shape. A cross-sectional area factor was included to modify the assumed spherical cross section area to allow for non-spherical particles. This factor was multiplied by the cross section area calculated assuming spherical particles. This affects the drag force calculated by ANSYS-CFX. Results indicate that the more the particles are sharped edged, the more they are erosive and result in an increase in erosion of the hydrocyclone walls. It is observed that an increase in the particle shape factor from 0.1 to 1.0 leads to a decrease in almost 70 % in the average erosion density of the hydrocyclone walls.

7 Conclusions

Hydrocyclones are designed to achieve a reasonably good separation between solid particles of various sizes or between solid particles and liquid flow. CFD simulations using ANSYS-CFX (14.5) were performed to investigate the slurry flow behavior and erosion rates inside a hydrocyclone. The consideration of an appropriate erosion model and specifically the erosion rate parameter is an important factor in the grid study. The number of elements of the final mesh to obtain mesh independent results was not identical with and without application of the erosion model in the CFD simulations. The grid study was initially carried out by simulating the hydrocyclone in ANSYS-CFX without coupling the erosion model. It was observed that taking the outlet velocity components at overflow and underflow as the grid study criterions, a fine mesh with around 150,000 elements would be sufficient to get mesh independent results. Afterwards, a grid study was carried out by coupling an erosion model with CFD simulations. The average erosion rate on hydrocyclone walls was considered here as the grid study parameter. It was observed that a fine mesh with around 700,000 elements, with specifically mesh refinement next to the wall surfaces, would be sufficient to get mesh independent results. In simulations for predicting the erosion, a much finer mesh at the vicinity of the wall surfaces is required in order to precisely model the particle–particle and particle–wall interactions near the surface.

The Euler–Euler approach was used to study the flow behavior and prediction of separation efficiency curves of the hydrocyclone. The Euler–Euler approach was capable of considering the particle–particle interactions and was appropriate for highly laden liquid–solid mixtures. The effect of solid mass loading on the separation efficiency of the hydrocyclone was also studied. The cut size d_{50} was predicted from the CFD simulations and was approximately 33 μm when the solid concentration was 14.4 wt% and the total input mass flow 1.19 kg/s. The cut size increased with an increase in the solid concentration. For example, the cut size with a solid concentration of 55.8 wt% was approximately 106 μm. Among the investigated turbulence models in the current study, the results obtained from the BSL-RSM matched better with the experimental results. However, the computational costs and simulation time were relatively higher in comparison with SST and k–ω turbulence models. The results of particle velocity profiles in distinct axial positions inside the hydrocyclone predicted by CFD simulations were validated with experimental results of Hsieh (1988) and showed a good agreement.

The Euler–Lagrange approach was applied to predicting the location and quantity of erosion on the hydrocyclone wall surfaces. The abrasive particles were traced in a Lagrangian reference frame as discrete particles. The most eroded part was at the intersection between the hydrocyclone cylindrical and conical body parts, which was in good accordance with the findings of Utikar et al. (2010). From a spatial analysis of the erosion rates, one can conclude that increases in the feed velocity, solid concentration, and particle size increased the erosion field at the upper part of the cylindrical body in front of the tangential inlet. The erosion density in the cylindrical to conical part transitional field in comparison to the other regions was also increased. Moreover, it was observed that an increase in the particle shape factor from 0.1 to 1.0 led to a decrease of almost 70 % in the average erosion density of hydrocyclone walls. Modeling of the change in geometry associated with erosion and the effects of erosion profiles on the fluid flow was postponed for subsequent works. Modeling the 3-D deformation of hydrocyclone walls due to erosion requires a finite element modeling (FEM) technique coupled to the CFD simulations. Such FEM-CFD coupling required long and costly transient simulations. Detailed hydroabrasion experiments in a broader range are planned to be undertaken.

Acknowledgments The authors would like to thank "Stiftung Rheinland-Pfalz für Innovation, Mainz, Germany," for financial support.

References

Chu L, Chen W. Research on the motion of solid particles in a hydrocyclone. Sep Sci Technol. 1993;28(10):1875–86. doi:10.1080/01496399308029247.

Clevenger WB, Tabakoff W. Erosion in radial inflow turbines—vol. IV: erosion rates on internal surfaces. Technical Report. University of Cincinnati; 1975.

Cullivan JC, Williams RA, Dyakowski T, Cross CR. New understanding of a hydrocyclone flow field and separation mechanism from computational fluid dynamics. Miner Eng. 2004;17(5):651–60. doi:10.1016/j.mineng.2004.04.009.

Desale G, Gandhi B, Jain S. Slurry erosion of ductile materials under normal impact condition. Wear. 2008;264(3–4):322–30. doi:10.1016/j.wear.2007.03.022.

Desale G, Gandhi B, Jain S. Particle size effects on the slurry erosion of aluminium alloy (AA 6063). Wear. 2009;266(11–12):1066–71. doi:10.1016/j.wear.2009.01.002.

Dwari R, Biswas M, Meikap B. Performance characteristics for particles of sand FCC and fly ash in a novel hydrocyclone. Chem Eng Sci. 2004;59(3):671–84. doi:10.1016/j.ces.2003.11.015.

Ghadirian M, Hayes RE, Mmbaga J, Afacan A, Xu Z. On the simulation of hydrocyclones using CFD. Can J Chem Eng. 2013;91(5):950–8. doi:10.1002/cjce.21705.

Grant G, Tabakoff W. An experimental investigation of the erosive characteristics of 2024 Aluminum alloy. Technical Report. Cincinnati University; 1973.

Grant G, Tabakoff W. Erosion prediction in turbomachinery resulting from environmental solid particles. J Aircr. 1975;12(5):471–8.

Hsieh K. A phenomenological model of hydrocyclone. Ph.D. Thesis. University of Utah; 1988.

Hsieh K, Rajamani R. Mathematical model of the hydrocyclone based on physics of fluid flow. AIChE J. 1991;37(5):735–46. doi:10.1002/aic.690370511.

Huang J, An L-S, Wu Z-Q. Study of application and operation optimization of hydrocyclone for solid–liquid separation in power plant. In: Proceedings of the world congress on engineering and computer science 2009, vol. I, 20–22 Oct 2009, San Francisco; 2009.

Ipate G, Căsăndroiu T. Numerical study of liquid–solid separation process inside the hydrocyclones with double cone sections. UPB Sci Bull. 2007;69(4):83–94.

Milin L, Hsieh K, Rajamani R. The leakage mechanisms in the hydrocyclone. Miner Eng. 1992;5(7):779–94. doi:10.1016/0892-6875(92)90246-6.

Monredon TC, Hsieh K, Rajamani R. Fluid flow model of the hydrocyclone: an investigation of device dimensions. Int J Miner Process. 1992;35(1–2):65–83. doi:10.1016/0301-7516(92)90005-H.

Morsi SA, Alexander AJ. An investigation of particle trajectories in two-phase flow systems. J Fluid Mech. 1972;55(2):193–208. doi:10.1017/S0022112072001806.

Panton RL. Incompressible flow. 4th ed. New Jersey: Wiley; 2013.

Pericleous K, Rhodes N. The hydrocyclone classifier—a numerical approach. Int J Miner Process. 1986;17(1–2):23–43. doi:10.1016/0301-7516(86)90044-X.

Process Group Pty Ltd. Hydrocyclones desander. Process Group Pty Ltd: Rowville; 2012.

Rajahram SS, Harvey TJ, Wood R. Electrochemical investigation of erosion–corrosion using a slurry pot erosion tester. Tribol Int. 2011;44(3):232–40.

Rietema K. Performance and design of hydrocyclones-III: separating power of the hydrocyclone. Chem Eng Sci. 1961;15(3–4):310–9. doi:10.1016/0009-2509(61)85035-5.

Romero J, Sampaio R, da Gama RS. Numerical simulation of flow in a hydrocyclone. Lat Am Appl Res. 2004;34:1–9.

Shook CA, Roco MC. Slurry flow: principles and practice. Stoneham: Butterworth-Heinemann; 1991.

Stephens DW, Mohanarangam K. Turbulence model analysis of flow inside a hydrocyclone. In: Seventh international conference on CFD in the minerals and process industries CSIRO, Melbourne; 2009.

Tabakoff W, Hamed A, Murugan DM. Effect of target materials on the particle restitution characteristics for turbomachinery application. J Propul Power. 1996;12(2):260–6. doi:10.2514/3.24022.

Thomas DG. Transport characteristics of suspension: VIII. A note on the viscosity of Newtonian suspensions of uniform spherical particles. J Colloid Sci. 1965;20(3):267–77. doi:10.1016/0095-8522(65)90016-4.

Utikar R, Darmawan N, Tade MO, et al. Hydrodynamic simulation of cyclone separators, Chap. 11. Croatia: InTech; 2010.

Vieira LG, Damasceno JJ, Barrozo MA. Improvement of hydrocyclone separation performance by incorporating a conical filtering wall. Chem Eng Process. 2010;49(5):460–7. doi:10.1016/j.cep.2010.03.011.

Vieira LG, Silverio BC, Damasceno JJ, Barrozo MA. Performance of hydrocyclones with different geometries. Can J Chem Eng. 2011;89(4):655–62. doi:10.1002/cjce.20461.

Wan G, Sun G, Xue X, Shi M. Solids concentration simulation of different size particles in a cyclone separator. Powder Technol. 2008;183(1):94–104. doi:10.1016/j.powtec.2007.11.019.

Wang B, Chu KW, Yu AB. Numerical study of particle–fluid flow in a hydrocyclone. Ind Eng Chem Res. 2007;46(13):4695–705. doi:10.1021/ie061625u.

Xu P, Wu Z, Mujumdar A, Yu B. Innovative hydrocyclone inlet designs to reduce erosion-induced wear in mineral dewatering processes. Dry Technol. 2009;27(2):201–11. doi:10.1080/07373930802603433.

18

Pore structure and tracer migration behavior of typical American and Chinese shales

Qin-Hong Hu[1,2] · Xian-Guo Liu[1] · Zhi-Ye Gao[3] · Shu-Gen Liu[4] · Wen Zhou[4] ·
Wen-Xuan Hu[5]

Abstract With estimated shale gas resources greater than those of US and Canada combined, China has been embarking on an ambitious shale development program. However, nearly 30 years of American experience in shale hydrocarbon exploration and production indicates a low total recovery of shale gas at 12 %–30 % and tight oil at 5 %–10 %. One of the main barriers to sustainable development of shale resources, namely the pore structure (geometry and connectivity) of the nanopores for storing and transporting hydrocarbons, is rarely investigated. In this study, we collected samples from a variety of leading hydrocarbon-producing shale formations in US and China. These formations have different ages and geologic characteristics (e.g., porosity, permeability, mineralogy, total organic content, and thermal maturation). We studied their pore structure characteristics, imbibition and saturated diffusion, edge-accessible porosity, and wettability with four complementary tests: mercury intrusion porosimetry, fluid and tracer imbibition into initially dry shale, tracer diffusion into fluid-saturated shale, and high-pressure Wood's metal intrusion followed with imaging and elemental mapping. The imbibition and diffusion tests use tracer-bearing wettability fluids (API brine or n-decane) to examine the association of tracers with mineral or organic matter phases, using a sensitive and micro-scale elemental laser ablation ICP-MS mapping technique. For two molecular tracers in n-decane fluid with the estimated sizes of 1.39 nm × 0.29 nm × 0.18 nm for 1-iododecane and 1.27 nm × 0.92 nm × 0.78 nm for trichlorooxobis (triphenylphosphine) rhenium, much less penetration was observed for larger molecules of organic rhenium in shales with median pore-throat sizes of several nanometers. This indicates the probable entanglement of sub-nano-sized molecules in shales with nano-sized pore-throats. Overall findings from the above innovative approaches indicate the limited accessibility (several millimeters from sample edge) and connectivity of tortuous nanopore spaces in shales with spatial wettability, which could lead to the low overall hydrocarbon recovery because of the limited fracture–matrix connection and migration of hydrocarbon molecules from the shale matrix to the stimulated fracture network.

Keywords Shale · Nanopore · Connectivity · Diffusion · Imbibition

✉ Qin-Hong Hu
water19049@gmail.com

[1] China University of Geosciences (Wuhan), Wuhan 430074, China

[2] The University of Texas at Arlington, Arlington, TX 76019, USA

[3] State Key Laboratory of Petroleum Resources and Prospecting, China University of Petroleum, Beijing 102249, China

[4] State Key Laboratory of Oil and Gas Reservoir Geology and Exploitation, Chengdu University of Technology, Chengdu 610059, China

[5] Nanjing University, Nanjing 210093, China

Edited by Xiu-Qin Zhu

1 Introduction

Since 2000, the technological development of horizontal drilling and hydraulic fracturing in US has led to a dramatic increase in hydrocarbon (gas and oil) production from shale formations, changing the fossil energy outlook in the US and worldwide (DOE 2009; Jarvie 2012; EIA 2014).

The production of shale hydrocarbon has recently increased significantly, and it is predicted that the production of shale gas will account for 50 % of total dry gas production in the USA by 2040, an increase from 34 % in 2011 (EIA 2014). Likewise, US tight oil production is 1.8 million barrels per day (accounting for 28 % of total US production) in 2012 and 2.8 million barrels (35 %) in 2013, and projected to be 4.8 million barrels in 2019 to provide more than a quarter of the 18.8 million barrels of oil that the US currently consumes daily.

Despite the increased gas and oil production from shale formations, analyses of data from 65,000 shale wells in 30 shale gas and 21 tight oil fields in the US led Hughes (2013a, b) to argue that the shale revolution will be hard to sustain. The data indicate that the production from a given well would decline rapidly within a few years. For example, wells in the top five US shale gas (Marcellus, Haynesville, Barnett, Fayetteville, and Eagle Ford) plays typically produce 80 %–95 % less gas after 3 years, and the productivity of new wells in two leading tight oil plays (Bakken and Eagle Ford) drops by about 60 % within the first year. The total gas recovery from the Barnett, the most developed shale play, is reported to be only 8 %–15 % of gas-in-place in 2002 (Curtis 2002), and 12 %–30 % in 2012 (King 2012). The recovery rate for tight oil is even lower at 5 %–10 % (Hoffman 2012), for example, the oil recovery rate in the middle member Bakken formation ranges from 2 % to 5 % to as high as 20 % (Jarvie 2012).

With estimated shale gas resources greater than that of US and Canada combined, China has been embarking on an ambitious shale development program. China has several types of shale (by area: 26 % marine, 56 % marine-terrestrial transitional, and 18 % terrestrial), whereas nearly all the US producing shales are of marine type (Zou et al. 2013). Sinopec (China) has recently reported that its first marine shale well (Jiao-Ye #1HF, drilled on Feb. 14, 2012 and completed on Nov. 24, 2012) initially produced 2.0×10^5 m^3 gas/day, and maintained a stable daily production of 6.6×10^4 m^3 over the next 7 months. This production behavior, though of limited duration, is consistent with the 60 % 1st-year decline observed in US shale wells.

Shale geology could be a bottleneck to its sustainable development. While Hughes' articles (2013a, b) mentioned this steep decline and low overall recovery, investigations into their root cause(s) are surprisingly scarce (Hu and Ewing 2013). This work highlights the studies of root causes from pore structure and fluid migration in the tight matrix of several leading US and China shales. China has so far drilled about 300 shale wells, dwarfed by the 80,000 US wells. A comparative analysis of pore structure and its resultant fluid migration behavior is critically important. Low recovery, coupled with the relatively high cost of

shale hydrocarbon development and the sharp drop in oil price that started in 2014, is the main barrier to sustainable development of US and China shale.

Fluid flow and solute transport in rock are macroscopic consequences of the pore structure, which integrates geometry (e.g., pore size and shape, pore size distribution; Bear 1972) and topology (e.g., pore connectivity; Dullien 1992). Especially when pore connectivity is low, topological factors outweigh the better known geometrical factors (Ewing and Horton 2002; Hu et al. 2012; Hunt et al. 2014). However, the prevalence of low pore connectivity in tight shales, and its impacts on fluid flow and chemical transport, is poorly documented and understood. This work examines the potential of low pore connectivity of typical American and Chinese shales, and its effects on fluid flow and chemical transport.

2 Materials and methods

2.1 Samples

The research objective of this work is to examine the pore structure (especially connectivity) characteristics of several shale samples from active and leading onshore US and China plays for hydrocarbon production (Table 1). These shale samples have different geologic ages, mineralogy, total organic carbon (TOC), kerogen types, and maturation levels. For example, the Chinese shales generally have much higher thermal maturation (with R_0 values at about 3 %) than American ones (Table 1). Various geologic controls on pore connectivity have not been systematically studied in the literature.

Located in the Fort Worth Basin in north-central Texas, the Barnett shale (the birth place of the "shale revolution") is a Mississippian-age marine shelf deposit, ranging in thickness from about 60 m (200 feet) in the southwest region to 305 m (1000 feet) to the northeast. The formation is a black, organic-rich shale (total organic carbon at 0.4 %–10.6 %, with an average of 4.0 %; Loucks and Ruppel 2007) composed of fine-grained, siliciclastic rocks with low permeability (70–5000 nD [nano-Darcy]) (Grieser et al. 2008; Loucks et al. 2009). The Barnett shale play currently has some 18,000 producing wells (Nicot et al. 2014). The reservoir produces at commercially viable levels only with hydraulic fracturing that establishes long and wide fractures, which connect large surface areas of the formation through a complex fracture network. However, gas production in such tight shale is still technically challenging, partly from the lack of the understanding of nanopore structure characteristics of the shale matrix.

The 120 foot long (37 m) Blakely #1core (API 42-497-33041), taken from southeastern Wise County within the

Table 1 Typical hydrocarbon-producing shales in China and USA tested in this work

Well/ sample ID	Formation	Geologic age (millions years ago)	Depth below ground, m	Location	TOC, wt%	R_o, %
EF outcrop	Eagle Ford (EF)	Late Cretaceous (99.6–65.5)	Surface outcrop	Comstock, Val Verde Co., TX	5.3	0.53
Leppard #1 (LP)			4143.9–4165.8	Bee Co., TX	NA	NA
Blakely #1 (BL)	Barnett (B)	Mississippi (359.2–318.1)	2166.8–2200.4	Wise Co., TX	3.08–6.62	~1.35
TZ-4H	Longmaxi (LMX)	Early Silurian (443.7–428.2)	Surface outcrop	Qilong Village, Xishui, Guizhou	6.23	NA
Jinye Well No. 1	Niutitang (NTT; also known as Qiongzhusi)	Early Cambrian (542.0–513.0)	136.7–220.5	Wazhi Village, Liujiazhen, Rong County, Sichuan	0.2–11.5	2.11–3.74
YC01-72D	Doushantuo (DST)	Sinian (800–570)	Surface outcrop	Jiulongwan, Yichang, Hubei	9.68	NA

Average of 31 Barnett core samples from 4 wells including Blakely #1 (Loucks and Ruppel 2007)

NA not available

Fort Worth Basin, includes part of the upper Barnett Shale (2166–2169 m below ground surface (bgs), the Forestburg Limestone (2169–218 mbgs), and the upper part of the lower Barnett Shale (2181–2202 mbgs) (Loucks and Ruppel 2007). We obtained core samples from this Blakely #1 well, courtesy of the core repository of the Texas Bureau of Economic Geology, representing five depths separated by at most 10 m: 2167 m (7109 ft; upper Barnett), 2175 m (7136 ft; Forestburg Limestone), and 2185 m (7169 ft), 2194 m (7199 ft) and 2200 m (7219 ft); all from the upper part of the lower Barnett) (Table 2).

Since around 2008, U.S. natural gas drillers, stung by decade-low gas prices from the over-supply of shale gas, have focused significant efforts on producing oil from the liquids-rich shale plays, such as Eagle Ford. The Eagle Ford shale of southeast Texas, covering 23 counties, is currently one of the most active shale hydrocarbon plays in the U.S. It trends across Texas from the Mexican border up into East Texas, roughly 80 km wide and 644 km long with an average thickness of 76 m, at depths ranging from 1220 to 3660 ft. The amount of technically recoverable oil in the Eagle Ford is estimated by the U.S. DOE to be 3.35 billion barrels of oil. The high percentage of carbonate makes the Eagle Ford shale brittle and "fracable" for the production of oil, condensate, and gas.

Eagle Ford shale samples in this work come from two sources: an outcrop and core samples from several depths. The Comstock West outcrop sample from Val Verde Co. was collected in Highway 90 about 50 km north-west of Del Rio, Texas (Slatt et al. 2012). The outcrop sample has an average TOC value of 5.3 %, and contains Type II kerogen, making it an excellent marine oil and gas source rock. However, at this location, the rocks are thermally immature with T_{max} values of 423–429 °C and average R_0

of 0.53 %. The sample is calcareous with an average $CaCO_3$ content of 64 %. Core samples from the J.A. Leppard #1 well (API 42-025-30389) in Bee County were obtained from the Texas Bureau of Economic Geology.

Samples from three typical shale formations in southwest China are studied in this work (Table 1): they are cores from an exploratory shale well (Jinye Well No. 1), and surface samples from two well-studied outcrops in Xishui Qilong Village for the Longmaxi Formation and Xiadong Jiulongwan for the Doushantuo Formation (Xie et al. 2008).

The Sichuan basin is an oil-bearing and gas-rich basin with an extensive development of the Lower Silurian Longmaxi Formation shale in southwestern China (southeastern Sichuan and western Hubei-eastern Chongqing) (Zou et al. 2013). The Longmaxi Formation contains 0.2 %–6.7 % of total organic carbon. The organic matter is over-mature, with R_o ranging from 2.4 % to 3.6 %, and dominated by Type II kerogen. Porosity measured on core samples of the shale from the Longmaxi Formation in exploratory wells ranges from 0.58 % to 0.67 % (Liu et al. 2011). The microporosity observed in thin sections of the shale is about 2 %, and dominated by intercrystalline and intragranular pores, with the pore size ranging from 100 nm to 50 µm. There are some differences between the Longmaxi Formation shale and the Barnett shale. The former is buried more deeply, with a higher degree of thermal evolution, lower gas content, denser, and more quartz of terrigenous origin (Liu et al. 2011).

In the Sichuan basin and its surrounding areas, shale gas in Lower Cambrian Niutitang Formation is distributed over all areas except for the eroded region in West Sichuan. Generally thicker than 40 m, its thickness increases from the central Sichuan to the north, the east, and the south. The

Table 2 Basic pore structure characteristics from MICP analyses

Sample ID[a]	Porosity, %	Total pore area, m²/g	Median pore-throat diameter (volume), nm[b]	Characteristic length, nm[c]	Permeability, nD	Effective Tortuosity[d]	L_e/L[d]
EF outcrop	7.85	6.38	22.7	7.2	258	80.5	2.51
EF LP4146[e]	3.28	7.16	6.5	4.1	2.72	1814	7.71
EF LP4155	1.56	3.14	9.5	3.2	1.60	NA	NA
EF LP4155[e]	2.58	5.49	8.0	3.2	2.46	2815	8,52
EF LP4157	3.11	5.58	17.4	6.4	2.97	2191	8.26
EF LP4166[e]	1.75	2.48	23.5	3.2	1.60	3867	8.23
B BL2167	1.98	4.87	5.9	4.8	1.35	2136	6.51
B BL2167	1.54	3.67	6.1	3.5	1.48	2026	5.58
B BL2175[e]	0.294	0.061	2698	13.7	3.24	1759	2.27
B BL2175	0.223	0.225	30.0	12.2	1.18	1784	1.99
B BL2185[e]	0.711	1.01	21.9	5.2	0.52	8699	7.86
B BL2185	0.555	1.02	7.0	4.6	1.05	2045	3.37
B BL2194	5.08	13.5	5.9	4.5	4.22	2273	10.7
B BL2194	4.84	13.0	5.9	4.0	3.71	2316	10.6
B BL2200	3.89	9.25	6.5	4.5	2.65	3845	12.2
B BL2200	3.27	7.51	6.7	3.6	2.78	3694	11.0
LMX TZ-4H	3.83	10.3	6.3	3.0	3.91	NA	NA
LMX TZ-4H	5.45	17.5	5.0	3.1	5.38	2214	11.0
NTT JY#1-137	1.49	1.01	65.9	11.0	9.58	NA	NA
NTT JY#1-137	1.18	1.10	20.9	6.5	1.99	1805	4.62
NTT JY#1-189	1.44	2.61	11.5	3.3	1.35	NA	NA
NTT JY#1-189	1.80	3.62	8.7	4.0	1.60	3683	8.13
NTT JY#1-220	3.91	9.07	8.3	3.3	3.66	NA	NA
NTT JY#1-220	4.39	7.86	10.3	5.7	6.64	1068	6.84
NTT JY#1-220	2.33	8.25	4.1	3.8	1.99	4103	9.77
DST YC01-72D	7.81	6.86	41.5	4.0	5.88	1413	10.5

[a] Numbers are the sample depths in a well (e.g., BL-2167 denotes sample location at 2167 m of the Blakely #1 well)

[b] Pore-throat diameter at which the volume of intruded mercury is 50 % of final value

[c] The characteristic (or threshold) length which is the pore-throat diameter corresponding to the threshold pressure P_t (psia); P_t is determined at the inflection point of the cumulative intrusion curve when mercury starts to percolate the whole sample

[d] *NA* not available. Some early measurements do not produce the effective tortuosity results, as the tortuosity calculation uses data points beyond the threshold pressure which is near 3-nm lower limit of the MICP instrument; recent measurements with more pressure measurement points consistently generate the tortuosity values

[e] These two sample cubes are detected to contain micron size fractures which lead to larger median pore-throat diameters than their non-fractured duplicate sample

Niutitang shale has a high organic carbon content, more than 2 % in most areas of Sichuan Basin. This formation is mature or over-mature. It is also deeply buried (more than 7 km in the west, the north and the east), and has evolved for a long time with complex preservation conditions which inhibit exploration and development. The most favorable exploration region lies in the south of Sichuan Basin, and the second most favorable areas are in western Hubei and eastern Chongqing (Liu et al. 2011; Sun et al. 2011; Guo 2013). Jinye well No. 1 is the first Sinopec shale gas well for obtaining geological parameters in southern

Sichuan, and was cored with a total core length of 101.3 m in the Niutitang Formation and used in this study (Table 1).

2.2 Mercury intrusion and pore characteristics

Pore structure characterization of shale samples includes measuring their porosity, particle and bulk density, total pore surface area, and pore size distribution with mercury injection capillary pressure (MICP). Porosity and pore-throat distribution were analyzed using a mercury intrusion porosimeter (AutoPore IV 9510; Micromeritics Instrument

Corporation, Norcross, GA). In addition, the MICP approach can also indirectly evaluate other pore characteristics, such as permeability and tortuosity (Webb 2001; Gao and Hu 2013).

Each shale sample (rectangular prisms with the largest linear dimension at either 10 or 15 mm) was oven-dried at 60 °C for at least 48 h to remove moisture, and cooled to room temperature ($\sim$23 °C) in a desiccator with less than 10 % relative humidity before the MICP test. Samples were then evacuated to 50 μm Hg (6.7 Pa, or 99.993 % vacuum). During a MICP test, each sample underwent both low-pressure and high-pressure analyses. The highest pressure produced by our MICP instrument is 60,000 psia (413 MPa), corresponding (via the Washburn equation) to a pore-throat diameter of about 3 nm. Under low-pressure analysis, the largest pore-throat diameter recorded by MICP is about 36 μm for a narrow-bore sample holder (called penetrometer) for samples with low (down to about 0.5 %) porosity. Equilibration time (minimum elapsed time with mercury volume change <0.1 μL, before proceeding to the next pressure) was chosen to be 50 s.

As reported by Gao and Hu (2013), permeability for the shale samples was calculated from the MICP data by the method of Katz and Thompson (1986, 1987). Effective tortuosity τ, another important parameter which indicates pore connectivity, can also be derived from MICP data (Hager 1998; Webb 2001).

$$\tau = \sqrt{\frac{\rho}{24k(1 + \rho V_{tot})} \int_{\eta=r_{c,\min}}^{\eta=r_{c,\max}} \eta^2 f_V(\eta) \mathrm{d}\eta}, \qquad (1)$$

where ρ is fluid density, g/cm^3; k is permeability, μm^2; V_{tot} is total pore volume, mL/g; $\int_{\eta=r_{c,\min}}^{\eta=r_{c,\max}} \eta^2 f_V(\eta) \mathrm{d}\eta$ is pore-throat volume distribution by pore-throat size. Effective tortuosity τ is related to the effective diffusion coefficient and travel distance of molecules by the following equation (Epstein 1989; Hu and Wang, 2003; Gommes et al. 2009):

$$\tau = \frac{D_0}{D_e} = \frac{1}{\phi}\left(\frac{L_e}{L}\right)^2, \qquad (2)$$

where D_0 is the aqueous diffusion coefficient and D_e effective diffusion coefficient in porous media, m^2/s; L_e is actual distance, m, traveled by a fluid particle as it moves between two points in a porous medium, which are separated by a straight line distance L, m; ϕ is porosity.

2.3 Spontaneous fluid imbibition and tracer migration

Because the concentration profile of various tracers provides a useful indication of the connectivity of a porous medium (Hu et al. 2012; Ewing et al. 2012), tracers were emplaced several different ways: by spontaneous

imbibition (Sect. 2.3), by diffusion (Sect. 2.4), and by injection under pressure (Sect. 2.5).

Spontaneous imbibition is a capillary force-driven process during which a wetting fluid displaces a non-wetting fluid only under the influence of capillary pressure. Because of the mathematical analogy between diffusion and imbibition, liquid imbibition can be used to probe a rock's pore connectivity (Hu et al. 2012). Imbibition tests, which are much faster than diffusion tests, involve exposing one face of a rock sample to liquid (for example, water, API brine or n-decane), and monitoring the fluid mass uptake over time (e.g., Hu et al. 2001; Schembre and Kovscek 2006). Using the network modeling results of Ewing and Horton (2002), we can probe pore connectivity, as indicated by the slope of log (imbibed liquid mass) versus log (time). The imbibition behavior—whether the imbibition slope is ¼, ¼ changing to ½, or ½ —conveniently classifies a rock's pore connectivity (Hu et al. 2012).

All sides of cm-sized cubes, except the top and bottom, were coated with quick-cure transparent epoxy to avoid evaporation of the imbibing fluid from, and reduce vapor transport and capillary condensation through, the side surfaces of the sample. The experimental procedure and data processing of imbibition tests were described in detail by Hu et al. (2001). Samples were first oven-dried at 60 °C for at least 48 h, in order to achieve a constant initial water saturation state, before being subjected to the imbibition experiments. During the imbibition tests with water or API brine, beakers of water were placed inside the experimental chamber to maintain a high and constant humidity level; these water beakers were removed for n-decane imbibition. The top of the side-epoxied samples was loosely covered with thin Teflon film, with a small hole left for air escape and co-current imbibition. The sample bottom was submerged to a depth of about 1 mm in the fluid reservoir. The imbibition rate was monitored by automatically recording the sample weight change over time.

In addition to replicate imbibition tests with de-ionized water being performed on the same sample (re-dried between runs), tracer imbibition experiment was conducted in API brine or n-decane fluid. Shales contain distinct phases of oil-wetting organic matter (e.g., kerogen) and water-wetting minerals, and tracers in two fluids (API brine and n-decane) are used to interrogate the wettability and connectivity of organic matter and mineral pore spaces. The organic fluid (n-decane) is expected to be preferentially attracted to the hydrophobic component (e.g., organic particles) of the shale matrix, with reported organic (kerogen) particle sizes ranging from less than 1 μm to tens of μm (Loucks et al. 2009; Curtis et al. 2012). Organic grains are found to be dispersed through the Barnett shale matrix (Hu and Ewing 2014), with their connection to mineral phases unknown. An understanding of the distribution and migration of the hydrocarbons stored in pores in these organic grains is critically needed.

API brine is made of 8 % NaCl and 2 % $CaCl_2$ by weight (Crowe 1969). Tracer imbibition test of API brine contains both non-sorbing (perrhenate) and sorbing (cobalt, strontium, cesium, and europium, with different sorption extents) tracers, and were prepared using ultrapure (Type 1) water and >99 % pure reagents ($NaReO_4$, $CoBr_2$, $SrBr_2 \cdot 2H_2O$, CsI, and $EuBr_3 \cdot 6H_2O$; Sigma-Aldrich Co., St. Louis, MO). Concentrations used were 100 mg/L $KReO_4$, 400 mg/L $CoBr_2$, 400 mg/L $SrBr_2$, 100 mg/L CsI, and 200 mg/L $EuBr_3$.

Dissolved in n-decane, organic phase tracer chemicals were >99 % pure tracer element-containing organic reagents: 1-iododecane {$CH_3(CH_2)_9I$, molecular weight of 268.18 g/mol} and trichlorooxobis (triphenylphosphine) rhenium(V){$[(C_6H_5)_3P]_2ReOCl_3$, molecular weight 833.14 g/mol}. The elements iodine (I) and rhenium (Re) of these organic chemicals are detected by LA–ICP–MS for the presence of these two organic tracers in shale. Trichlorooxobis (triphenylphosphine) rhenium (referred as organic-Re in this paper) was first dissolved in acetone solvent until over-saturation, filtered to obtain the saturated liquid, and then added to n-decane at a volume ratio of 2 %. Organic-I (1-iododecane) was directly added to n-decane fluid at a volume ratio of 1 %.

We use jmol software (http://jmol.sourceforge.net/) to estimate the structural dimensions of these organic tracers. The organic-I has a diameter at 1.39 nm length (L) × 0.29 nm width (W) × 0.18 nm height (H), while the structural dimensions of organic-Re molecules are 1.27 nm L × 0.92 nm W × 0.78 nm H, with its width and height larger than organic-I.

After tracer imbibition tests of 12–24 h in either API brine or n-decane fluids, the shale samples were lifted out of the fluid reservoir, flash-frozen with liquid nitrogen, kept at −80 °C in a freezer, freeze-dried as a batch at −50 °C and near-vacuum (less than 1 Pa) for 48 h, then stored at <10 % relative humidity prior to the analyses by laser ablation inductively coupled plasma-mass spectrometry (LA-ICP-MS).

2.4 Saturated diffusion

Chemical diffusion within fluid-saturated shale was measured following Hu and Mao (2012). Dry shale samples were evacuated for nearly 24 h at 99.99 % vacuum, and saturated by allowing non-traced fluid to invade the evacuated sample. Samples were then exposed (1 face only) to a tracer-containing fluid (either API brine or n-decane) in a reservoir. At fixed diffusion times (generally 1 day for 1 cm-sided cubes, and 2–4 days for 1.5-cm cubes),

individual shale samples were removed from the reservoir and processed (liquid nitrogen freezing, freeze-drying, and dry-storing) for tracer mapping.

2.5 Wood's metal intrusion

Wood's metal intrusion, and consequent SEM imaging and LA-ICP-MS mapping, provides a direct assessment of surface-accessible pore structure. Rock pore networks are examined by injecting the sample with molten Wood's metal alloy (50 % Bi, 26.7 % Pb, 13.3 % Sn, and 10 % Cd; melting point around 78 °C), a method pioneered by Swanson (1979) and Dullien (1981). Because of its high bismuth (Bi) content, Wood's metal alloy does not shrink as it solidifies (Hildenbrand and Urai 2003). Because this metal alloy is solid below 78 °C, pore structures filled by Wood's metal alloy can be readily imaged and mapped for the presence of alloy component elements. Injecting molten Wood's metal alloy also offers the possibility of "freezing" the invaded network at any stage of the injection, allowing micro-structural studies to be made on the iteratively filled pore networks (Kaufmann 2010).

Wood's metal alloy impregnation on shales was carried out at Kaufmann's laboratory at Empa (Zurich, Switzerland), which is capable of injecting the alloy at pressures up to 6000 bars (600 MPa or 87,023 psi). This corresponds to a pore-throat diameter of 2.35 nm; this estimation is based on the Washburn equation (1921), using a surface tension of 0.4 N/m and a contact angle of 130° for Wood's metal (Darot and Reuschle 1999). Shale samples (about $5 \times 5 \times 5$ mm^3) were first dried at 110 °C for 1 week, then glued to the bottom of a cell, which was then filled with solid Wood's metal pieces. The cell was closed by a piston and heated to 85 °C under vacuum (<5 Pa); O-rings connecting the piston to the cell wall are able to bear pressures of more than 6,000 bars (Kaufmann 2010). Once the metal was molten, a load was applied to the piston by means of a press in controlled load mode. Pressure in the cell was increased by 7 MPa/min until the maximum pressure was reached; the maximum pressure was then held for 15 min. After that, the temperature was decreased at a cooling rate of about 1 °C/min to solidify Wood's metal in the intruded pore spaces. During this solidifying step, the load was regulated to maintain the desired pressure. Once cool, the alloy-impregnated sample was taken out of the cell and cut horizontally in the middle of the sample height, to allow tracing Wood's metal alloy intrusion from the sides. One piece was cut to 150 μm thick, polished, and mounted on a glass slide, and imaged with an environmental SEM (Quanta 200, FEI, Hillsboro, OR) using back-scattered electrons (Dultz et al. 2006).The other cut piece was used for LA-ICP-MS elemental mapping, using the distribution of Wood's metal alloy component (e.g., Pb) to identify surface-accessible connected pore spaces.

2.6 LA-ICP-MS elemental mapping

For tracer imbibition, saturated diffusion, and Wood's metal intrusion tests, tracer concentrations at spatially resolved (mostly at 100 μm for cm-sized samples) locations were measured using laser ablation followed by LA-ICP-MS. The laser ablation system (UP-213, New Wave; Freemont, CA) used a 213 nm laser to vaporize a hole in the shale sample at sub-micron depths; elements entrained in the vapor were analyzed with ICP-MS (PerkinElmer/ SCIEX ELAN DRC II; Sheldon, CT). For various tests described here, this LA-ICP-MS approach can generate 2-D and 3-D maps of chemical distributions in rock at a spatial resolution of microns, and a concentration limit of low mg/kg (Hu et al. 2004; Hu and Mao 2012; Peng et al. 2012).

First we spot-checked for the presence of tracers at the sample's bottom (reservoir) and top faces, then the sample was cut dry in the middle from the top face, transversewise with respect to the imbibition/diffusion direction. A grid of spot analyses were then performed by LA-ICP-MS on the saw-opened interior face to map the tracer distribution within the shale sample, with a two-grid scheme to capture the tracer penetration into the samples from the edge. The first grid used an area of about 10 mm × 0.3 mm (in the direction of imbibition/diffusion), close to the sample bottom, with 100 μm laser spot size and 100 μm spacing between spots in the imbibition/ diffusion direction. This fine gridding scheme was performed in the area close to the bottom of the sample in the flow direction, because this area can have a steep tracer concentration gradient over a short distance. A second grid was then used to the right of the first grid at ∼800 μm spacing among laser spots to capture the possible presence of tracers away from the bottom edge.

3 Results and discussion

3.1 Mercury intrusion

As measured by MICP, pores in American and Chinese shales are predominantly in the nm size range with a measured median pore-throat diameter of 4.1–65.9 nm (Table 2). Excluding a few samples with fractures, 80 %–95 % of matrix pore-throat sizes by volume are smaller than 100 nm. For example, duplicate measurements of Longmaxi Formation samples (LMX TZ-4H in Table 2) show a porosity of 3.83 % and 5.45 %, median pore-throat sizes of 6.3 and 5.0 nm, with nearly 50 % of the pore-throat sizes located in 3–5 nm range and 90 %–95 % smaller than 100 nm. This is consistent with other literature values. For the Longmaxi shale samples from the

Changxin#1 well drilled in southeastern region of the Sichuan Basin, Chen et al. (2013) reported that the average porosity is 5.68 %. The micropores in the shale observed by the SEM are dominated by intergranular (dissolution) mineral pores, intergranular gaps, intragranular pores, the organic matter micropores, and micro-fractures. The mercury intrusion porosimetry results show that the maximum radius of the pore-throat is 33 nm, with the average at 10 nm (Chen et al. 2013).

While the presence of nanopores in shales has been well recognized since the first application of Ar-ion milling and field-emission SEM imaging (Loucks et al. 2009) to indicate the dominant organic nanopores in Barnett shales, larger pores cannot be discounted in their roles of mass transport and connecting nanopore regions. Pores less than 3 nm can be quantified with approaches such as low-pressure gas sorption isotherm, but such small pores probably do not play critical roles in hydrocarbon movement (Javadpour et al. 2007). When small pores or narrow pore-throats have diameters in the same order of magnitude as the chemicals, the steric hindrance effect can be significant (Hu and Wang 2003). Most hydrocarbon species of interests have a molecular diameter from 0.5 to 10 nm (Nelson 2009), and therefore probably the pore systems that contribute to hydrocarbon movement are these connected ones with pore sizes of 5–100 nm.

Using SEM/field-emission SEM methods to determine porosities of a range of shale, Slatt and O'Brian (2011) and Slatt et al. (2013) reported that micropores (>1 μm in pore length and nanopores (<1 μm) are subequal. The micropores are commonly porous floccules (clumps of electrostatically charged clay flakes arranged in edge–face or edge–edge card house structure) of up to 10 μm in diameter, which are not often seen or identified in ion-milled shale surfaces, perhaps due to the collapse of floccules during milling (Slatt et al. 2013). For the Eagle Ford shale with a high carbonate content, Slatt et al. (2012) reported the presence of μm-sized pore types from coccospheres (internal chambers and hollow spines are up to 1 μm in diameter and several μm in length) and foraminifera (their internal chambers can be 10 μm in diameter). Our MICP analyses, performed on individual 1 cm sized cubes, observed appreciable μm-sized pores, with >1 μm sized pores account for 10 %–20 % for American and Chinese shales studied in this work.

Since core samples are not readily available, we compare the outcrop with core samples for Eagle Ford Lower Shale, and observe a quite different behavior. The outcrop sample has a unimodal pore-throat size (at 95 %) located at 10–50 nm, compared to 30 %–44 % for three core samples. Pore-throats for the core samples are widely distributed from 3 nm to 36 μm, though most of them (70 %–80 %) are smaller than 100 nm. The disparity between

Table 3 Experimental imbibition results for Eagle Ford Leppard #1 well samples

Sample ID	Sample dimension	Height/width	Imbibition slope[a]
EF LP-4144	1.10 cm L × 1.00 cm W × 1.56 cm H	1.49	0.266 ± 0.073 ($N = 5$)
EF LP-4146	1.03 cm L × 1.38 cm W × 1.35 cm H	1.12	0.350 ± 0.059 ($N = 3$)
EF LP-4155	1.50 cm L × 0.97 cm W × 4.06 cm H	3.29	0.312 ± 0.042 ($N = 4$)
EF LP-4157	1.12 cm L × 1.14 cm W × 2.43 cm H	2.14	0.295 ± 0.018 ($N = 4$)
EF LP-4166	1.19 cm L × 1.44 cm W × 1.83 cm H	1.39	0.256 ± 0.008 ($N = 3$)

[a] Average ± standard deviation for replicate measurements on the same sample

outcrop and core samples is also reflected in median pore-throat sizes, as well as permeability with a difference of a factor of 100. However, caution needs to be extended when using data from outcrop samples to infer underground conditions.

Matrix permeabilities for these twelve American and Chinese shales, at a total of 24 MICP measurements, are 0.52–9.6 nD (Table 2), as estimated by the method of Katz and Thompson (1986, 1987). This is consistent with reported permeabilities of 1–10 nD by Heller and Zoback (2013), who used a pulse-decay permeameter with helium (more labor- and instrumentation-intensive) on 1–2 mm sized Barnett chips. The matrix permeabilities for Lower Eagle Ford shale from a well at different depths (4053.2 to 4087.8 m) range from 0.44 to 23.3 nD (5.82 ± 6.89, $N = 14$) (David Maldonado, 2012, personal communication), measured by the Gas Research Institute (GRI) (crushed rock) method (Luffel et al. 1993; Guidry et al. 1996). Our MICP technique, with cm-sized samples under confined pressure conditions and at an analytical cost (and sample mass) required of less than 10 % of amount of sample required for the GRI method, consistently produces comparable permeability values of 1.60–2.97 nD for the Leppard #1 core samples (Table 2).

MICP data can also be used with Hager's (1998) method to estimate effective tortuosity values, which characterize the convoluted pathways of fluid flow through porous systems (e.g., Epstein 1989; Hu and Wang 2003; Gommes et al. 2009). Following Gommes et al. (2009) approach (Eq. 2) of relating geometrical tortuosity to the travel paths that molecules will need to follow through a porous medium, the L_e/L ratios for the American and Chinese shales are on the order of 2–11 (Table 2). These relatively large values of tortuosity imply that fluid within the tight shale formations will need to make its way through some tortuous pathways in order to migrate from one location to another. For example, a tortuosity L_e/L of 11 for Longmaxi Formation sample means that it will take 11 cm for fluid to travel a linear distance of 1 cm in that formation. This is consistent with the findings from other approaches presented in this paper, which indicate that the nanopores of the tight shale formations are poorly connected so that

fluids require much time to find connected pathways to travel a limited distance.

3.2 Spontaneous fluid imbibition and tracer migration

Imbibition experiments for American shales consistently produce imbibition slopes of approximately ¼. The results for Eagle Ford core samples are presented in Table 3, with a typical one shown in Fig. 1; similar results are presented for Barnett core samples in Hu and Ewing (2014). The ¼ imbibition slope was consistently observed across different sample shapes, and imbibition directions (parallel/horizontal versus transverse/vertical to the bedding plane) examined for Barnett samples, and for all sample depths (e.g., Table 3). From this, we conclude that pores in the Eagle Ford and Barnett shales are poorly connected for water movement, with a correlation length (imbibition distance beyond which behavior becomes Fickian) greater than the sample height (approximately 15 mm) used. On other rocks, we have experimentally observed all three types of imbibition slopes (¼, ¼ changing to ½, and ½), consistent with percolation theory (Stauffer and Aharony 1994; Hunt et al. 2014). For example, Hu et al. (2012)

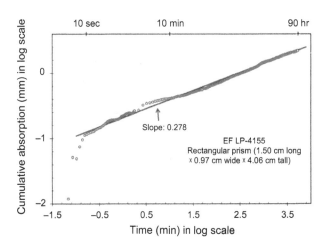

Fig. 1 Water imbibition results for Leppard #1 well 4155 m core sample of Eagle Ford shale

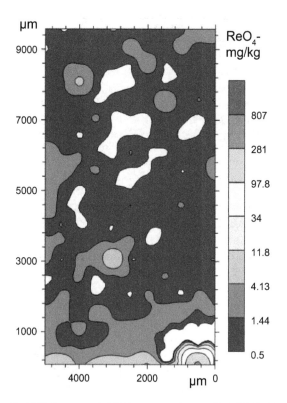

Fig. 2 API brine-tracer imbibition penetration profile in LMX TZ-4H sample of Longmaxi shale, mapped on the interior face for non-sorbing ReO_4^-; imbibition time 47.5 h. This 5000 μm × 9575 μm grid nearly covers whole sample, with a mapping routine of 100 μm laser spot size and 500 μm spacing between spots. The *bottom* (tracer imbibing) *face* is shown to the *bottom* of the figure. The LA-ICP-MS detection limit and background levels in the clean sample for Re are about 0.05 and 0.5 mg/kg (used as the smallest value for 8-scale color scheme), respectively. The largest value of color scale (807 mg/kg) is the averaged Re concentration of 81 measurements obtained on the imbibing face (807 ± 391 mg/kg)

observed the ½ slope across all aspect ratios for Berea sandstone samples, indicating its well-connected pore spaces.

In addition to fluid imbibition tests to probe the pore connectivity, some tracer imbibition tests were conducted and resultant tracer penetration distance was mapped using LA-ICP-MS. The mapping result for API brine-tracer imbibition into the interior face of a Longmaxi shale sample (LMX DZ-4H) is shown in Fig. 2. When imbibition proceeds from this initially dry sample, capillary-driven advective flow produces a relatively steep imbibition front for non-sorbing ReO_4^-. At about 1 mm into the matrix from the edge, the tracer is only distributed at about 1–4 mg/kg (compared with a measured mean concentration of 807 mg/kg within 2 μm of the edge on the imbibing face; in other words, the connected porosity away from the edge is only about 0.5 %–2 % of the total pore space), from the imbibition test of 47.5 h. Such a result indicates the limited pore connectivity for brine fluid flow and tracer

movement, with the connected pore spaces limited to the sub-mm range from the sample edge in Longmaxi shale; similar results are observed for Barnett shale (Hu and Ewing 2014).

3.3 Saturated diffusion

Figure 3 presents 2-D tracer concentration profiles in n-decane fluid for a Niutitang shale (NTT JY#1-137) after 25 h diffusion time. Two molecular tracers in n-decane fluid have the sizes of 1.39 nm × 0.29 nm × 0.18 nm for organic-I (Fig. 3a) and 1.27 nm × 0.92 nm × 0.78 nm for organic-Re (Fig. 3b). Much less diffusive penetration was observed for larger molecules of organic-Re inside a sample with 47 %–66 % pore-throat sizes between 3–50 nm (measured characteristic pore-throat sizes from MICP are 6.5–11.0 nm, when fluid mercury starts to percolate across the whole sample; Table 2). This result indicates the entanglement of nano-sized molecules in nanopore spaces of tight samples. Comparatively, the narrower sized organic-I is present nearly throughout the whole sample, indicating that this shale sample has oil-wetting characteristics and possesses well-connected pore spaces for hydrophobic molecules to move, barring the very sensitive molecular size effect. The practical implication is that the out-diffusion of nm-sized hydrocarbon is expected to be slow, of limited quantity, and largely limited to a small distance from rock matrix into a fracture; this will affect the hydrocarbon recovery in stimulated shale reservoirs.

3.4 Limited edge-accessible porosity from Wood's metal intrusion

The Barnett shale sample from 2185 m (B BL-2185) was injected with Wood's metal alloy at 1542 bars (154 MP). At this pressure, pore-throats of diameter greater than 9.2 nm are invaded. For this sample, SEM images show that there is little connected matrix porosity farther than about 60 μm from the sample's edge (figure not shown); Wood's metal mainly occupies small cracks and matrix pores connected to the sample surface. As in the case for mercury intrusion, the sample was externally and uniformly surrounded by non-wetting molten Wood's metal alloy, with even pressurization to reduce experiment-related fractures.

The sample was also mapped by LA-ICP-MS for more sensitive detection of the presence of Wood's metal component elements (Bi, Cd, Pb, and Sn). Distribution of all these elements inside the shale sample was similar, so only the plot for Pb is presented (Fig. 4). Wood's metal alloy has 26.7 % Pb (267,000 mg/kg), consistent with the

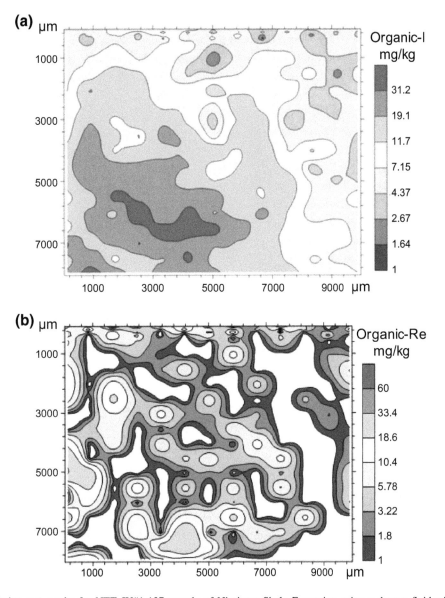

Fig. 3 Saturated diffusion test results for NTT JY#1-137 sample of Niutitang Shale Formation using n-decane fluid with tracers of organic iodine and organic rhenium of different molecular sizes. The total grid size is 10,000 μm × 8050 μm to cover whole sample, and mapped with a 100 μm laser spot size and two-grid scheme of variable spacing between spots; the white areas indicate the regions of concentration below the LA-ICP-MS detection limit

concentration detected by LA-ICP-MS, but its natural abundance in the shale matrix is only about 0.5 mg/kg. Lead is, therefore, an excellent proxy for the presence of Wood's metal inside the sample, identifying porous regions that are connected to the sample's exterior. LA-ICP-MS has a higher chemical resolution and larger observation scales than SEM, though lower spatial resolution, and reveals connected pore spaces that are not visible with SEM imaging. Wood's metal penetrates all pore spaces (at the resolution of 100-μm laser spot) within this sample with a dimension of 7.0 mm × 7.75 mm, as seen from the

concentrations uniformly across the sample at above the 0.5 mg/kg background level. However, the surface-accessible connected pores constitute a very low percentage of the interior volume. Detected Pb concentrations in the interior area are only in the range of 10,000 mg/kg, compared to 267,000 mg/kg for the Pb content in Wood's metal itself. Even under a high pressure of 1542 bars, a mean connected porosity therefore comprises about 4 % of interior pore volume of this sample, consistent with other results of this work with respect to limited pore connectivity for tight shales.

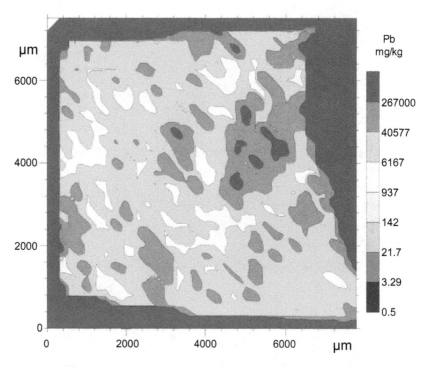

Fig. 4 2-D LA-ICP-MS mapping of Pb (lead) distribution in B BL-2185 m sample following Wood's metal alloy injection at a pressure of 1542 bars; laser spot size is 100 μm and spacing between spots is 250 μm

4 Conclusions

We investigated pore geometry and topology in several typical American and Chinese shales, and their implications for fluid movement and tracer migration. Our experimental approaches included mercury porosimetry, spontaneous imbibition and tracer migration, tracer diffusion, and Wood's metal intrusion under high pressures. Consistent with other reports, we found that these shale pores are predominantly in the nm size range, with a median pore-throat diameter of 4.1–65.9 nm. It is expected that the shale's geometrical properties—low porosity and small pore-throat size—will yield extremely low diffusion and imbibition rates.

Using multiple characterization approaches to infer the shale's topology, we also found consistent evidences that pore connectivity in these shales could be quite low for a particular probing fluid of either API brine or n-decane. Tracer diffusion inside the shale matrix was not well described by classical Fickian behavior; the anomalous behavior suggested by percolation theory gave a better description. The spontaneous imbibition tests saw an imbibition slope characteristic of low-connectivity material. Our different methods assessing edge-accessible porosity—following spontaneous imbibition, diffusion, and high pressure intrusion—consistently showed a steep decline in accessible porosity with distance from the

sample exterior. In fact, Wood's metal concentration in the interior is only about 4 % under high pressures, which indicates that only a limited number of the interior pores are connected to the outside by a continuous pathway.

Because shale has both mineral (water-wetting) and organic (kerogen; water-repelling) pores, a fluid-based approach such as spontaneous imbibition and diffusion are expected to give lower accessible porosity profiles than will pressure-injected Wood's metal, which accesses both the water-wet and oil-wet pores. But despite Wood's metal accessing both kinds of pores, it consistently shows the steep decline in accessible porosity with distance that is characteristic of low pore connectivity. All results converge into findings that pores more than a few mm inside a sample are unlikely to be connected to the exterior, with additional implications of molecular size effects, as shown from our tracers of different sizes. In terms of hydrocarbon production, matrix pores more than a few mm from an induced fracture are only sparsely connected to fracture network and the producing borehole.

The consequences of low pore connectivity, specifically distance-dependent accessible porosity, will lead to an initial steep decline in production rates and low overall recovery. Once the hydrocarbons residing in the fracture network are exhausted, hydrocarbon replenishment from shale matrix (i.e., "matrix feeding") is limited by the slower anomalous diffusion through the shale matrix, and reduced by the fraction of the matrix that is actually accessible.

This study addresses knowledge gaps regarding pore connectivity in typical American and Chinese shales. We show that pore connectivity may be a dominant constraint on diffusion-limited hydrocarbon transport. Pore connectivity at the nanoscale may be the cause of underlying steep first-year declines in hydrocarbon production, as well as the low overall recovery observed in hydraulic fractured shales.

Acknowledgments Funding for this project was partially provided by the following three State Key Laboratories in China: State Key Laboratory of Oil and Gas Reservoir Geology and Exploitation, Chengdu University of Technology, Chengdu (PLC-201301); State Key Laboratory of Organic Geochemistry, Chinese Academy of Sciences, Guangzhou (No. OGL-201402); the Foundation of State Key Laboratory of Petroleum Resources and Prospecting, China University of Petroleum, Beijing (No. PRP/open-1403). The authors would like to thank Harold Rowe and Roger Slatt for providing Barnett and Eagle Ford samples. Laboratory assistance from Troy Barber, Yuxiang Zhang, Golam Kibria, and Francis Okwuosa is also much appreciated.

References

Bear J. Dynamics of fluid in porous media. New York: Dover Publications; 1972.

Chen WL, ZhouW LuoP, et al. Analysis of the shale gas reservoir in the Lower Silurian Longmaxi Formation, Changxin 1 well, Southeast Sichuan Basin, China. Acta Petrol Sin. 2013;29(3): 1073–86 (in Chinese).

Crowe CW. Methods of lessening the inhibitory effects to fluid flow due to the presence of solid organic substances in a subterranean formation. U.S. Patent Office 3482636, patented on 9 Dec 1969.

Curtis JB. Fractured shale-gas systems. AAPG Bull. 2002;86(11): 1921–38.

Curtis ME, CardottBJ SondergeldCH, et al. Development of organic porosity in the Woodford shale with increasing thermal maturity. Int J Coal Geol. 2012;103:26–31.

Darot M, Reuschle T. Direct assessment of Wood's metal wettability on quartz. Pure Appl Geophys. 1999;155(1):119–29.

DOE (Department of Energy). Modern shale gas development in the United States: A primer. 2009.

Dullien FAL. Wood's metal porosimetry and its relation to mercury porosimetry. Powder Tech. 1981;29:109–16.

Dullien FAL. Porous media: fluid transport and pore structure. 2nd ed. San Diego: Academic Press; 1992.

Dultz S, Behrens H, Simonyan A, et al. Determination of porosity and pore connectivity in feldspars from soils of granite and saprolite. Soil Sci. 2006;171(9):675–94.

EIA (Energy Information Administration). Annual energy outlook 2014: with projections to 2040. U.S. Department of Energy, DOE/EIA-0383, released on May 7, 2014. Available at http://www.eia.gov/forecasts/AEO/. Accessed 28 Jan 2015.

Epstein N. On tortuosity and the tortuosity factor in flow and diffusion through porous media. Chem Eng Sci. 1989;44(3):777–9.

Ewing RP, Horton R. Diffusion in sparsely connected pore spaces: temporal and spatial scaling. Water Resour Res. 2002;38(12):1285. doi:10.1029/2002WR001412.

Ewing RP, Liu CX, Hu QH. Modeling intragranular diffusion in low-connectivity granular media. Water Resour Res. 2012;48:W03518. doi:10.1029/2011WR011407.

Gao ZY, Hu QH. Estimating permeability using median pore-throat radius obtained from mercury intrusion porosimetry. J Geophys Eng. 2013;10:025014. doi:10.1088/1742-2132/10/2/025014.

Gommes CJ, Bons AJ, Blacher S, et al. Practical methods for measuring the tortuosity of porous materials from binary or gray-tone tomographic reconstructions. AIChE J. 2009;55:2000–12.

Grieser WV, Shelley RF, Johnson BJ, et al. Data analysis of Barnett shale completions. SPE J. 2008. doi:10.2118/100674-PA.

Guidry FK, Luffel DL, Curtis JB. Development of laboratory and petrophysical techniques for evaluating shale reservoirs. GRI (Gas Research Institute) Final Report GRI-95/0496, 1996.

Guo TL. Evaluation of highly thermally mature shale-gas reservoirs in complex structural parts of the Sichuan Basin. J Earth Sci. 2013;24(4):863–73.

Hager J. Steam drying of porous media. Ph.D. Thesis. Department of Chemical Engineering, Lund University, Sweden. 1998.

Heller R, Zoback M. Laboratory measurements of matrix permeability and slippage enhanced permeability in gas shales. Soc Pet Eng, SPE 168856. 2013.

Hildenbrand A, Urai JL. Investigation of the morphology of pore space in mudstones-first results. Mar Pet Geol. 2003;20(10):1185–200.

Hoffman T. Comparison of various gases for enhanced oil recovery from shale oil reservoirs. This paper was presented for presentation at the eighteenth SPE improved oil recovery symposium held in Tulsa, OK, pp. 14–18, SPE 154329. 2012.

Hu QH, Wang JSY. Aqueous-phase diffusion in unsaturated geological media: a review. Crit Rev Environ Sci Technol. 2003;33(3):275–97.

Hu QH, Mao XL. Applications of laser ablation-inductively coupled plasma-mass spectrometry in studying chemical diffusion, sorption, and transport in natural rock. Geochem J. 2012;46(5):459–75.

Hu QH, Ewing RP. Comment on "Energy: a reality check on the shale revolution" (Hughes, J.D., Nature, Vol. 494, p. 307–308. doi:10.1038/494307a. http://www.nature.com/nature/journal/v494/n7437/full/494307a.html. Accessed 21 Feb 2013.

Hu QH, Ewing RP. Integrated experimental and modeling approaches to studying the fracture-matrix interaction in gas recovery from Barnett Shale. Final Report, Research Partnership to Secure Energy for America (RPSEA), National Energy Technology Laboratory, Department of Energy. 2014.

Hu QH, Persoff P, Wang JSY. Laboratory measurement of water imbibition into low-permeability welded tuff. J Hydrol. 2001;242(1–2):64–78.

Hu QH, Ewing RP, Dultz S. Pore connectivity in natural rock. J Contam Hydrol. 2012;133:76–83.

Hu QH, Kneafsey TJ, Wang JSY, et al. Characterizing unsaturated diffusion in porous tuff gravels. Vadose Zone J. 2004;3(4):1425–38.

Hughes JD. Drill, Baby, Drill: can unconventional fuels usher in a new era of energy abundance? Post Carbon Institute. 2013a.

Hughes JD. Energy: a reality check on the shale revolution. Nature. 2013;494:307–8.

Hunt AG, Ewing RP, Ghanbarian B. Percolation theory for flow in porous media. 3rd ed., Lecture notes in physicsHeidelberg: Springer; 2014. p. 880.

Jarvie DM. Shale resource systems for oil and gas: part 1—shale-gas resource systems. In: Breyer JA, editor, Shale reservoirs—giant resources for the 21st century. AAPG Memoir; 2012, 97: 69–87.

Javadpour F, Fisher D, Unsworth M. Nanoscale gas flow in shale gas sediments. J Can Pet Technol. 2007;46(10):55–61.

Katz AJ, Thompson AH. A quantitative prediction of permeability in porous rock. Phys Rev B. 1986;34:8179–81.

Katz AJ, Thompson AH. Prediction of rock electrical conductivity from mercury injection measurements. J Geophys Res. 1987;92(B1):599–607.

Kaufmann J. Pore space analysis of cement-based materials by combined nitrogen sorption: Wood's metal impregnation and multi-cycle mercury intrusion. Cement Concr Compos. 2010;32(7):514–22.

King GE. Hydraulic fracturing 101. SPE 152596. 2012.

Liu SG, Ma WX, Luba J, et al. Characteristics of the shale gas reservoir rocks in the Lower Silurian Longmaxi Formation, east Sichuan Basin, China. Acta Petrol Sin. 2011;27(8):2239–52 (in Chinese).

Loucks RG, Reed RM, Ruppel SC, Jarvie DM. Morphology, genesis, and distribution of nanometer-scale pores in siliceous mudstones of the Mississippian Barnett Shale. J Sediment Res. 2009;79(11–12):848–61.

Loucks RG, Ruppel SC. Mississippian Barnett Shale: lithofacies and depositional setting of a deep-water shale-gas succession in the Fort Worth Basin, Texas. AAPG Bull. 2007;91(4):579–601.

Luffel DL, Hopkins CW, Schettler PD. Matrix permeability measurement of gas productive shales. SPE 26633. 1993.

Nelson PH. Pore-throat sizes in sandstone, tight sandstone, and shale. AAPG Bull. 2009;93(3):329–40.

Nicot J-P, Scanlon BR, Reedy RC, Costley RA. Source and fate of hydraulic fracturing water in the Barnett shale: a historical perspective. Environ Sci Technol. 2014;48(4):2464–71.

Peng S, Hu QH, Ewing RP, et al. Quantitative 3-D elemental mapping by LA-ICP-MS of basalt from the Hanford 300 area. Environ Sci Tech. 2012;46:2035–42.

Schembre JM, Kovscek AR. Estimation of dynamic relative permeability and capillary pressure from countercurrent imbibition experiments. Transp Porous Media. 2006;65:31–51.

Slatt RM, O'Brien NR. Pore types in the Barnett and Woodford gas shales: contribution to understanding gas storage and migration pathways in fine-grained rocks. AAPG Bull. 2011;95(12):2017–30.

Slatt RM, O'Brien NR, Miceli R, et al. Eagle Ford condensed section and its oil and gas storage and flow potential. AAPG Annual Convention and Exhibition, Long Beach, California, 22–25 April 2012. Search and Discovery Article #80245. 2012.

Slatt RM, O'Brien NR, Molinares-Blanco C et al. Pores, spores, pollen and pellets: small, but significant constituents of resources shales. SPE 168697/URTeC 1573336. 2013.

Stauffer D, Aharony A. Introduction to percolation theory. 2nd ed. London: Taylor and Francis; 1994.

Sun W, Liu SG, Ran B, et al. General situation and prospect evaluation of the shale gas in the Niutitang Formation of Sichuan Basin and its surrounding area. J Chengdu Univ Technol (Science and Technology Edition). 2011;39(2):170–75 (in Chinese).

Swanson BF. Visualizing pores and nonwetting phase in porous rock. J Petrol Technol. 1979;31:10–8.

Washburn EW. Note on a method of determining the distribution of pore sizes in a porous materials. Proc Natl Acad Sci USA. 1921;7:115–6.

Webb PA. An introduction to the physical characterization of materials by mercury intrusion porosimetry with emphasis on reduction and presentation of experimental data. Micromeritics Instrument Corporation. 2001.

Xie GW, Zhou CM, McFadden KA, et al. Microfossils discovered from the Sinian Doushantuo Formation in the Jiulongwan section, East Yangtze Gorges area, Hubei Province, South China. Acta Palaeontologica Sinica. 2008;47(3):279–91 (in Chinese).

Zou CN, Tao SZ, Yang Z, et al. Development of petroleum geology in China: discussion on continuous petroleum accumulation. J Earth Sci. 2013;24(4):796–803.

Permissions

List of Contributors

Yong-Zhang Huang, Bao-Sheng Zhang and Ren-Jin Sun
School of Business Administration, China University of Petroleum, Beijing 102249, China

Xin-Qiang Wei
Overseas Investment Environment Research Department, CNPC Economics and Technology Research Institute, Beijing 100724, China

Azizollah Khormali and Dmitry G. Petrakov
Department of Oil and Gas Field Development and Operation, Oil and Gas Faculty, National Mineral Resources University (Mining University), Saint Petersburg, Russia 199106

Zahra Sadat Mashhadi, Ahmad Reza Rabbani and Ahmad Khajehzadeh
Petroleum Engineering Department, Amirkabir University of Technology, Hafez Street, 15875-4413 Tehran, Iran

Mohammad Reza Kamali and Maryam Mirshahani
Research Institute of Petroleum Industry (RIPI), West Blvd. Azadi Sport Complex, 14665-37 Tehran, Iran

Dong-Kun Luo and Liang-Yu Xia
School of Business Administration, China University of Petroleum (Beijing), Beijing 102249, China

Hao-Ran Zhang, Yong-Tu Liang, Meng-Yu Wu and Qi Shao
Beijing Key Laboratory of Urban Oil and Gas Distribution Technology, China University of Petroleum, Beijing 102249, China

Qiao Xiao
CNPC Trans-Asia Gas Pipeline Company Ltd., Beijing 100007, China

Ying Li and Gang Wang
College of Computer Science and Technology, Jilin University, Changchun 130012, Jilin, China

Zhi-Qi Guo and Cai Liu
College of Geo-Exploration Science and Technology, Jilin University, Changchun 130026, Jilin, China

Xiang-Yang Li
British Geological Survey, Edinburgh EH9 3LA, UK
CNPC Key Laboratory of Geophysical Prospecting, China University of Petroleum, Beijing 102249, China

Zhi-Gang Wei, Chao-Yu Yan, Meng-Da Jia, Jian-Fei Song and Yao-Dong Wei
State Key Laboratory of Heavy Oil Processing, China University of Petroleum, Beijing 102249, China

Xing-Yao Yin, Xiao-Jing Liu and Zhao-Yun Zong
School of Geosciences, China University of Petroleum (Huadong), Qingdao 266580, Shandong, China

Chuan-Zhi Cui and Jian-Peng Xu
College of Petroleum Engineering, China University of Petroleum, Qingdao 266580, Shandong, China

Duan-Ping Wang, Zhi-Hong Liu and Ying-song Huang
Exploration and Development Research Institution, Shengli Oilfield of SINOPEC, Dongying 257015, Shandong, China

Zheng-Ling Geng
Center for Educational Development, China University of Petroleum, Qingdao 266580, Shandong, China

Yi-Jin Zeng, Xu Zhang and Bao-Ping Zhang
Sinopec Research Institute of Petroleum Engineering, Beijing 100101, China

Selin Ozdemir and Işil Akgul
Department of Econometrics, Faculty of Economics, Marmara University, Istanbul, Turkey

Shao-Ying Huang, Ke Zhang, Zhong-Yao Xiao, Bao-Shou Zhang and Qing Zhao
Research Institute of Petroleum Exploration and Development, Tarim Oilfield Company, PetroChina, Korla 841000, Xinjiang, China

Mei-Jun Li, T.-G. Wang, Rong-Hui Fang, Dao-Wei Wang and Fu-Lin Yang
State Key Laboratory of Petroleum Resources and Prospecting, College of Geosciences, China University of Petroleum, Beijing 102249, China

Zhuoheng Chen
Geological Survey of Canada, Calgary, AB, Canada

Kirk G. Osadetz
University of Calgary, Calgary, AB, Canada

Xuansha Chen
ZLR Valeon, Beijing, China

Ahmed H. Kamel and Essam A. Ibrahim
College of Business & Engineering, University of Texas of the Permian Basin, 4901 E University, Odessa, TX 79762-0001, USA

Ali S. Shaqlaih
Department of Mathematics & Information Sciences, University of North Texas at Dallas, 7400 University Hills Blvd, Dallas, TX 75241, USA

M. Enamul Hossain and Mohammed Wajheeuddin
Department of Petroleum Engineering, College of Petroleum Engineering and Geosciences, King Fahd University of Petroleum & Minerals (KFUPM), Box: 2020, Dhahran 31261, Kingdom of Saudi Arabia

Zheng-Song Qiu, Xin Zhao, Jun-Yi Liu and Wei-Ji Wang
School of Petroleum Engineering, China University of Petroleum, Qingdao 266580, Shandong, China

Wei-An Huang
School of Petroleum Engineering, China University of Petroleum, Qingdao 266580, Shandong, China
School of Mechanical and Chemical Engineering, The University of Western Australia, Crawley 6009, Australia

Ming-Lei Cui
Shandong Shengli Vocational College, Dongying 257097, Shandong, China

Melike Bildirici
Department of Economics, Yıldız Technical University, Istanbul, Turkey

Özgür Ersin
Department of Economics, Beykent University, Istanbul, Turkey

Mehdi Azimian and Hans-Jörg Bart
Department of Mechanical and Process Engineering, University of Kaiserslautern, 67663 Kaiserslautern, Germany

Xian-Guo Liu
China University of Geosciences (Wuhan), Wuhan 430074, China

Qin-Hong Hu
China University of Geosciences (Wuhan), Wuhan 430074, China
The University of Texas at Arlington, Arlington, TX 76019, USA

Zhi-Ye Gao
State Key Laboratory of Petroleum Resources and Prospecting, China University of Petroleum, Beijing 102249, China

Shu-Gen Liu and Wen Zhou
State Key Laboratory of Oil and Gas Reservoir Geology and Exploitation, Chengdu University of Technology, Chengdu 610059, China

Wen-Xuan Hu
Nanjing University, Nanjing 210093, China

Index

Printed in the USA
CPSIA information can be obtained
at www.ICGtesting.com
JSHW052022301024
72690JS00004B/133